TRAFFIC FLOW THEORY AND CONTROL

TRAFFIC FLOW THEORY AND CONTROL

DONALD R. DREW
HEAD, DESIGN AND TRAFFIC DIVISION
TEXAS TRANSPORTATION INSTITUTE
AND
ASSOCIATE PROFESSOR OF CIVIL ENGINEERING
TEXAS A&M UNIVERSITY

LIANG YU HSIA

1971

發行人：卓　劉　慶　弟
　　地址：台北市重慶南路一段一三三號
發行所：新月圖書股份有限公司
　　地址：台北市重慶南路一段一三三號
印刷所：盛昌印製廠有限公司
　　地址：三重市正義北路二十八號
　　電話：九七二六八四號
　中華民國六十年　月　日

登記證內版台業字第一〇四一號

PREFACE

How well our highways are planned, the safety and efficiency of their design and operation, and in the final analysis, the climate for future growth of highway transportation depend on a continuing supply of competent engineering manpower. Efforts to furnish an adequate supply of such talent have not been successful. In spite of the fact that traffic problems constitute one of the great continuing challenges facing the nation, about one out of every five job opportunities in the profession goes vacant.

This shortage of manpower is due primarily to a lack of interested engineering students at the various universities offering specialized graduate training in this field. This shortage stems from the fact that traffic engineering lacks the glamour appeal of some of the newer branches of engineering such as aerospace, electronics, and computer science. This is ironic in light of the number of aerospace organizations and operations research groups who, without enough defense contracts to go around, have turned to traffic and transportation research.

At the same time, Don Quixotes are springing up all over the country, jousting with the gigantic windmill in Detroit. Readers anticipating yet another attempt to tame these Detroit tigers will be let down. In this book, the auto is not regarded as a weapon. It is treated as a component in the man-machine-road system—a "black box," though not necessarily a coffin.

The primary purpose of this book is to present traffic engineering as the science it is, so as to challenge and convert young minds. A second purpose, equally ambitious, is to provide practicing traffic engineers with an insight into contemporary traffic research approaches.

Although it takes theory into account, this is not a highly theoretical book. Only a knowledge of mathematics through differential equations is required. This is an applied book in the same sense that traffic engineering is an applied science. It is the application of experimental psychology, mathematics, physics, and statistics.

This book stems from two sources. In part, it has been developed from lecture notes prepared for two courses in the Civil Engineering Department's transportation engineering and transportation science curriculum at Texas A&M University. However, to a much greater extent, this book is an outgrowth of those phases of the Texas Transportation Institute's research program in which the author was directly involved. I am deeply indebted to my students and associates in the Texas Transportation Institute and the Civil Engineering Department.

Sincere appreciation is expressed to Mr. Charles J. Keese, director of research for the Texas Transportation Institute and professor of civil engineering, who inspired this undertaking and who collaborated on many of the publications from which this material was gathered. Chapter 14 is taken from the Highway Research Board's award-winning paper, "Freeway Level of Service as Described by an Energy-acceleration Noise Model," coauthored with Professor Keese and Mr. Conrad L. Dudek, assistant research engineer.

There are other obligations for material. Chapters 9 and 16 make extensive use of freeway surveillance and control research conducted with Mr. W. R. McCasland, head of the Texas Transportation Institute's Freeway Surveillance and Control Department, Dr. Joseph A. Wattleworth, head of the Traffic Systems Department, Dr. Johann H. Buhr, assistant research engineer, and Dr. Charles Pinnell, associate dean of Texas A&M University's Graduate College.

Gratitude is expressed to the entire T.T.I. staff for their assistance in many phases of the work. Special thanks to Mr. Kenneth Brewer of T.T.I. and to Dr. Bob Smith of Kansas State University for their critical reviews of the manuscript.

Donald R. Drew

CONTENTS

Preface vii

CHAPTER ONE INTRODUCTION
1.1 A Nation on Wheels *1*   **1.2** The Problem *2*   **1.3** The Challenge *3*   **1.4** Role of Traffic Engineering *4*   **1.5** Organization of the Book *5*

CHAPTER TWO THE VEHICLE
2.1 Introduction *7*   **2.2** Static Characteristics *8*   **2.3** Vehicle Kinematics *8*   **2.4** Nonuniform Acceleration Theory *10*   **2.5** Vehicle Dynamics *12*   **2.6** Braking *13*   **2.7** Speed Computations *14*   **2.8** Impact and Collision *15*   **2.9** Control *17*   **2.10** Turning *17*   **2.11** Stability *19*   **2.12** Vehicle Power Systems *20*   **2.13** Challenges *21*

CHAPTER THREE THE DRIVER
3.1 Learning *25*   **3.2** Motivation *26*   **3.3** Attitude *27*   **3.4** Vision *28*   **3.5** Night Visibility *29*   **3.6** Driver Eye Height *31*   **3.7** Human-factors Engineering *32*   **3.8** Driver Behavior *33*   **3.9** Driver Response *34*   **3.10** Driver Error *36*   **3.11** Driver Simulation *36*   **3.12** Perception-Reaction Time *37*   **3.13** The Fallacy of the Average Man *40*   **3.14** The Rule of Thumb *40*   **3.15** Challenges *42*

CHAPTER FOUR THE ROAD

4.1 Introduction *46*   **4.2** Roadway Surface and Lighting *47*   **4.3** Geometric Design *50*   **4.4** The Cross Section *52*   **4.5** Vertical Alignment *55*   **4.6** Horizontal Alignment *64*   **4.7** Control Devices *69*   **4.8** Project Skyhook *70*   **4.9** The Road as a System *72*   **4.10** Challenges *75*

CHAPTER FIVE THE CIRCULATION SYSTEM

5.1 Introduction *79*   **5.2** Functional Classification *80*   **5.3** The Freeway Concept *84*   **5.4** Measuring Traffic Volumes *85*   **5.5** Volume Survey Devices *88*   **5.6** Volume Criteria *89*   **5.7** Future Traffic *100*   **5.8** Capacity *102*   **5.9** Level of Service *104*   **5.10** Challenges *108*

CHAPTER SIX TRAFFIC RESEARCH

6.1 Introduction *111*   **6.2** The Systems Approach *112*   **6.3** The Mathematical Model *113*   **6.4** Experiment and Evaluation *115*

CHAPTER SEVEN TRAFFIC EVENTS

7.1 Introduction *120*   **7.2** The Binomial Distribution *121*   **7.3** Moments of a Discrete Distribution *123*   **7.4** The Poisson Distribution *123*   **7.5** Statistical Tests for Randomness *125*   **7.6** The Probability-generating Function *126*   **7.7** The Negative Binomial Distribution *127*   **7.8** Analysis of Arrivals *128*   **7.9** The Geometric Distribution *131*   **7.10** Poisson Approximation to the Binomial *132*   **7.11** Application to Signalization *134*   **7.12** Illustration of Capacity-Design Procedure *140*

CHAPTER EIGHT CONTINUOUS VARIABLES

8.1 Introduction *147*   **8.2** The Moment-generating Function *148*   **8.3** The Rectangular Distribution *149*   **8.4** The Normal Distribution *150*   **8.5** Development of Traffic Variables *152*   **8.6** The Negative Exponential Distribution *153*   **8.7** The Pearson Type III Distribution *155*   **8.8** The Erlang Distribution *158*   **8.9** The Pearson Type I Distribution *161*   **8.10** Distribution Relationships *164*   **8.11** Functions and Distributions *166*   **8.12** Summary *167*

CHAPTER NINE GAP ACCEPTANCE

9.1 Introduction *173*   **9.2** Merging Parameters *177*   **9.3** Delay Models *180*   **9.4** Critical-gap Distributions *187*   **9.5** Gap Acceptance Functions *190*   **9.6** The Binomial Response *193*   **9.7** The Probit Method *194*   **9.8** Gaps and Lags *196*   **9.9** Multiple Entries *197*   **9.10** The Ideal Merge *198*   **9.11** Speed of the Merging Vehicle *199*   **9.12** Angular Velocity Model *202*   **9.13** Two-variable Probit Analysis *204*   **9.14** Effects of Geometrics *208*   **9.15** Angle of Convergence *209*   **9.16** Acceleration-lane Length *211*   **9.17** Accepted-gap Number *214*   **9.18** Application to Design *215*

CHAPTER TEN QUEUEING PROCESSES

10.1 Introduction *223*   **10.2** The Single Channel *224*   **10.3** Application of the Generating Function *227*   **10.4** Finite Queues *229*   **10.5** Waiting Times *230*   **10.6** Multiple Channels *233*   **10.7** Moving Queues *236*   **10.8** Entrance Ramps *240*   **10.9** The Markov Process *244*   **10.10** Significance of Queueing Service Rate *246*   **10.11** Nonexponential Distributions *248*   **10.12** Status of Queueing *251*

CHAPTER ELEVEN SIMULATION

11.1 Introduction *255*   **11.2** Monte Carlo Methods *256*   **11.3** Random Walks *258*   **11.4** Simulated Sampling (Simulation) *261*   **11.5** Random Numbers *263*   **11.6** Generation of Pseudorandom Numbers *263* **11.7** Number Systems *264* **11.8** Power-residue Method *265* **11.9** Random Sequences Satisfying Desired Distributions *266* **11.10** The Method of Inversion *267* **11.11** Point-distribution Method *269* **11.12** Discrete Random Deviates *270* **11.13** Special Conversion Methods *272* **11.14** Scanning Techniques *274* **11.15** Steps in Simulation *276* **11.16** The Simulation Program *278* **11.17** Model Calibration *283* **11.18** Why Simulate *284* **11.19** An Example—Freeway Merging *286*

CHAPTER TWELVE DETERMINISTIC RELATIONSHIPS

12.1 Introduction *298*   **12.2** Curve Fitting *300*   **12.3** The Boundary-condition Approach *302*   **12.4** The Heat-flow Analogy *304*   **12.5** The Fluid-flow Analogy *305*   **12.6** The Moving-vehicle Method *312*   **12.7** Shock Waves *314*   **12.8** The Bottleneck-control Approach *315*   **12.9** Stream Measurements *319*   **12.10** System Models *324*   **12.11** Challenges *326*

CHAPTER THIRTEEN THE MAN-MACHINE SYSTEM

13.1 Introduction *330*   **13.2** Feedback Control *331*   **13.3** The Laplace Transform *332*   **13.4** System Stability *332*   **13.5** The Human Servomechanism *333*   **13.6** Time Lags *336*   **13.7** Linear Car-following *337*   **13.8** Nonlinear Car-following Models *341*   **13.9** Car maneuvering—A New Concept *342*   **13.10** The Analog Simulator *346*   **13.11** Determination of Car-following Variables *350*   **13.12** Challenges *352*

CHAPTER FOURTEEN ENERGY-MOMENTUM APPROACH TO LEVEL OF SERVICE

14.1 Introduction *355*   **14.2** The Parameter *359*   **14.3** Mathematics of Acceleration Noise *362*   **14.4** Momentum–Kinetic Energy *364*   **14.5** Internal Energy *368*   **14.6** Study Methods *372*   **14.7** Evaluation of Geometrics *374*   **14.8** Acceleration Noise Due to Traffic Interaction *374*   **14.9** Verification of Energy Model *376*   **14.10** Capacity and Level of Service *378*   **14.11** Relationship of Energy and Fuel Consumption *383*   **14.12** Challenges *384*

CHAPTER FIFTEEN CREATIVE DESIGN

15.1 Introduction *388*   **15.2** Freeway Design *389*   **15.3** The Freeway System *393*   **15.4** Interchange Requirements *393*   **15.5** Main-lane Requirements *399*   **15.6** Reverse-flow Freeway Schematics *404*   **15.7** Downtown Distribution System *408*   **15.8** Geometrics of the Reverse-flow Interchange *411*   **15.9** Summary *416*

CHAPTER SIXTEEN SURVEILLANCE AND CONTROL

16.1 Introduction *420*   **16.2** Freeway Operations *420*   **16.3** Freeway Surveillance and Control *422*   **16.4** The Gulf Freeway Project *424*   **16.5** Surveillance Projects as Research Facilities *424*   **16.6** The Evolution of Ramp Control Criteria *427*   **16.7** Merging Control *434*   **16.8** Merging Control System *435*   **16.9** Prototype Merging Control Installation *438*   **16.10** Application to Bus Rapid Transit *443*

Name Index *449*

Subject Index *453*

CHAPTER ONE
INTRODUCTION

*I have a great subject to write upon, but feel keenly my
literary incapacity to make it easily intelligible without
sacrificing accuracy and thoroughness.*

Sir Francis Galton

1.1 A NATION ON WHEELS

There are many indications of the complexity of living in today's world.
The population of the world has doubled in the last century. There is
even a greater increase in the number of city dwellers as opposed to those
who stay on the land. We travel about 15 times faster and more often
than we did 100 years ago. Fortunately, our productivity has increased
at a rate commensurate with the problems we have created. We have
developed more and more complex solutions to the problems that perplex
us. Scholars who have attempted to document our rampaging expansion
tell us that we are living in the Golden Age of Science.

Although this complexity manifests itself in virtually every aspect of
modern life, nowhere is it as dramatically exhibited in our society as on
our streets and highways. Among America's 200 million people, half are
licensed to operate motor vehicles. During 1965, they drove 90 million
autos, trucks, and buses on almost 4 million miles of roads and streets,
traveling 800 billion miles. There is a vehicle for every two persons in the

United States, 25 vehicles for every mile of road, and a mile of road for every square mile of land.   More pertinently, the car population has risen by almost 50 million since World War II, growing an average 5.7 percent a year while population increased by 1.7 percent a year.   Millions of families have bought their first car or their first second car or their first third car. Great fleets of new cars will continue to cascade onto United States highways, but eventually, a point of saturation comes —it is hoped at a ratio not exceeding one car for every person who can drive.   Then the growth of the auto population will be tied to, and limited by, the growth of human population.   Building roads for this controlled total becomes a definable, if enormous, job.

The private car is, and for scores of years to come will be, the most used form of transportation.   Nowhere is this acknowledged so dramatically as in the Interstate System.   This system represents the biggest public work of all history.   Pavement area of the Interstate System, assembled in one huge parking lot, would be 20 miles square and could accommodate two-thirds of all motor vehicles in the United States.   One and one-half million acres of right-of-way will be used, and the total excavation will move enough material to bury the whole state of Connecticut knee-deep in dirt.   Enough concrete will be poured to lay a two-lane highway to the moon.   The steel will take 30 million tons of iron ore, 18 million tons of coal, and 6.5 million tons of limestone.

The effect of highway transportation on our lives is indeed profound. Transportation costs rank, for example, along with food and lodging as the principal expense for the average American family.   The President, recognizing the specialty of transportation, has created a Department of Transportation Cabinet post.   Indeed, the countries of the world can be ranked according to their transportation capabilities.

1.2 THE PROBLEM

Any one driver's traffic experience on a bad day can make it seem that the United States is well on its way to hell on wheels.   And indeed, maybe we are!

Fifty thousand people were killed on our roads and streets in 1965; another 2 million, a 10 percent increase over 1964, were injured.   An equivalent death toll at sea would require the sinking of 15 to 20 giant ocean liners with all hands aboard.   In the air it would require around 500 jet airplane crashes, or over one a day.   It is a gruesome coincidence that the number injured in 1965, according to the National Safety Council, exactly equaled the total number of hospital beds in the United States.

The total cost of these accidents was $8.5 billion —exactly the same as the year's total outlay for highway construction and improvement.

Actually, the accident record is more a symptom, a "side effect," of the highway transportation problem. Fortunately, most of the improvements in highway and vehicle design, over the years, have benefited both transportation efficiency and safety; otherwise the macabre statistics would have been worse. Thus, using the best relative measure of the accident problem—the number of deaths per 100 million miles of travel— some consolation may be taken in the fact that from a horrifying 15 in 1935 it has been reduced to about 5 in 1965.

Another symptom of the problem is congestion. The expanses of concrete built to unclog traffic often are jammed almost from the moment they open. The Long Island Expressway and Hollywood Freeway, both designed to carry an anticipated 100,000 vehicles per day in 1970, now move about twice that number daily. There is a kind of Parkinson's law that causes any new highway facility to overfill the instant it opens.

Everyone knows that the urban transportation problem exists because of the expansion of auto transportation far exceeding the provisions for highway and terminal facilities—because of population growth, migration to metropolitan areas, increased ownership and use of the auto, suburban movement, decline of public transportation, concentration of auto trips in time (the peak hours) and space (freeways), and the lack of funds and facts. Everyone knows we have a problem!

1.3 THE CHALLENGE

The transportation problem described in the previous section is, we must remember, a product of progress, prosperity, and personal freedom. In a country where jobs and housing are assigned and citizens are neither permitted nor can afford to buy automobiles, there is no urban transportation problem. There are, by the same token, no civic groups in such a society clamoring over the decline of a central business district. There are no save-our-trees aesthetes and no handwringing over the loss of customers by public transit.

In attacking the problem, one must consider an important constraint. Contemporary American society is truly mobile. This mobility inherent in this society is so obvious and has been discussed so often that mention of it seems trite. Yet, in attempting to understand the plight of urban public transit, one must appreciate the high value Americans have placed on this personal mobility. It is for this reason that small time-cost savings seldom entice a motorist to abandon his auto in favor of public transit.

The question, in metropolitan areas where automated public transit systems at up to $15 million per mile are being contemplated to woo motorized commuters away from their cars, is whether anybody wants to make the switch.   Most drivers view it as a solution to the problem to the extent that other drivers, not they, will leave their cars in their garages. The average American views transit as part of the cold city and his car a part of his own cozy home.

We must remember that even at its worst, city traffic in most places is moving better than it ever has. Peak-hour speeds in such freeway-oriented cities as Los Angeles and Houston are up from an average of 20 mph in 1955 to over 30 mph 10 years later.

The solution lies in giving urban dwellers the mode of transportation they want.   The facts are that the trend is toward even larger increases in crosstown trips and relatively sharp decreases in downtown trips, which will effect substantial changes in the nature of the downtown area. Employment there will not keep pace with increases in the suburban areas. New commercial and industrial developments will be in relatively low-density areas.   These trends will produce densities that will not permit heavy reliance on public transit.   A flexible highway system with its free-ways and arterial streets must of necessity be the backbone of the trans-portation system.   Transit service to downtown areas may have to be improved in some cases in frank recognition that transit has an essential, if limited, role to play in the solution of the urban transportation problem.

The transportation system of the future will be a rapid transportation system; however, it will not be a public rapid transit system.   The system referred to will consist of motorists; their private vehicles; and multilevel surface streets, underground highways, and super freeways spanning and tunneling commercial and industrial areas.   By A.D. 2000, over 300 million wealthy Americans living in a completely automated pleasure-oriented society will thank us for providing them with a choice in transportation mode, thus ensuring their personal mobility.

1.4   ROLE OF TRAFFIC ENGINEERING

Through the centuries, and into the early period of the automobile, the engineering approach to highways was almost exclusively structural.   With the advent of the modern motorcar, the functional requirements of the new type of traffic gradually became recognized.   Emphasis began to shift from the static aspects of road design to the dynamic factors so important to efficient movement.   However, since the majority of our key highway and street systems were built under the old philosophy, it is not surprising that we seem to face a perpetual traffic crisis.

Because the street plan of cities is impotent to cope with the traffic upon them, there have been various methods devised to control operation. The installation and administration of these controls, so essential to maintaining the movement of vehicles, is the popular conception of *traffic engineering*. Actually traffic engineering is much more. It is the art and science of estimating traffic demand and highway capacity and measuring and determining the relationships between traffic variables—and the application of this knowledge to planning, designing, operating, and administering highway facilities in order to achieve the safe and efficient movement of persons and goods. Note the use of both the words *art* and *science*. Art is knowledge made efficient by skill; science is systematized knowledge. This text accentuates the latter in the belief that the traffic engineer is a planner, not a deviser; a builder, not a repairman; a surgeon, not a first-aid traffic corpsman.

The choice of *systematized* above is significant. One is impressed with the use of such terms as *systems, components, analyses, networks,* and *optimization* in discussing traffic problems. Because of the complexity of the traffic phenomena and the importance of highway transportation, researchers from widely varying disciplines have become interested in traffic engineering and have contributed immeasurably to the understanding of traffic interaction and movement. Unfortunately, these scientists— predominately mathematicians, psychologists, statisticians, and physicists —often have little appreciation for the practical constraints placed on the traffic engineer. There is a need for more communication between these theorists and practitioners.

Traffic engineering is a dynamic human activity with the purpose of benefiting and serving mankind. It is a growing and changing discipline that becomes more and more complex as new knowledge is gained.

1.5 ORGANIZATION OF THE BOOK

There are five distinct bases on which a book on traffic engineering can be organized. First are the chronological phases through which the traffic-engineering effort passes, such as characteristics, planning, design, and operation—in that order. Second are the logical research steps, such as theory formulation, theory verification, and application. Third are the parts of traffic analyses, such as traffic demand and highway capacity. Fourth are the tools of traffic engineering, such as probability distributions, queueing theory, control theory, and physical analogies. Fifth are the elements of the traffic system, the driver, the vehicle, the road, the intersection, etc.

Actually all these bases—elements, tools, parts, steps, and phases— are employed here. Chapters 2 to 4 develop the basic elements of any traffic situation. Chapters 5 and 6 establish the systems concept and research approach, including organization of traffic theory and the construction of the mathematical model. Chapters 7 and 8 develop the basic tools of probability; followed by Chaps. 9 and 10, the powerful and important tools of queueing theory and simulation. Whereas these general tools are referred to as stochastic models, Chap. 10 treats a particular application of the stochastic approach peculiar to traffic —that is, the concept of gap acceptance. Gap acceptance is basic to almost every traffic maneuver and is therefore fundamental to the description and understanding of traffic interaction.

Chapters 12 to 14 illustrate what is called the deterministic approach to traffic problems and include the utilization of such tools as curve fitting, differential equations, control-system theory, linear programming, and physical analogies. Chapters 15 and 16 are truly application chapters dealing with design and operations. The phases of traffic characteristics and highway planning are discussed in Chap. 5 rather briefly. The parts of traffic analyses —traffic demand and highway capacity —are introduced in Chap. 5 and are emphasized throughout the text.

CHAPTER TWO
THE VEHICLE

And the gray mare will prove the better horse.

Daniel Defoe

2.1 INTRODUCTION

The traffic engineer, paradoxically, has done relatively little to study and understand the very element that is responsible for his professional existence. In this chapter, we wish to discuss the mechanics of the vehicle. Classical mechanics is usually subdivided into three parts: statics, kinematics, and dynamics. The static properties of the vehicle are its size and weight. Kinematics deals with motion without reference to its cause and is, therefore, practically a branch of geometry. Thus, it is not surprising that the science of proportioning the visible elements of a road to the limitations and performance characteristics of the vehicle is called *geometric design*. Finally, dynamics considers the forces which effect this motion or changes in motion. The basic motions or maneuvers to be considered in the case of a moving vehicle are starting, turning, and stopping.

2.2 STATIC CHARACTERISTICS

Vehicle size affects traffic-lane width, shoulder width, length and width of parking spaces, vertical curve, sight distance, and channelization geometrics. The weight of the vehicle, in addition to being of importance in pavement design, affects fuel consumption and speed-change ability.

Although the average length of domestic passenger cars has increased from 165 in. in 1930 to 210 in. in 1960, the range of 50 in. between the longest and shortest car has not changed during the same period.[1]* A space of 22 ft is usually allowed for parallel parking.[2]

The trend in overall width of domestic cars since 1930 shows an increase up to World War I. Since 1945 an average width of 77 in. with a range of about 12 in. between the widest and narrowest has been maintained. To accommodate vehicles at present speeds, 8-ft parking lanes and 10-ft shoulders are desirable minimums.

The average height of autos has been steadily reduced from 72 in. in 1930 to 55 in. in 1960. The range of 6 in. has not changed during this period. This reduction in the overall height of the vehicle has permitted a proportionate reduction in the center-of-gravity height from 25 to 21 in. and a corresponding increase in stability. The minimum ground-clearance value seems to be established at 5 in.[3]

The average weight of cars has increased from 3,500 lb in 1930 to 4,000 lb in 1960; the range between lightest and heaviest was 2,000 lb in 1960 as against 3,000 lb in 1930. Although fuel economy is related closely with the vehicle weight, streamlining in design and efficiency in operation have effected a 25 percent average reduction in fuel consumption over the three decades considered.[1]

2.3 VEHICLE KINEMATICS

All problems in which force results in motion necessarily involve acceleration. It is therefore necessary to examine some aspects of accelerated motion before considering the forces on a moving vehicle.

If we assume that acceleration is constant, then the following relationships are readily obtainable:

$$\frac{dv}{dt} = a \tag{2.3.1}$$

$$\int_{v_0}^{v} dv = \int_{0}^{t} a \, dt$$

$$v = v_0 + at \tag{2.3.2}$$

* Raised numbers indicate the References at the end of each chapter.

$$\int_0^x dx = \int_0^t (v_0 + at)\, dt$$

$$x = v_0 t + \tfrac{1}{2}at^2 \tag{2.3.3}$$

where a = acceleration
$\quad\quad v$ = speed
$\quad\quad v_0$ = initial speed
$\quad\quad x$ = distance
$\quad\quad t$ = time

Equations (2.3.1) to (2.3.3) are the acceleration-time, speed-time, and distance-time relationships respectively for uniformly accelerated motion. The expression for distance as a function of speed may be obtained from (2.3.1) as follows:

$$a = \frac{dv}{dx}\frac{dx}{dt} = \frac{dv}{dx}v$$

$$\int_0^x a\, dx = \int_{v_0}^v v\, dv$$

$$x = \frac{1}{2a}(v^2 - v_0^2) \tag{2.3.4}$$

These relationships are illustrated in Fig. 2.1.

2.4 NONUNIFORM ACCELERATION THEORY

If, instead of assuming that acceleration is constant, we assume that acceleration is allowed to vary inversely with speed, the equations for the fundamental acceleration-, speed-, and distance-time relationships may be derived as in the previous section. The following differential equation expresses the inverse relationship between acceleration and speed:

$$\frac{dv}{dt} = \alpha - \beta v \tag{2.4.1}$$

where α and β are constants. The units of the constants α and β^{-1} are those of acceleration and time, respectively, and by setting first $v = 0$ then $dv/dt = 0$ it is apparent that α is the maximum acceleration and α/β the maximum possible speed attainable.

If the vehicle is moving at a speed v_0 at time $t = 0$, then the limits of integration for (2.4.1) are

$$\frac{1}{-\beta}\int_{v_0}^v \frac{-\beta\, dv}{\alpha - \beta v} = \int_0^t dt$$

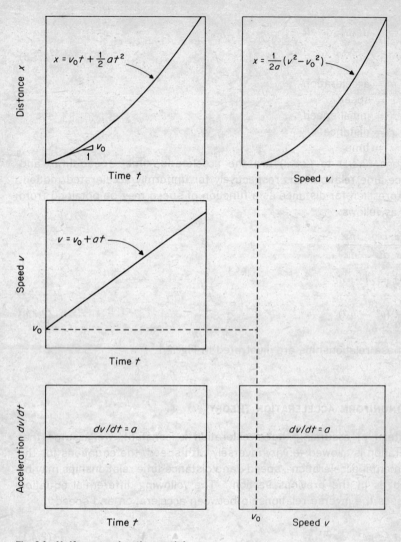

Fig. 2.1 Uniform acceleration model.

and the speed-time relationship is

$$\frac{\ln (\alpha - \beta v)}{-\beta} \Big|_{v_0}^{v} = t$$

$$\frac{\alpha - \beta v}{\alpha - \beta v_0} = e^{-\beta t}$$

$$v = \frac{\alpha}{\beta} (1 - e^{-\beta t}) + v_0 e^{-\beta t} \tag{2.4.2}$$

Since $v = dx/dt$, integration of (2.4.2) provides the equation of distance as a function of time,

$$x = \frac{\alpha}{\beta} t - \frac{\alpha}{\beta^2} (1 - e^{-\beta t}) + \frac{v_0}{\beta} (1 - e^{-\beta t}) \qquad (2.4.3)$$

Finally, substitution of (2.4.2) in (2.4.1) gives the acceleration-time relationship for a speed-change maneuver,

$$\frac{dv}{dt} = (\alpha - \beta v_0)e^{-\beta t} \qquad (2.4.4)$$

The forms of Eqs. (2.4.1) through (2.4.4) are illustrated in Fig. 2.2.

Probably the most appreciated feature of the domestic passenger car is its flexibility in the traffic stream.   This flexibility is manifest in the accelerating capability of the vehicle.   It is a significant factor in the use of entrance ramps, in passing on two-way two-lane highways, and in using modern traffic facilities in general.   The behavior of a motor vehicle, starting from an initial speed and accelerating as rapidly as possible to its ultimate maximum speed, can be rather accurately expressed by the non-uniform acceleration model derived in this section.

2.5 VEHICLE DYNAMICS

The forces acting on the vehicle are illustrated in Fig. 2.3.   The forces resisting the movement of the vehicle are air resistance,[4] rolling resistance, grade resistance,[5] and friction resistance.[6]   Rolling resistances are those forces inherent in the vehicle itself that tend to retard its motion.   The grade resistance is simply that component of the weight of the vehicle acting in the plane of the roadway.   The tractive-effort force is equal to the force supplied by the engine less certain internal-friction losses; it is the force available to overcome the resistance forces and to move the vehicle forward.   The accelerating performance of the vehicle is determined by the excess of tractive effort over these resistance forces.[7]

All the forces considered above are assumed for most purposes to be acting on the center of gravity of the vehicle.   The force of friction, however, acts between the tires of the vehicle and the pavement surface of the roadway.

It is friction that makes it possible for a driver to start, stop, and maneuver his vehicle.   Friction may be defined as a force which opposes motion.   There are two principal kinds to be considered: sliding friction and rolling friction; and each may be divided into static and kinetic varieties.

Sliding friction is represented by countless phenomena, including slowing down our automobiles when the brakes are applied.   Static

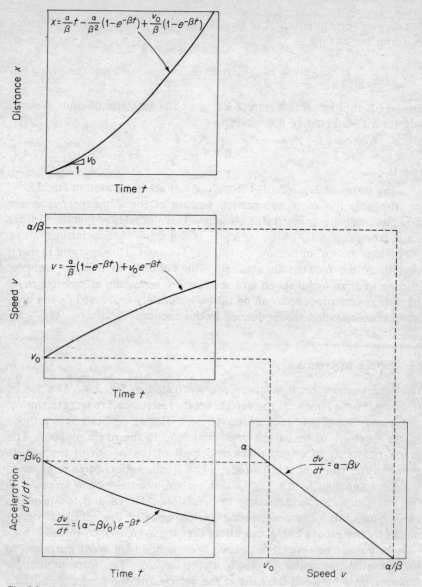

Fig. 2.2 Nonuniform acceleration model.

sliding friction is greater than the kinetic form. Because of the slight roughness, the two bodies seem to settle into more intimate contact when at rest. We know from elementary physics experiments that after motion begins just as much force is needed to pull a body rapidly as slowly over a surface. This means that, theoretically, sliding friction is independent of

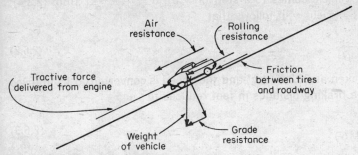

Fig. 2.3 Forces on a moving vehicle.

the relative speed of the surfaces in contact—a fact that tends to justify application of the uniform-acceleration model described in Sec. 2.3 as a model for braking.

2.6 BRAKING

The conditions for braking for a vehicle traveling uphill are illustrated in Fig. 2.4, where W is the weight of the vehicle, v_0 is the initial speed in feet per second at the beginning of braking, f is the coefficient of friction between the pavement and tires, γ is the angle of incline, G is the percent grade divided by 100 (equal to $\tan \gamma$), g is the acceleration due to gravity, x is the distance on the incline, and D_b is the horizontal braking distance in feet. The braking distance is obtained by solving the equation of the forces on the vehicle in the plane of the incline:

$$\frac{W}{g} a + Wf \cos \gamma + W \sin \gamma = 0 \qquad (2.6.1)$$

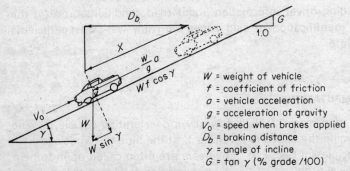

W = weight of vehicle
f = coefficient of friction
a = vehicle acceleration
g = acceleration of gravity
V_0 = speed when brakes applied
D_b = braking distance
γ = angle of incline
G = $\tan \gamma$ (% grade /100)

Fig. 2.4 Conditions for a vehicle braking on a grade.

Because the vehicle comes to a stop $(v = 0)$, $a = -v_0^2/2x$ from (2.3.4), and since $D_b = x \cos \gamma$, it follows that

$$D_b = \frac{v_0^2}{2g(f + G)}$$ (2.6.2)

Where g is taken as 32.2 ft/sec² and the speed is converted to V_0 in miles per hour, the braking distance in feet reduces to   *1 mile = 5280'*

$$D_b = \frac{V_0^2}{30(f + G)}$$ (2.6.3)

Actually the coefficient of friction between the tires and pavement is not constant during deceleration. In applying the equations for braking distance the f factor is used as an overall value that is representative for the whole of the speed change. Actual measurements show that f is not the same for all speeds but varies inversely with speed. In stops from high speeds the actual braking distance is dependent upon the vehicle braking system rather than the skidding of the tires on the pavement. Brake fade—the temporary reduction in braking effectiveness because of heat generated—and such physical elements as tire pressure, tire tread, type and condition of pavement, and the presence of moisture, mud, or ice affect the coefficient of friction considerably.[8]

There is evidence of considerable research concerning braking and skid prevention. The braking performance of motor vehicles in everyday traffic—new vehicles in optimum condition and used vehicles before and after brake-system maintenance and improvement—has been reported.[9] The effect on brake performance of such factors as speed, weight, vehicle type, brake type, and the braking testing device itself must be considered in the intelligent selection of the values of performance to be used.

When a skid is developed at high speeds, the auto becomes very difficult to control, often resulting in accidents. Investigators[10] studying skidding and its prevention recommend (1) having all jurisdictions report skidding on accidents, (2) having skid instruction included in driver education, (3) providing driver information about road conditions, and (4) the isolation and identification of skid zones with low wet-skid coefficients within each state.

2.7 SPEED COMPUTATIONS

The length of skid marks D_b laid down by a decelerating vehicle may be used to determine the speed of a vehicle V at the time the wheels locked, using the principles of Sec. 2.6. Engineers are often called upon to help police in technical problems of this type associated with accident analyses.

The skid marks for each tire are measured and divided by 4 to determine the braking distance D_b.   Then the coefficient of friction f may be determined by a trial skid from a known speed V_t under similar conditions:

$$f = \frac{V_t^2}{30D_t} - G \tag{2.7.1}$$

The unknown speed is obtained by substituting (2.7.1) in (2.6.3) and solving for V:

$$V = \left(\frac{D_b}{D_t}\right)^{\frac{1}{2}} V_t \tag{2.7.2}$$

The results obtained by computing speeds from (2.7.1) and (2.7.2) will always be conservative; that is, the estimated speed will always be lower than the actual speed.   This is true because any reduction in speed before skidding and any speed at impact, in the case of an accident ending in impact, will not be reflected in the length of skid marks.

2.8 IMPACT AND COLLISION

As suggested in the previous section, it is often important to be able to determine the velocity of a vehicle immediately before impact.   Moreover, physical data on the forces and deformations developed in collisions are important in automobile design and geometric designs.   Many serious and fatal accidents result from vehicles striking rigid structures along the highway right-of-way.

Human tolerance to deceleration is the controlling factor in the design specifications for automobile seat belts and shoulder harnesses.   Permissible forces to the human body are related to time (approaching $40g$ for periods of a few hundred milliseconds) and how fast the force builds up (not to exceed $500g$ per second).[11]   Theoretically, the force in g on a driver is given by F, average force of the vehicle at impact, divided by W, the weight of the vehicle:

$$a = \frac{F}{W} g \tag{2.8.1}$$

The average force at impact may be obtained by equating the loss in kinetic energy of the vehicle to the work done in deforming the vehicle on impact:

$$F = \frac{Wv^2}{2gd} \tag{2.8.2}$$

where d is the total deformation.

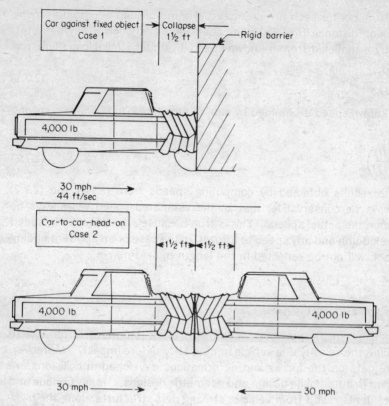

Fig. 2.5 Impact and collision characteristics.

It is known that high-speed impacts do not necessarily mean intolerable decelerating forces on human occupants, because the vehicle structure absorbs large amounts of energy. Impact tests comparing the case of a head-on collision between two cars show that the decelerating forces are about equal (see Fig. 2.5). Note that each car in the second case absorbs the same kinetic energy as if it were hitting a fixed object at 30 mph. This is contrary to the popular conception that the force on each driver is a function of the sum of the speeds of the opposing vehicles.

Tests[12] have shown a need for a closer look into the frame design of automobiles. The front end of some cars tested collapsed upon collision without contributing appreciably to the deceleration of the car. It is believed that the severity of a crash could be reduced significantly if the vehicle frame were designed with an energy-absorbing section.

By the same token, it is apparent from Fig. 2.5 that the severity of single-car accidents in which the vehicle strikes a highway structure can be greatly reduced by increasing the energy-absorbing characteristics of the barrier.   Of great significance are the development of criteria in the design. of guardrails and sign supports based on the reduction of highway fatalities and serious injuries, with property damage to the high-way structures a secondary consideration (see Sec. 4.8).

2.9 CONTROL

Automotive engineers have made great strides in finding what factors are variables in car control, measuring these variables objectively, determining their relative importance, and finding the laws describing their variation.[13] The *slip angle* and *cornering force* are two such variables.

The behavior of vehicles is determined by the slip angles of the front and rear wheels and the relation between them.[14]   A side force due to centrifugal acceleration acting at the center of gravity of the vehicle makes it necessary that the required steering direction be different from the desired path of motion by an angle equal to the slip angle.   If the front-wheel slip angle is greater than the rear-wheel slip angle, the vehicle will move in a gradual curved path with the centrifugal force.[15]   This is called *understeer.*   If the rear-wheel slip angle is greater, the vehicle will move in a curved path against the centrifugal force.   Understeering cars are found to handle better than oversteering cars, requiring less width of lane on curves, for example.[16]   Tire pressures and the distribution of weight between front and rear wheels are factors affecting steering properties.

In order to establish equilibrium, a force to counteract the centrifugal (slip angle) force is needed.   This is called the cornering force and it is achieved by side friction between the tires and pavement.   It is apparent then that the cornering force varies with the slip angle and is dependent on the design of the tire and characteristics of the pavement.   Ironically, cornering power decreases with the depth of tread, so that a worn tire will show higher stability than when it was new.[17]

2.10 TURNING

The problems associated with steering described above are compounded in turning. For example, the rear wheels cannot follow the same arcs as the front wheels in turning, an effect referred to as *off-tracking.*   Since the turning radii for the front wheels are larger than those corresponding to the rear wheels, the radius path of the outer front wheel controls. The *minimum turning radius* of a vehicle, then, refers to the radius of the outer front wheel.   These turning radii may be established by the actual maneu-vering of representative test vehicles,[18] by mathematical analyses, or by

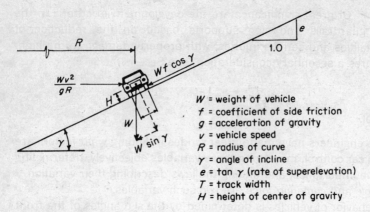

W = weight of vehicle
f = coefficient of side friction
g = acceleration of gravity
v = vehicle speed
R = radius of curve
γ = angle of incline
e = tan γ (rate of superelevation)
T = track width
H = height of center of gravity

Fig. 2.6 Conditions for a vehicle traveling around a curve.

scale models.[19] These vehicle characteristics are important in the design of urban roadways relating to channelization, the designs of parking facilities, and the design of driveways.[20]

The forces on a turning vehicle are a centrifugal force (a radial force acting outward) opposed by a centripetal force provided by the friction between the tires and road. Where the highway is superelevated, the vehicle-weight component parallel to the roadway surface also opposes the centrifugal-force component. These conditions are illustrated in Fig. 2.6.

By equating the components of centrifugal force, centripetal force, and vehicular weight acting parallel to the roadway surface, we obtain

$$\frac{W}{g} a \cos \gamma = Wf \cos \gamma + W \sin \gamma \tag{2.10.1}$$

The fundamental relationship between acceleration a, velocity v, and radius of curvature R for curvilinear motion is

$$a = \frac{v^2}{R} \tag{2.10.2}$$

Substituting (2.10.2) in (2.10.1) and letting $e = \tan \gamma$, where e is the superelevation in feet per foot of horizontal width, yields the basic equation relating superelevation, transverse coefficient of friction, radius of curvature, and vehicle speed:

$$R = \frac{v^2}{g(e + f)} \tag{2.10.3}$$

Taking g at 32.2 ft/sec² and converting the vehicle speed to miles per hour yields the more conventional form

$$R = \frac{V^2}{15(e + f)} \tag{2.10.4}$$

The similarities in the forms of the control equations for the basic maneuvers of stopping and turning are apparent in comparing Figs. 2.4 to 2.6 and Eqs. (2.6.2) and (2.6.3) to (2.10.3) and (2.10.4).

2.11 STABILITY

The stability of a vehicle on a curve may be analyzed by taking moments about the outside wheels (Fig. 2.6).   Since there is no weight on the inside wheels when a vehicle is in danger of overturning, the equation becomes

$$\left(\frac{W}{g} a \cos \gamma - W \sin \gamma\right) H = W \cos \gamma \frac{T}{2} \tag{2.11.1}$$

Dividing by $W \cos \gamma$ yields

$$\frac{a}{g} - e = \frac{T}{2H} \tag{2.11.2}$$

The right side of (2.11.2) is the *stability factor*, where T is the track width and H is the center of gravity height of the vehicle.
From Eq. (2.10.1) we can obtain

$$\frac{a}{g} - e = f \tag{2.11.3}$$

Combining (2.11.2) and (2.11.3) we get

$$f = \frac{T}{2H} \tag{2.11.4}$$

The interpretation is this: A vehicle will slide if its dimensions are such that $T/2H$ is greater than the coefficient of friction f developed.   The amount of friction developed depends on the motion of the vehicle a/g, the amount of superelevation e, and the characteristics of the tires and surface.   Thus, theoretically, even a fast-moving vehicle on an unsuperelevated curve would not overturn if the surface were icy because not enough friction could be developed to resist sliding.

Stonex[21] compares the effect on the stability factor $T/2H$ of varying the track width T and the center of gravity height H throughout the practical ranges of both variables.   This comparison can be visualized by taking the partial derivative of (2.11.4) with respect to both T and H, obtaining $1/2H$ and $-T/2H^2$, respectively.   The effect of lowering the height of the center of gravity is of more importance on this variation than is the effect of changing the track. It is not surprising that the relative stability of the current cars has been achieved largely by virtue of a lower center of gravity height.

The stability concepts of this section are not limited to horizontal-curve and superelevation design applications. About one-third of the 12,000 to 13,000 highway fatalities in the United States each year occur in noncollision accidents, most of which involve vehicles leaving the highway. The forces on a car sliding down a side slope of a highway embankment are indicated in Fig. 2.7. In contrast to the conditions on a superelevated curve (Fig. 2.6), the fundamental equation becomes

$$\frac{a}{g} + e = \frac{T}{2H} = f \tag{2.11.5}$$

where e is the slope ratio of the side slope. If we assume that the car leaves the road at some angle ϕ and travels along the side slope and curvilinear ditch at this angle, the radius of curvature of the path of travel may be expressed as a function of ϕ and $\tan^{-1} e$. Using the fundamental expression for curvilinear motion (2.10.2) to solve for a, we have in Eq. (2.11.5) the relationship among the stability factor of the vehicle, the roadside slope, and the roadside surface as a means of predicting whether a car leaving the highway at a known angle and speed will roll over. This provides the designer a rational approach to roadside design.

2.12 VEHICLE POWER SYSTEMS

The power source and application method determine to a great extent the

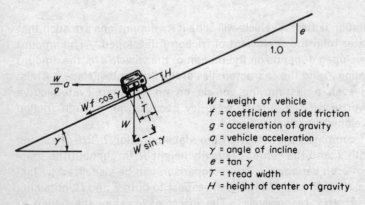

W = weight of vehicle
f = coefficient of side friction
g = acceleration of gravity
a = vehicle acceleration
γ = angle of incline
e = tan γ
T = tread width
H = height of center of gravity

Fig. 2.7 Conditions for a vehicle sliding down a side slope.

vehicle's design, operating cost, and performance characteristics. The basic power and propulsion system used today is the self-contained gasoline or diesel engine which propels the vehicle by means of a mechanical drive. The gasoline reciprocating engine used in automobiles is characterized by a relatively low power output per thermal unit of energy input. The reciprocating diesel engine used in trucks has a somewhat higher thermal efficiency and the advantage of using lower-cost fuels. There is relatively little opportunity for major improvement in either except in efforts to eliminate smog-producing unburned hydrocarbon and noxious fumes.

Traffic engineers take great pains to show how a reduction in traffic congestion, an increase in flow, or improved geometric design can save millions of dollars but tend to overlook how a 25 percent decrease in fuel consumption would save billions.

Economy in vehicle operation is manifest in reduction in fuel consumption, travel time, maintenance, and depreciation. In the usual analysis these variables are expressed as functions of profile grade, pavement condition, and vehicle classification.[5] Fuel consumption and travel time, for example, have been measured on nearly every possible type of truck and trailer combination, as well as urban and intercity buses, under varying conditions of grade, surface speed, and weight. On an even tonnage basis, trucks show more savings in fuel costs and travel time for grade reductions than cars (the savings being much greater for reductions from, say, 9 to 6 percent than a reduction in grade from 6 to 3 percent). Gasoline consumption seems to be related to gross weight, and travel time to the weight/power ratio.[22]

The purpose of these studies in the field of vehicle performance is to establish better geometric-design[23] policies, to program highway improvements based on benefits to road users,[24] and to pioneer advances in automotive design.[25] With reference to the latter, improved power systems seem to offer the greatest potential economies made possible by (1) smaller, lighter-weight, and more reliable power plants; (2) initial capital-cost savings in dollars per horsepower; and (3) reduced cost of maintenance and fuel.

One field of great interest to power engineers is that of turbine-powered cars. The largest application to date has been turbojet-aircraft power plants. Someday these new turbines may well be combined with generators to provide a new, efficient power package.

2.13 CHALLENGES

There is a need to analyze the tire-road interface at the transition between rolling and sliding friction in order to define the tire and pavement characteristics required to reduce sensitivity to lowered coefficients of friction. There is a need to predict and control skidding. If the factors determining incipient skidding can be measured, it becomes possible to predict a skid and hence prevent it.

One attack on the accident problem should be based on minimizing collisions; a second should be based on reducing their destruction. Since a collision between two vehicles or between a single vehicle and a barrier is essentially an energy-transfer and -conversion problem, there is a basic need to define the nature of this process not only in terms of collision forces but also in terms of the energy transfer through the vehicle structure to the occupant.

There is a need for special-purpose vehicles—automobiles especially designed for commuting, for example. Such vehicles might be designed to seat two people, to be considerably smaller than even today's compacts, and to be powered by high-efficiency batteries or chemical cells. Another special-purpose vehicle might be equipped with automatic guidance and control for high-speed (150 mph) intercity freeway driving.

PROBLEMS

2.1. A driver traveling at 30 mph behind another car decides to pass it and presses the accelerator to the floor. Assuming that the accelerating behavior of the vehicle may be described by

$$\frac{dv}{dt} = 4.0 - .05v$$

where v is speed in ft/sec and t is time in sec, find the rate at which the vehicle is accelerating after 3 sec.

2.2. How long would it take the vehicle in Prob. 2.1 to reach a speed of 80 mph?

2.3. In exiting from a freeway, a car is expected to decelerate from 50 to 20 mph on the exit ramp. Assuming $f = 0.2$ and $g = 32.2$ ft/sec² compare the braking distance required on a +4.0 percent upgrade exit ramp from the depressed freeway with that required on an at-grade freeway (0 percent grade).

2.4. The driver of a vehicle traveling 50 mph up a grade requires 28 ft less to stop after he applies the brakes than a driver traveling at the same initial speed down the same grade. If the coefficient of friction between the tires and the pavement is 0.55, what is the percent grade and what is the braking distance on the downgrade?

2.5. One vehicle is following another on a two-lane two-way highway at night according to the rule of thumb which allows one car length of spacing per 10 mph of speed. If both vehicles are traveling at 70 mph and the lead car crashes into the rear of an unlighted parked truck, at what speed will the rear vehicle hit the wreckage? (Assume a reaction time of 0.5 sec and a coefficient of friction of 0.65.)

2.6. Why are friction factors used for calculating braking distances higher than those used for analyzing turning?

2.7. Because of location restrictions, a turning roadway is limited to a radius of 500 ft. If the maximum rate of superelevation is 0.06 and the coefficient of side friction is taken as 0.14, what would be considered a safe speed?

2.8. Two superelevated concentric roadways are separated by a distance of 40 ft between their center lines. If the radius and rate of superelevation for the inside roadway are 1,000 ft and 0.10 ft/ft and the design speeds and friction coefficients are the same for both roadways ($f = 0.16$), find the rate of superelevation for the outside roadway.

2.9. How would you determine appropriate information on tractive resistance based on field tests using your own vehicle?

2.10. How would you determine power requirements for your vehicle for speeds between 20 and 50 mph?

2.11. An undergraduate civil engineering student upon completion of a highway engineering course, calling on his background of the laws of motion from mechanics and stopping-sight distance from highway engineering, hypothesizes that the braking distance of a vehicle is described by the model

$$x = a\dot{x}^2 + b$$

where x = distance from point where braking starts to point where vehicle comes to rest

$\dot{x}$ = vehicle velocity at start of braking

a, b = factors relating x and $\dot{x}$ constant for given vehicle under given conditions

Test and evaluate this model by working in teams and utilizing your own vehicles. Determine a coefficient of friction for the test conducted.

2.12. The behavior of a motor vehicle, starting from a standstill and accelerating as rapidly as possible to its ultimate maximum speed, may be expressed by the experimental equation

$$v = \frac{a}{b}\,(1 - e^{-bt})$$

where t = time, sec

v = speed, ft/sec after t sec

e = base of Naperian system of logarithms

a, b = normalizing factors constant for any particular vehicle

Working in groups and using your own vehicles, perform appropriate experiments to test the above model.

REFERENCES

1. Stonex, K. A.: Review of Vehicle Dimensions and Performance Characteristics, *Highway Res. Board Bull.* 195, 1958.

2. Froehlich, W. R.: Parking Facility Design, *Highway Res. Board Bull.* 195, 1958.

3. McConnell, W. A.: Passenger Car Overhand and Underclearance as Related to Driveway Profile Design, *Highway Res. Board Bull.* 195, 1958.

4. Conrad, L. E., and E. R. Pawley: Air Resistance of Motor Vehicles, *Highway Res. Board Proc.*, vol. 15, pp. 70–80, 1935.

5. Moyer, R. A.: Motor Vehicle Requirements on Highway Grades, *Highway Res. Board Proc.*, vol. 14, pp. 147–186, 1934.

6. Moyer, R. A.: Roughness and Skid Resistance Measurements of Pavements in California, *Highway Res. Board Bull.* 37, pp. 1–35, 1951. (See also Institute of Transportation and Traffic Engineering, University of California, Berkeley, Special Report, "Summary of 1953 Skid Resistance Tests.")

7. Schmidt, R. E.: High Speeds vs. Horsepower, *Traffic Quart.*, July, 1954, pp. 339–350.
8. Normann, O. K.: Braking Distances of Vehicles from High Speeds, *Public Roads,* vol. 27, no. 8, pp. 159–18 ', June, 1953.
9. "Braking Performal ze of Motor Vehicles," U.S. Bureau of Public Roads, 1954, 170 pp. (Conclusions 9–10, F ocedure and Results 15–50.)
10. Skid Prevention Research Committee Reports, *Highway Res. Board Bull.* 219, 1959.
11. Platt, F. N.: "A New Approach to the Design of Safer Guard Structures for Highways," Traffic Safety Department, Ford Motor Company, Dearborn, Mich.
12. Severy, D. M., and J. H. Mathewson: Automobile-Barrier Impacts, *Highway Res. Board Bull.* 91, 1954.
13. Stonex, K. A.: Car Control Factors and Their Measurement, *J. Soc. Automotive Engrs.,* vol. 48, pp. 81–93, 1941.
14. Segal, L.: Research in the Fundamentals of Automobile Control and Stability, *SAE Trans.,* vol. 65, p. 527, 1957.
15. Olley, M.: Road Manners of Modern Cars, *J. Inst. Automobile Engrs. (London),* vol. 15, p. 147, February, 1947.
16. Fox, M. L.: Relations between Curvature and Speed, *Highway Res. Board Proc.,* vol. 17, pp. 202–214, 1937.
17. Bull, A. W.: Tire Behavior in Steering, *SAE Trans.,* vol. 34, pp. 344–350, 1939.
18. Young, J. C.: Truck Turns, *Calif. Highways Public Works,* March–April, 1950, pp. 14–31.
19. Foxworth, De V. M.: Determination of Oversized Vehicle Tracking Patterns by Adjustable Scale Models, *Highway Res. Board Proc.,* vol. 39, p. 479, 1960.
20. Bus Turn Radii, *Mass Transportation,* February, 1953.
21. Stonex, K. A.: Roadside Design for Safety, *Highway Res. Board Proc.,* vol. 39, pp. 120–157, 1960.
22. Time and Gasoline Consumption in Motor Truck Operation, *Highway Res. Board Repts.* 9A, 1950, 75 pp.
23. Willey, W. E.: Uphill Speeds of Trucks on Mountain Grades, *Highway Res. Board Proc.,* vol. 38, pp. 304–310, 1953. (See 1953 *Proc.* for downhill speeds.)
24. "Road User Benefit Analyses for Highway Improvements," pp. 89–112, American Association of State Highway Officials, 1952. (Costs of fuel, tires, oil.)
25. Carmichael, T. J.: Motor Vehicle Performance and Highway Safety, *Highway Res. Board Proc.,* vol. 33, pp. 414–421, 1953.

CHAPTER THREE
THE DRIVER

Americans are always moving on . . .
I think it must be something in the blood.

Stephen Vincent Benét

3.1 LEARNING

Man has certain unique characteristics. These characteristics must be considered if the safe and efficient movement of vehicular traffic is to be achieved. Only by understanding driver behavior is it possible to protect drivers through the rational utilization of traffic control devices (engineering), to train prospective drivers adequately (education), and to improve the behavior of violators (enforcement).

Man can learn in all or one of the following ways. He can learn by practicing a task over and over until he becomes proficient in its execution. He can learn by trial and error, attempting various solutions to a problem and adopting the one that succeeds. Man can also learn by the transfer of previous training.[1]

Man acquires most of his driving skills through practice and the transfer of previous training.[2] The rate at which he learns is mainly dependent on simplicity, understanding, and feedback of results. The simpler the problematical situation with which a driver is faced, the more likely he is to respond correctly.   A good rule is to operate a highway facility so that a driver is faced with but one decision at a time.   This is an application of the first of the three E's, that is, engineering.

The second factor affecting the rate of learning is understanding, which corresponds to the second E, education.   The more a driver is able to understand or acquire insight into a situation, rather than depend on repetitive practice or trial and error, the more rapidly he can learn.[3]   On our early expressways many drivers were slow to learn to use acceleration lanes as an aid to merging because they did not understand their purpose.

The third factor affecting the rate at which man learns is feedback of results.   The more a person gets feedback of results about whether he is correct or incorrect, the less likely he is to acquire bad habits.   For example, to improve at gunnery, a person must know if he is hitting the target. In driving, the feedback of results is manifest in the third of the three E's, enforcement, and lately in formal driver-education classes.

Man is a creature of habit.   Through the impact of past experience, the driver is conditioned by his learning processes, responding to similar situations with predictable patterns of acquired behavior.   Learned responses are powerful factors which may be beneficial if properly developed or harmful if formed irrationally.

3.2   MOTIVATION

People enter the traffic stream for diverse purposes.   Once in the traffic stream, the driver is motivated by the desire for time-distance economy on the one hand and for safety, comfort, and convenience on the other. Fear of accident motivates many driver decisions resulting, for example, in jamming the brakes when confronted with any danger.   Drivers also tend to follow the path of least change of speed and least discomfort resulting, for example, in the cutting of corners at unchannelized intersections.

An operator's capabilities depend on whether he is motivated to perform as effectively as possible.   Motivation is a combination of vigilance and attitude.   Vigilance can be defined as the prolonged ability to detect certain environmental signals.[4]   Attitude is a personality aspect; it is a subjective element.

Several factors can reduce human vigilance and thus impair human performance.[5]   Among these are fatigue and boredom.   Man gets fatigued or bored if he must do the same thing for a long time, and in most tasks his performance deteriorates accordingly.   The principal effect of fatigue and boredom is to reduce vigilance as well as the ability to do one's best.

Fatigue does not necessarily impair the ability to drive well; lack of sleep greatly impairs performance because it lowers motivation.   With fatigue or boredom, a person's attention wanders and the likelihood of having serious accidents increases.   Drivers become more careless when tired and take chances they would not take when rested.   Prolonged uneventful driving and cramped or unchangeable positions are important conditions contributing to fatigue and boredom.[6]

Mental and emotional instability affect motivation.[7]   A man drives as he lives; no driver ceases to be a parent, a hater, a dreamer, or a bully merely because he gets behind a steering wheel.[8]   Emotions can be killers.   Emotional instability or confused emotions caused by many things such as anger, fear, hate, and worry often replace rationality behind the wheel.[9]   Poor driving seems to be a problem of emotional makeup and social inadequacy.[10]   Emotions can motivate driving above the safe speed for prevailing conditions, unnecessary or dangerous weaving, following too closely, or other careless acts.

3.3 ATTITUDE

Although a certain level of intelligence is required to maneuver a vehicle, attentiveness is believed to be more important than superior intelligence.[11] In research conducted to attempt to relate specific personal characteristics to patterns which clearly and consistently distinguish accident repeaters from accident-free drivers, no statistically significant differences were found in physiological reactions, coordination, reaction time, anxiety, IQ, or in most objective personality inventories.[12]   In another study,[13] however, such biographical characteristics as youth, unmarried status, little education, high job turnover, and low income appeared to be important.

Without a doubt, our cultural climate, aspirations, and myths find reflection in the prevailing attitude toward the automobile and contribute in no small degree to the accident rate.   Evidence is mounting that many drivers, because of either ignorance or immaturity, have no desire to stay within the law or obey the rules of the road.   This might be attributed to a lack of public opinion against reckless driving.   There is no law against one person handing another a gun with his finger on the trigger, yet there seem to be customs or conventions developed over centuries about handling guns that rule out this sort of behavior.   On the other hand, many drivers have the attitude they can drive the way they want to provided they are insured.   Much attention[14] has been focused on the under-25-years age group, with three areas of needed action being pointed out: (1) improved driver-education programs to teach good driving techniques, (2) stricter police enforcement of traffic rules and stricter licensing requirements, and (3) attitude changes through proper driving examples by parents.

3.4 VISION

Of the several senses of man, vision is undoubtedly the most important from the point of view of informing him about the world around him. Many traffic operational and design problems require a knowledge of the general characteristics of human vision.

The sense of vision consists of a number of different abilities of the eye. The eye is in many respects something like a camera. Light passes through the pupil, is reflected by the lens, and is brought to a focus on the retina. The retina receives the light stimulus and transmits an impulse to the brain through the optic nerve. Important for driving are visual acuity, peripheral vision, glare recovery, color perception, and depth perception. This is to say that a driver should be able to identify objects; when looking straight ahead, he should be able to detect motion at the sides. He should be able to see his way at night when there is little light — and in the face of glare. Finally, he should be able to distinguish colors on signs and signals and the relative distances to different objects.

One of the most important kinds of data about the eye is visual acuity— the size of detail that the eye is capable of resolving. It is usually specified as the reciprocal of the minimum visual angle subtended by the viewed object. A person with normal 20/20 vision can read standard Snellen chart letters about ⅓ in. high at a distance of 20 ft. Maximum acuity, or the clearest seeing, takes place in a relatively small portion of the visual field. For distinct vision the subtended angle should not exceed 3°, or 1½° on each side of the center line of regard. The limit of fairly clear seeing is 10°; this is the point at which visual acuity drops off rapidly.[15] Traffic signs should be designed and located so that the message falls within this 10° cone of vision.

A driver's visual acuity is affected by the amount of light on the traffic sign, its brightness contrast, its shape, and the amount of time the driver has to look at it. Another variable that affects the ability of people to make visual discriminations is that of movement. Movement of objects in the visual field may occur when the viewer is moving (as in a vehicle), when the object is moving (as a crossing pedestrian), or when both the viewer and object are moving (the usual traffic situation). Goodson and Miller[16] define this deterioration of visual acuity with speed as *dynamic visual acuity.*

The ability to see objects outside the cone of clearest vision is called *peripheral vision.* Driver studies have shown that the angle of peripheral vision varies from 120° to 160°, reducing as the speed of the driver's vehicle increases. The peripheral area is sensitive to motion and light and therefore can serve to alert the driver to events at either side of him. The sensitivity is a function of the angle from the center of vision.[15]

The human eye has extraordinary capacities for seeing small differences between things but is very poor at estimating absolute values. This characteristic of the eye is particularly true in the case of light and shows up in the estimation of size, distance, speed, and acceleration in absolute or numerical terms. Forbes[17] relates the speed-judgment factor to the increasing frequency of rear-end collisions on high-speed highways. The estimates of size, distance, speed, and acceleration are interrelated. Thus, uniformity in the size of control devices aids distance estimation. Standardization of sign shapes and traffic-signal sequences also aids the color-blind driver.

Estimates of the speed of moving vehicles by both pedestrians and other drivers are poor. Little is known about human ability to estimate speed changes except that it is inaccurate and unreliable.

3.5 NIGHT VISIBILITY

Traffic at night has always had accident rates which average approximately twice that of day rates. Night driving involves not only the problems of darkness but hazards of fatigue, drowsiness, and other factors. Good visibility on roadways at night results from lighting which provides adequate pavement brightness with good uniformity and appropriate illumination of adjacent areas, together with reasonable freedom from glare.

The predominant method of discernment on roadways at night is by silhouette.[18] This occurs when the general level of brightness of all or a substantial part of the object is higher or lower than the brightness of its background. Another method of discernment is by surface detail. This occurs when the level of illumination of the sides of the object toward the driver are illuminated enough to make contrasts notable on the surfaces themselves. Usually distant objects are seen by silhouette, and closer objects are seen by surface detail.

When the eye is viewing a field of brightness, light enters the eye. Some of this light is reflected, and stray light in the eye occurs. The ideal seeing conditions are achieved when the whole field of view is as uniform as possible, but even this condition produces stray light in the eye. The effect of stray light in the eye is to superimpose a veiling brightness upon the object viewed, and thereby to decrease the brightness contrast needed for discernment.[19] When a relatively bright light source (direct glare) or its reflected image (specular glare) appears in the visual field, decreased visibility and visual discomfort result.   This phenomenon must be taken into consideration in both vehicle and highway lighting systems.

Vehicle headlights have provided a constant challenge through the years to automobile designers desiring improved visibility without increased glare.   In 1939, the sealed-beam head lamp was introduced as the result of combined work by vehicle and lamp manufacturers.   In 1955, in response to the need for a better low beam, a *fog cap* was added in order to block off light rays emanating forward and upward from the low-beam filament.   The dual head-lamp system, introduced on some 1957 cars, featured increased wattage and improved lens design.[20]   Automotive engineers stress the importance of periodic inspection of head lamps to check proper aim in order to ensure maximum visibility with minimum glare.[21]

Between 1955 and 1959, vehicle manufacturers, in their zeal to increase passenger comfort, equipped millions of automobiles with heat-absorbing tinted windshield glass.   Although eliminating half of the radiant solar energy in the infrared range, the corresponding reduction in light transmission decreased visibility, thus creating a potential safety hazard.[22] Specifically, tests on tinted windshields showed no improvement in glare protection, because of a corresponding reduction in the visibility of the test target.   Moreover, acuity was reduced slightly and depth perception was reduced appreciably.

The fact that tinted windshields do not reduce the effects of glare can be explained by the possibility that the discomfort and reduced visibility during headlight glare may be an adaptation problem instead of a brightness problem.[23]   This opinion is supported by the fact that there is no discomfort when lighted head lamps approach during daylight.   The danger of wearing tinted contact lenses for night driving has been found to be the same as tinted windshields since both prevent light from reaching the sensitive retina of the eye.[24]

Forbes[25] summarizes some of the factors affecting highway efficiency at night.   The increased complexity and difficulty of the driver's task can increase his reaction time by a factor of 2 or 3.   Such increases may be critical in relation to time headways in night traffic on high-volume highways.   Pupillary response time may be a factor in some accidents.   Data from studies[26] of pupil diameter changes show that the dilation response may be up to 4 times as slow as the contraction to light.   It may take up to 9 sec for the pupil to dilate after being exposed to the headlights of an opposing vehicle.   This slow rate of dilation may lead to an almost continuous lowered visibility for dark objects while passing a succession of headlights or streetlights.

Drivers should be made to realize that night driving is a more difficult task than daylight driving and that they must avoid driving under sleep-deprived conditions.   It is important that all lighting, signs, and markings be designed with the characteristics of the eye taken into consideration and with a view to reducing fatigue- and drowsiness-inducing effects as much as possible.[25]

3.6 DRIVER EYE HEIGHT

The trend in recent years toward lower vehicles has been documented in the previous chapter.   It has resulted in a gradual reduction of the height of the driver's eyes above the surface of the roadway, reducing sight distances in many situations.   Representative dimensions for driver eye height are important in geometric design in calculating safe sight distances on vertical curves.[27]   Driver eye height should also be considered in traffic operation.   For drivers whose eyes are less than 4 ft from the roadway surface, horizontal-sight distance may be restricted by guardrails, wing walls, and parked vehicles at many locations.[28]   By the same token, even moving vehicles may obstruct the horizontal visibility of the lower driver in certain traffic-merging areas.

The variability in driver eye height is a function of vehicle character- istics as well as driver characteristics.   Stonex[29] has pointed out that a fair estimate of average eye height can be made from overall vehicle height. During the period from 1930 to 1960 the average height of the vehicle decreased from 67 to 55 in., with a corresponding change in average driver eye height from 59 to 47.5 in.

Because this decrease in eye height meant a decrease in sight distance over crests, AASHO (American Association of State Highway Officials) standards were changed in 1961.   The driver's eye height was changed from 4.5 to 3.75 ft, and the height of object was increased from 4 to 6 in. The latter, justified by improvements in vehicle performance, tended to offset the increase in crest vertical-curve lengths due to lower eye height. A 3.75-ft driver eye height represents a lower limit for about 95 percent of current drivers and as such is a realistic value for design purposes.

Horizontal-sight distance is restricted at many locations by a lower driver eye height.   Lee[28] shows one situation in which a wing wall is the critical factor and another in which the driver's view of traffic is obstructed by a guardrail.   In both cases, from a 4.5-ft height the driver could see traffic plainly, but from a 3.5-ft height he could see nothing but the obstruction.

The view to the front, also, has been reduced with respect to driving in hilly terrain.   With distance constant, the probability of detecting obstacles in the roadway just beyond the crest of a hill has been reduced generally.   The reduction in the height of the vehicle has been accompanied by a lowering of the front seat, making it necessary to consider the operator's position in relation to the controls he must manipulate efficiently as well as his eye height in relation to the amount of environment he must observe.

3.7 HUMAN-FACTORS ENGINEERING

The efficient operation of vehicles depends on human perception and response and on the degree to which the vehicle designer considers the physiological, structural, and functional factors of the driver. The cabs of various types of vehicles illustrate the need for integration of space relations in work. Since truck and bus drivers spend a high percentage of their lives in the cabs of their vehicles, the design of the cabs is of some consequence to both their comfort and safety.

In the case of truck cabs, the two most important factors seem to be space relations and standardization.[30] Accelerators and brakes on some models are positioned so that when a driver has his foot on the accelerator, part of his shoe is under the brake. Similar conflicts between the gear shift and the steering wheel on other models preclude some drivers from applying the brake without first shifting the gear to a position on the right-hand side of neutral. Many models have the steering wheel in such a position that tall drivers cannot get their knee directly under the steering wheel, making it necessary to apply the brake at an angle.

Detailed measurements of the internal spatial dimensions of stock cars have also been compared to human sizing criteria.[31] It was found, for instance, that the areas cleaned by windshield wipers range between 37 and 54 percent of the total area of any given windshield. In some models had the designers been aware that a choice could have been made in favor of the operator, the front rather than the rear seat could have been designed with more generous proportions. The distance from the top of the brake pedal to the lower edge of the steering wheel on most vehicles does not meet the recommended distance. Few vehicles have seats that can be adjusted vertically although differences in the physique of drivers suggest 4 in. of vertical adjustment should be built into all passenger cars.

The ability of the driver to see what is happening about him is obviously vital to vehicle operation and driver safety. All driver decisions resulting in vehicle maneuvers depend entirely on the act of visual perception. Fosberry and Mills[32] have studied the factors that determine the extent to which the driver can see out of his vehicle. In experiments in which two lamps located $2\frac{1}{2}$ in. apart were placed in vehicles and served as the eyes of the driver, shadows cast by the vehicle structure (windshield posts, wiper blades, and perimeter of the windshield) upon a background were used for developing a 180° visibility pattern. Blind spots were easily identified.

The importance of visual perception has been recognized by automobile manufacturers who have increased the area of effective vision in the front, side, and rear windows; introduced tinted glass; and improved the headlight systems.[33]   In spite of the efforts of the automobile manufacturers, research reported by the Harvard School of Public Health[34] suggests that much more can be done.

Since the widest possible range of vision is mandatory, tinted windshields are not recommended for night driving. It is suggested that windshield and windshield-wiper design be modified and augmented in order to approximate 100 percent clearing during inclement weather. Some mode of clearing side windows should also be devised. More careful control of the curvature and quality of glass to reduce distortion is recommended.

3.8 DRIVER BEHAVIOR

Probably the weakest link in the chain of knowledge regarding the traffic phenomenon is the description of driver behavior. Researchers and practitioners still rely heavily on assumptions and theoretical formulations regarding the driver in a traffic situation.

Perchonok[35] describes a stimulus-oriented approach and a response-oriented approach to traffic analyses. The former is based on determining the effect of some preselected stimulus on driver behavior; the latter is concerned with isolating those stimuli that cause a certain response. The stimulus-oriented approach is the conventional one. It is the one taken by the traffic engineer who wants to determine the effect of speed zoning on spot speeds,[31] the effect of lane-closure signals upon driver decision making and traffic flow,[37] or the effect of ramp metering on the merging maneuver.[38]

The response-oriented approach treats causes rather than symptoms. Examples are an investigation of factors (short headways and high relative speeds) that are conducive to lane changing;[35] an investigation to define relevant stimuli (time lag, space lag, relative speed) in gap acceptance;[39] or a study of perceptual factors (offset distance, visual-offset angle, speed) causing the lateral displacement of a vehicle with respect to a fixed object alongside the road.[40]

An increased knowledge of driver behavior is fundamental to rational traffic flow theory and practice. Methodological barriers which hampered progress in this area in the past seem to have been overcome, as is evidenced in general by the literature and in particular by the approach in the following section.

3.9 DRIVER RESPONSE

When an object is placed near the path of a driver, a lateral movement away from the object often occurs as the vehicle passes the object. Michaels and Cozan[40] offer an explanation of the process that the driver carries out in order to locate his vehicle relative to some object.

In isolating the variables involved in this location process (see Fig. 3.1), it is apparent that the lateral distance w is significant in exerting the displacement response. It is equally apparent that w is not the only stimulus, since if it were, all drivers would displace at the instant the object came into sight. We know, however, that this is not true. For example, the longitudinal distance x between the driver and the object enters into the problem, the fundamental relationship being

$$\tan \theta = \frac{w}{x} = \theta \qquad \theta \to 0 \tag{3.9.1}$$

in which the approximation $\tan \theta = \theta$ is true for small angles. According to the model described in (3.9.1.) the driver evaluates the location of the object as defined by only the lateral and longitudinal distances. If this were true, lateral displacement should begin at some fixed distance from the object independent of the speed at which the driver is traveling.

It is apparent that the driver's speed v must be included as a stimulus. Mathematically this is accomplished by writing (3.9.1) in terms of x:

$$x = w \cot \theta \tag{3.9.2}$$

and then differentiating with respect to time t:

$$V = \frac{dx}{dt} = -w \csc^2 \theta \, \frac{d\theta}{dt} = -w \, \frac{w^2 + x^2}{w^2} \, \frac{d\theta}{dt} = -\frac{w^2 + x^2}{w} \, \frac{d\theta}{dt} \tag{3.9.3}$$

The quantity $d\theta/dt$ is the rate of the change of the driver's visual angle. Rearranging (3.9.3) gives
不管是號, 低误 Rate,

$$\frac{d\theta}{dt} = \frac{vw}{w^2 + x^2} \tag{3.9.4}$$

This states in effect that if the driver tracks the object over a period of time, estimating the rate at which the angle is changing, he is subjectively taking into account the lateral distance w, the longitudinal distance x, and his own speed v. If the driver can detect a rate of change in his visual angle, he knows the object cannot be in his path. If there is no such angular velocity, the object is perceived as an obstruction, and the driver must displace.

It is easily shown how the visual-angle model can be generalized to include the cases in which the object is another vehicle, moving in the same direction, the opposite direction, or diagonally (see Fig. 3.1). In these cases, v becomes the relative speed between the two vehicles.

Important applications[40,41] of the generalized model are in the placement of objects along the road (signs, lateral clearance for bridges); speed control (in temporary construction areas through the judicial use of cones, in channelized areas by gradually varying the offset distances to delineators); lane widths on two-lane two-way roads with wide truck loads and the effects of wide truck loads; effects on lane capacity of shoulder objects; and gap acceptance and intersection operation.

Fundamental equation $\dfrac{d\theta}{dt} = \dfrac{wv}{w^2 + x^2}$ where $v = v_a \pm v_b$

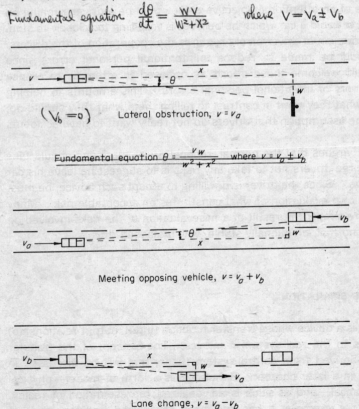

$(v_b = 0)$ Lateral obstruction, $v = v_a$

Fundamental equation $\theta = \dfrac{v\,w}{w^2 + x^2}$ where $v = v_a \pm v_b$

Meeting opposing vehicle, $v = v_a + v_b$

Lane change, $v = v_a - v_b$

Fig. 3.1 Human-factors approach.

3.10 DRIVER ERROR

Many traffic investigations are based on determining what is best for the driver as a means of improving safety and on determining what is best for the traffic stream as a means of improving efficiency.   Driver errors contribute to accidents as well as inefficiency and must be considered in research.

Perchonok[42] defines driver errors as follows: A *personal* error is the difference between observed driver behavior and that which the driver, himself, intended to do, whereas an *objective* error is the difference between observed driver behavior and that which an observer (the traffic engineer or enforcement officer) says the driver should have done.   An *unintentional* error is due to the failure of the driver to perceive the environmental conditions correctly, whereas an *intentional* error is the result of a decision to disregard an environmental condition.   Thus, every error is classifiable as personal or objective and as intentional or unintentional. An example of an intentional objective error would be a driver running an amber light to avoid a car which he believes is following too closely to stop.

The reduction of driver errors requires the cooperation of the drivers. Efforts should be made to reduce unintentional personal errors since drivers would welcome these efforts as opposed to attempts to reduce objective errors or intentional personal errors.   This amounts to helping drivers do what they want in contrast to telling them what they should do, based on the assumption that drivers do not really want to have accidents at all.

Hurst[43] argues that the normal, rational driver must and should take risks.   To urge drivers not to take any risks is to suggest he leave his car in the garage.   Since the driver is unwilling to accept such advice, he must develop his own evaluation of what constitutes an acceptable risk.   Many accidents are simply the result of a misevaluation of the risks involved on the part of normal, emotionally stable, well-motivated drivers.

3.11 DRIVING SIMULATION

A simulator is a device whose transfer function (input/output relationship) is analogous to the real system it mimics.   The concept of the transfer function is derived from control-system theory and will be discussed in more detail in a later chapter.   Simulation is a form of model—physical or mathematical—and as such is an idealized representation of reality. In aviation, for example, it has long been recognized that the study of human performance in the system provides the key to safe operation.

Flight simulators make possible experimental control of flight situations, thus providing the essential conditions for controlled experimental research. The success of simulators in the fields of aviation and space suggests the practicality of the driving simulator.

Hulbert and Wojcik[44] visualize the driving simulator as consisting of an actual automobile with its rear wheels on the steel rollers of a chassis dynamometer, operated in response to two motion pictures. One picture would present the forward view and the other the rearward scene as would be viewed in the rearview mirror. The forward picture would be shown on a curved screen 8 ft from the driver by a projector located 1 ft above the car. The rear picture would be projected from inside the car onto a screen which just covers the rear window. Steering-wheel movements would displace the forward picture from side to side giving the driver the impression of controllability of his lateral position on the highway. Engine noise would be real, and a flywheel attached to the rollers would simulate vehicle momentum, thus contributing to the realism of engine performance and speed changes.

A driving simulator may be regarded as consisting of five components: the vehicle proper, the associated mechanical system for simulating the force environment, the optical system for simulating the visual environment, the feedback system, and a means of measuring results.[45]

The advantages of a driving simulator lie in the ability to acquire facts which would tend to improve both safety and efficiency.[46] It permits research which is either unsafe, impractical, or impossible to do otherwise. It will permit extensive inquiries into human behavior, experimental design, and statistical analyses. The effect of grades and of vehicle weight can be simulated through the introduction of a tilting chair and of suitable inertia into the dynamometer. Thus the driver would experience the sensations that are felt when a vehicle climbs up a hill.

Conclusions obtained from driving-simulation research are still quite modest. Hulbert and Wojcik[47] state that it was possible to conclude that drivers responded differently to straight-road scenes than to winding-road scenes both as to speed and steering-wheel movements. A testimonial to its infancy is that the majority of literature devoted to driving simulation deals with the methodology,[48,49,52] suggested research priorities,[50,51] and possible applications.[52,53]

3.12 PERCEPTION-REACTION TIME

Reaction time in man-machine systems is the time interval elapsing from the beginning of the stimulus signal, including its perception, to the completion of the operator's response.   Some authorities define reaction time as ending when the operator starts to make his response rather than when he finishes it.   For driving situations the former definition is the more logical since the response time is often appreciable.

In driving, as in most dynamic human situations, the dangers of impact have to be perceived, and motion must be continuously evaluated and adjusted.   The driver's response of the next moment is controlled by the position of his vehicle with respect to the traffic environment at the present moment.   Included in this evaluation is some margin of safety, which varies from driver to driver.   For example, there exists at every instant a minimum braking distance, which is fixed by several determinants —most notably, speed.   At every instant there is a zone of free travel, which is fixed by the road surface and geometrics.   If the braking distance is ever allowed to exceed this zone of free travel by the slightest amount, the driver is in danger of crashing.

Perception may fail for a variety of reasons.   Most notably, the physical-stimulus information may be compromised under conditions of poor highway illumination.   Secondly, the available physical-stimulus information may be unreliable or undependable, as it is for the gap selection by a ramp driver for merging into freeway traffic.   The remedy for both is to provide the information by artificial means, by illuminating the environment (see Sec. 4.2), or by providing substitute information in the form of signs or signals (see Secs. 4.7 and 16.10).

Reaction time is probably the most important human parameter in such traffic problems as highway and intersection capacity, safe gaps for maneuvers, placement of control devices, and geometric design.   For example, on a two-lane two-way highway, a driver pulls into the left lane to pass; at the instant he starts the passing maneuver, he detects an approaching vehicle, and his eyes move to center to focus on it (approximate time, 0.1 sec); he sees the object clearly and interprets the image to be a pickup truck (0.3 sec); he selects a course of action, electing to complete the passing maneuver rather than to back off (0.4 sec); and so he presses down hard on the accelerator (0.2 sec).   A total of 1.0 sec has transpired — enough to move the opposing vehicle 176 ft closer at a speed of 60 mph.

Reaction time includes the times required by the operator to perceive, to interpret, to make a decision, and to respond.   Perception and intellection times depend on the number and type of stimuli.   Some traffic-control devices such as railroad-crossing signals appeal to the sense of hearing as well as sight.   Decision time varies widely, depending on the complexity of the decision to be made, but even simple decisions can significantly increase reaction time. For example, it is pointed out that voluntary eye blinks take approximately 20 percent longer than involuntary ones. Response time is a function of the complexity of the response and of the part of the body responding.   Very simple responses, such as pushing a button, involve a few hundredths of a second; but more complex responses take a few tenths of a second.

No single value can adequately represent the reaction time for a given driving task.   Not only are there variations among individuals, but the reaction time of any one person will vary on occasion.   Measurements of reaction time are difficult to make, and published results are few, as far as traffic applications are concerned.   The Ohio State Highway Department[54] tested over 1,000 persons in order to illustrate the effect on stopping distance, at various speeds, of the time taken to apply the brakes after the driver sees a hazard.   The average reaction time for men drivers was 0.57 sec and for women drivers, 0.62 sec.   These times were believed to be less than expected under ordinary conditions, since the drivers were anticipating some kind of stimulus.

The extent of variation in reaction time depends on the particular environmental conditions and age of the person involved as well as sex. In the study reported,[54] the average reaction times were lowest for the age group from 20 to 29 years and highest for ages 60 and over.   It has been found that quick-reacting motorists tend to drive at higher speeds, essentially nullifying their advantage.   Therefore motorists' susceptibility to accidents is independent of their average reaction time.[55]   Although differences between individuals may not be a factor, some authorities believe that the variability in the reaction time of a given driver is an important factor.

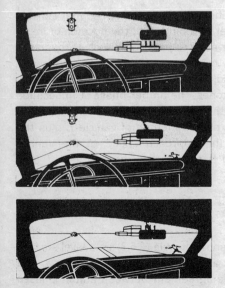

Fig. 3.2 View from eye position of three drivers (5th, 50th, 95th percentile in height).

3.13 THE FALLACY OF THE AVERAGE MAN

Designing and operating to accommodate the average driver is not good practice since by definition about half the population could conceivably suffer. In Fig. 3.2, it is apparent that although the average driver can see both the child and the traffic signal, neither the short driver nor the tall driver can see both.[56]

Few drivers are average in all respects or even in several. If one takes the middle third of a group of drivers in any variable (say eye height when seated, as depicted in Fig. 3.2), then the middle of that third with respect to an independent or uncorrelated variable (say visual acuity), and then repeats the process for a third independent variable such as reaction time, one obtains $\frac{1}{3} \times \frac{1}{3} \times \frac{1}{3} = \frac{1}{27}$, or less than 4 percent of the initial population of drivers who are "average" in only three traits.

3.14 THE RULE OF THUMB

Traffic engineering, like other disciplines, is not without its "rules of thumb." As everyone knows, a driver should follow another vehicle no closer than one car length for each 10 mph of speed. The commonly used rule of thumb for determining the duration of amber is to allow 1 sec of amber for each 10-mph legal approach speed, giving 3 sec for 30-mph approach speeds, 4 for 40, and 5 for 50. Because of its simplicity, the amber-phase rule will be discussed at this time; the safe-following rule will be explored in a later chapter.

Theoretically, the amber phase should be long enough so that a driver approaching the intersection will never find himself in the position of being too close to stop safely and yet too far away to pass completely through the intersection before the red phase commences. Gazis, Herman, and Maradudin[57] refer to these intersection approach areas as *dilemma zones*. In order to eliminate such dilemma zones, the amber phase A in seconds should be set equal to the sum of the following distances expressed in feet (the vehicle braking distance D_b plus the length of vehicle L plus the intersection width W), divided by the approach speed of the vehicle expressed in feet per second, added to the driver's reaction time T (see Fig. 3.3).

$$A = \frac{D_b + W + L}{v_0} + T \qquad (3.14.1)$$

Using the standard braking-distance formula (2.6.3) on a level surface with $f = 0.5$ and allowing $T = 1$ sec and representative values for L and W, it is seen that there is reasonable agreement between amber phases

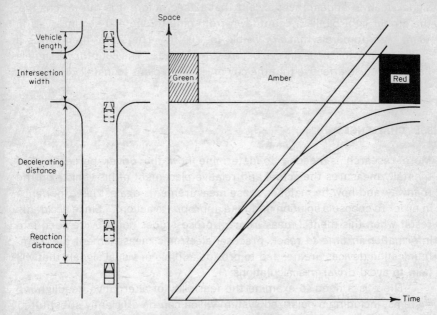

Fig. 3.3   Theoretical conditions for determining duration of amber.

obtained by the rule and those obtained by (3.14.1).   It would seem at first glance that the rule of thumb is therefore justified.   However, closer examination of (2.6.3) shows that the equation behind the rule-of-thumb method assumes an average deceleration during braking of 16 ft/sec^2. Actual observation of motorist response to the amber phase showed that drivers brake with this high a deceleration less than 1 percent of the time. Olson and Rothery[58] concluded that a constant amber phase of about 5.5 sec would be practical for a wide range of approach speeds.   This amber phase would be of such a length that only 5 percent of the vehicle operators who do not choose to stop when confronted with the amber light would fail to clear the intersection before the red phase.

If a moral is to be gained from this, it should be to the effect that stopping distances and deceleration rates used as the basis for design should reflect the behavior of the road user rather than the limitations of the vehicle.   For many motorists, even a properly timed amber phase is more of a challenge than a warning: instead of slowing down, they speed up, trying to beat the light before it changes.   The problem is that there is no way of knowing precisely when the light will turn red.

The answer may lie in a new "countdown" traffic light that has been tested in Abilene, Texas, and in several intersections in Houston, cutting accidents by 50 percent.   It is much the same as an ordinary traffic-light signal head in appearance, except that 12 sec before it is due to change, the amber light blinks a countdown from nine to one at 1-sec intervals in 10-in. high numerals that are visible for 200 ft.   The signal then burns a steady amber for the last 3 sec before turning red.   The countdown signals have proved so effective that the city of Houston plans to install 73 of them.

3.15 CHALLENGES

More research is needed to determine how the driver perceives and mentally measures the actual and relative placement of his vehicle on the roadway and how he matches these measurements against his experience in order to choose a solution and take appropriate action.   Since accidents result when this mental measurement process does not provide adequate information in time to react, practical electronic measurement and communication devices are needed to provide audible or visual signals that will help to avoid driver miscalculations.

There is a need to examine the feasibility of alternative headlighting systems, including a polarized system which can be efficiently substituted for the present system without a long transition period.   Highway illumination should also be studied to determine its compatibility with vehicle lighting in order to develop an optimum night driving environment.

PROBLEMS

3.1. Assuming that a driver with a visual acuity of 20/20 can read a road sign within his area of vision at a distance of 50 ft for each inch of letter height, prepare a table with required letter heights for signs 100 ft, 500 ft, and 1,000 ft for drivers with acuities of 20/20, 20/40, and 20/70.

3.2. A driver with 20/20 vision traveling at 60 mph in the outside lane desires to leave the freeway at exit ramp A.   If 4-in. letters are used, what is the minimum distance the sign should be placed ahead of the ramp?

3.3. If the driver in Prob. 3.2 has 20/40 vision and the sign is placed as computed, what travel speed would allow him to safely turn at the exit ramp?   Assume all conditions such as the driver's perception-reaction time are identical.

3.4. Would you expect your answers to Probs. 3.2 and 3.3 to be different under night-time conditions?   If so, show changes.

3.5. An intersection approach has a 3 percent downgrade and a 4-sec amber phase. The street has a 45 mph speed limit and the cross street is 40 ft wide.   Assuming a driver's reaction time of 1 sec, calculate the average deceleration during braking that can be expected for a vehicle 20 ft long traveling at the legal speed.

3.6. Tests indicate that drivers consider a stop which develops an $f = 0.3$ a comfortable stop.   Define any dilemma zone which may exist for the approach and conditions described in Prob. 3.5.

3.7. What minimum length of amber for a signal on a level tangent will allow a semi-trailer combination (see Table 4.1) that is too close to make a comfortable stop $(f = 0.26)$ to proceed through the intersection described in Prob. 3.5 at a constant speed without violating the subsequent red signal?

3.8. A vehicle traveling at 60 mph was observed to first displace laterally at a distance of 350 ft from a sign located 7 ft to the right of his path. How close would you expect a vehicle traveling at the same speed and driven by the same driver to get to a sign located 15 ft from his path before it displaces?

3.9. Assuming the same threshold rate of change of visual angle as for the driver in Prob. 3.8, how close together would two rows of road cones have to be placed in order to ensure a speed of 30 mph through a school zone?

3.10. The Tri-state Turnpike Authority has received complaints that drivers using the left lane often find it impossible to maneuver to the right lane in time to use the Mid-state exit ramp. As engineer for the authority, you must find what percentage of the left lane drivers who are headed for Mid-state during the peak hour could be expected to be prevented by right lane traffic from making the maneuver, assuming all strangers will not attempt the maneuver until the sign is legible. In order to solve this problem, what information would have to be known and how would you go about getting it?

1. McCormick, E. J.: "Human Factors Engineering," McGraw-Hill Book Company, New York, 1957.
2. Malfetti, James L.: Critical Incidents in Behind-the-wheel Instruction in Driver Education, *Highway Res. Board Bull.* 330, August, 1962.
3. Rainey, Robert V., John J. Conger, and Charles R. Walsmith: Personality Characteristics as a Selective Factor in Driver Education, *Highway Res. Board Bull.* 285, April, 1961.
4. Dobbins, D. A., D. M. Skordahl, and A. A. Anderson: Human Factors Research Report—AASHO Road Test: II. Prediction of Vigilance, *Highway Res. Board Bull.* 330, August, 1962.
5. Dobbins, D. A., D. M. Skordahl, and A. A. Anderson: Prediction of Vigilance: AASHO Road Test, *APO Tech. Res. Note* 119, December, 1961.
6. Lauer, A. R., and Virtus W. Suhr: Effect of Rest Pauses and Refreshment on Driving Efficiency, *Highway Res. Board Bull.* 152, pp. 15–22, 1957.
7. Selling, L. S.: Personality Traits Observed in Automobile Drivers, *J. Criminal Psychopath.*, vol. 1, pp. 258–263, 1940.
8. Van Lennep, D. J.: Psychological Factors in Driving, *Traffic Quart.*, October, 1952, pp. 484–498.
9. Uhlaner, J. E., L. G. Goldstein, and W. Van Steenberg: Development of Criteria for Safe Motor-vehicle Operation, *Highway Res. Board Bull.* 60, pp. 1–16, 1952.
10. Brody, Leon: Personal Characteristics of Chronic Violators and Accident Repeaters, *Highway Res. Board Bull.* 152, pp. 1-2, 1957.
11. Alligaier, Earl: Some Road User Characteristics in the Traffic Problem, *Traffic Quart.*, January, 1950, pp. 59–77.
12. Rainey, Robert V., John J. Conger, Herbert S. Gaskill, Donald D. Glad, William L. Sawery, Eugene S. Turrell, Charles R. Walsmith, and Leo Keller: An Investigation of the Role of Psychological Factors in Motor Vehicle Accidents, *Highway Res. Board Bull.* 212, March, 1959.
13. Heath, Earl Davis: Relationships between Driving Records, Selected Personality Char-

acteristics, and Biographical Data of Traffic Offenders and Non-offenders, *Highway Res. Board Bull.* 212, March, 1959.

14. Wallace, Ralph: Kid Killers at the Wheel, *Colliers,* May 28, 1949.

15. Bartlett, Neil R., Albert E. Bartz, and John V. Wait: Recognition Time for Symbols in Peripheral Vision, *Highway Res. Board Bull.* 330, August, 1962.

16. McCormick, E. J.: "Human Factors Engineering," McGraw-Hill Book Company, New York, 1957.

17. Forbes, T. W.: Driver Characteristics and Highway Operations, *Inst. Traffic Engrs. Proc.,* 1953, pp. 68–73.

18. "IES Lighting Handbook," 3d ed., chap. 20, p. 2, Illumination Engineering Society, N.Y., 1959.

19. Fry, Glenn A.: A Re-evaluation of the Scattering Theory of Glare, *Illum. Eng.,* vol. 49, no. 2, p. 98, February, 1954.

20. Gilgour, T. R.: Some Results of Cooperative Vehicle Lighting Research, *Highway Res. Board Bull.* 255, 1960.

21. Roper, Val J., and G. E. Meese: More Light on the Headlighting Problem, *Highway Res. Board Record* 70, pp. 29–40, 1964–1965.

22. Wolf, Ernst, Ross A. McFarland, and Michael Zigler: Influence of Tinted Windshield Glass on Five Visual Functions, *Highway Res. Board Bull.* 255, pp. 30–46, 1960.

23. Peckham, Robert H., and William M. Hart: The Association between Retinal Sensitivity and the Glare Problem, *Highway Res. Board Bull.* 225, pp. 57–60, 1960.

24. Richards, Oscar W.: Vision at Levels of Night Road Illumination. VIII Literature 1962, *Highway Res. Board Record* 25, pp. 76–82, September, 1963.

25. Forbes, T. W.: Some Factors Affecting Driver Efficiency at Night, *Highway Res. Board Bull.* 255, pp. 61–70, 1960.

26. Forbes, T. W., and M. S. Katz: Driver Behavior and Highway Conditions as Causes of Winter Accidents, *Highway Res. Board Bull.* 161, pp. 18–29, 1957.

27. Loutzenheiser, D. W., and E. R. Haile: Vertical Curve Design, *Highway Res. Board Bull.* 195, pp. 4–8, 1958.

28. Lee, Clyde R.: Driver Eye Height and Related Highway Design Features, *Highway Res. Board Proc.* 39, pp. 46–60, 1960.

29. Stonex, K. A.: Vehicle Data, *Highway Res. Board Bull.* 195, pp. 1–4, 1958.

30. Morgan, C. T., J. S. Cook, A. Chapanis, and M. W. Lund: "Human Engineering Guide to Equipment Design," McGraw-Hill Book Company, New York, 1963.

31. McFarland, Ross A., and Richard C. Domey: Park Adaptation Threshold, Rate and Individual Prediction, *Highway Res. Board Bull.* 298, pp. 3–17, 1961.

32. Fosberry, R. A. C., and B. C. Mills: Measurement of Driver Visibility and Its Application to a Visibility Standard, *Inst. Mech. Engrs. (London) Proc. Automobile Div.,* 1959–1960, no. 2, pp. 50–81.

33. Brody, Leon: The Role of Vision in Motor Vehicle Operation, *Intern. Rec. Med. & Gen. Pract. Clins.,* June, 1954.

34. McFarland, R. A., et al.: "Human Body Size and Capabilities in the Design and Operation of Vehicular Equipment," Harvard School of Public Health, Boston, 1964.

35. Perchonok, K.: "The Determination of Environmental Factors Controlling Driver Behavior," Highway Research Board–Singer, Inc., State College, Pa., 1965.

36. Rowan, N. J., and C. J. Keese: A Study of Factors Influencing Traffic Speeds, *Highway Res. Board Bull.* 341, 1962.

37. Perchonok, K., and P. M. Hurst: "The Effect of Lane Closure Signals upon Driver Decision-making and Traffic Flow," Division of Highway Studies, Institute for Research, State College, Pa.

38. Drew, D. R.: Gap Acceptance Characteristics for Ramp-Freeway Surveillance and Control, *Highway Res. Board Record* 157, 1966.

39. Hurst, P. M., K. Perchonok, and E. L. Sequin: "Measurement of Subjective Gap Size," Division of Highway Studies, Institute for Research, State College, Pa.
40. Michaels, R. M., and L. W. Cozan: Perceptual and Field Factors Causing Lateral Displacement, *Highway Res. Board Record* 25, pp. 1–13, 1963.
41. Taragin, A.: Driver Behavior as Related to Shoulder Type and Width on Two-lane Highways, *Highway Res. Board Bull.* 170, 1958.
42. Perchonok, K.: "The Measurement of Driver Errors," Division of Highway Studies, Institute for Research, State College, Pa.
43. Hurst, P. M.: "Errors in Driver Risk-taking," Division of Highway Studies, Institute for Research, State College, Pa.
44. Hulbert, Slade, and Charles Wojcik: Driving Simulator Research, *Highway Res. Board Bull.* 261, pp. 1–13, 1960.
45. Hulbert, S. F., and J. H. Mathewson: The Driving Simulator, *Inst. Traffic Transportation Eng.*, Reprint 65, University of California, Berkeley, 1965.
46. Fox, B. H.: Engineering and Psychological Uses of a Driving Simulator, *Highway Res. Board Bull.* 261, pp. 14–38, 1960.
47. Hulbert, Slade, and Charles Wojcik: Human Thresholds Related to Simulation of Inertia Forces, *Highway Res. Board Record* 25, pp. 106–109, 1963.
48. Michaels, R. M., and B. W. Stephens: Part-task Simulation in Driving Research, *Highway Res. Board Record* 25, pp. 87–94, 1963.
49. Braunstein, M. L., W. J. White, and R. C. Sugarman: Use of Stress in Part-task Driving Simulators: A Preliminary Study, *Highway Res. Board Record* 25, pp. 95–101, 1963.
50. Fox, B. H., and M. W. Fox: Some Criteria for Priorities of Research in Driving Simulation: Difficulties in Their Measurement and Application, *Highway Res. Board Record* 55, pp. 36–53, 1964.
51. Silver, C. A.: Performance Criteria: Direct or Indirect, *Highway Res. Board Record* 55, pp. 54–63, 1964.
52. Kobayashi, M., and T. Matsunaga: Development of the Kaken Driving Simulator, *Highway Res. Board Record* 55, pp. 29–35, 1964.
53. Fox, B. H.: Some Technical Considerations in Driving Simulation, *Highway Res. Board Bull.* 261, pp. 38–43, 1960.
54. Neal, H. E.: Reaction Time, *Highway Res. Abstr.* 134, pp. 6–7, 1946.
55. Olmstead, F. R.: "Studies of Factors Influenced by Automobile Brake-reaction Time," 22d *Mich. Highway Conf. Proc.*, 1936, pp. 16–27.
56. McCormick, E. J.: "Human Factors Engineering," McGraw-Hill Book Company, New York, 1957.
57. Gazis, D. C., R. Herman, and A. A. Maradudin: The Problem of the Amber Signal Light in Traffic Flow, *Operations Res.*, vol. 8, pp. 112–132, 1960.
58. Olson, P. L., and R. Rothery: Driver Response to Amber Phase of Traffic Signals, *Highway Res. Board Bull.* 330, 1962.

CHAPTER FOUR
THE ROAD

There's a cement octopus sits in Sacramento, I think
Gets red tape to eat, gasoline taxes to drink
And it grows by day and it grows by night
And it rolls over everything in sight
Oh, stand by me and protect that tree
From the freeway misery.

Malvina Reynolds

4.1 INTRODUCTION

The practice of road building dates back to the discovery of the wheel almost 6,000 years ago. By the time of Christ, extensive road systems existed on the continents of Europe, Asia, Africa, and South America. For example, the Appian Way, built from Rome to the sea in preparation for the invasion of Greece, was of such elaborate construction that it is still in existence today after 2,000 years. A trench was excavated, and the pavement was placed in three courses so that the finished surface would be at ground level. The Great Wall of China, which rises 20 ft high and is 10 ft wide at the top, was built not only to defend against the Mongolian hordes, but also to link cities along the 1,000-mile frontier. As such, it marked civilization's first elevated freeway, inspiring the use of the term *Chinese wall* by modern sociologists to describe the barrier effect of elevated urban freeways.

The mating of the vehicle to the needs of man has been a challenge throughout the ages. History tells us that Julius Caesar forbade vehicles from entering the business districts of large cities in the Roman Empire during certain hours of the day to reduce congestion. Europe's early roads charged stiff tolls to pay for improvements, such as sufficient widening "to let a man pass with a dead corpse on a cart."   Early Saxon laws imposed on all lands an obligation to repair roads and bridges and required abutting property owners to maintain the right-of-way. In this manner such present-day concepts as the government's responsibility for highways, the obligations of owners of abutting property, and the rights of the road user were established.

Between the invention of the wheel and the invention of the automobile, the primary concern of road builders was "getting the road user out of the mud."   Only the structural aspects of design were considered. With the 1920s came the concept of traffic engineering, providing gradual curves, smooth highway surface, flat grades, and route markers. This proportioning of the visible elements of the road is called *geometric design.* The third phase of road history is the present period of making the road fit the environment. Land use, the natural setting, social conditions, and human psychology are its concerns. Its application might be called *functional design.* Structural design is related to vehicular load, geometric design to vehicle capabilities and driver requirements, and functional design to the demands of traffic and travel.

4.2 ROADWAY SURFACE AND LIGHTING

The type of pavement is determined by the volume and composition of traffic, the availability of materials, and available funds. The choice of pavement type has some influence on traffic operation and geometric highway design.

There are several factors relating to road surfaces to which the engineer should give special attention in the design, construction, and maintenance of highways for the safe and economical operation of modern high-speed traffic. Some of these factors are road roughness, tire wear, tractive resistance, noise, light reflection, and electrostatic properties. Unfortunately, it is impossible to build a road surface which will provide the best possible performance for all these conditions. A highly skid-resistant surface is usually abrasive and contributes to tire wear. This same surface may have excellent light-reflecting qualities for night driving, but it may have a coarse open surface texture that causes an irritating tire noise at high speeds.

Roughness, raised surface, and color contrast may be used to designate traffic lanes. Drivers tend to seek the smoother surface when given a choice. On four-lane highways where the texture of the surface of the inside lane is rougher than that of the outside lane, passing vehicles tend to return to the outside lane after execution of the passing maneuver. Shoulders or even speed-change lanes may be deliberately roughened as a means of delineation.

Drivers become accustomed to a pavement color, and having driven some distance on a light or dark surface, begin to associate the color with the route. At an interchange the surface color may be used to direct traffic into through lanes. Shoulders and medians should also be chosen to contrast with the traffic lanes. For this purpose, a light aggregate surface treatment is often used on the shoulders of a black asphaltic concrete pavement.

Pavements may be considered as three general types —high, intermediate, and low. For heavy traffic volumes, a smooth riding surface with good all-weather antiskid properties is desirable. The surface should be chosen to retain these qualities so that maintenance cost and interference to traffic operations are kept to a minimum. With high-type pavements there is no raveling at the edges nor loss of effective pavement width due to unequal settlement. Since a smooth surface offers little frictional resistance to the flow of surface water, high-type pavements are designed with minimum cross slopes. On the other hand, low-type rough surfaces must be crowned enough to drain well. Low-type surfaces tend to reduce operating speeds. However, the choice of design speed, on which many features of geometric design depend, should not be unduly influenced by the type of surface to be used. A low design speed should not be assumed solely because of an initial low-type surface, which ultimately may be improved or replaced with a higher type.

Of utmost concern are night traffic operations since the night accident rate is approximately double the day experience. Nightfall brings increased hazards to users of streets and highways primarily because of reduced visibility, although such factors as distraction due to extraneous background lighting, lack of environmental clues, and increased driver fatigue undoubtedly contribute.

Nighttime visibility is provided primarily by contrast. In the case of signing, this contrast is attained by reflectivity or external illumination, depending on whether the sign is a ground-mounted roadside sign or an externally illuminated overhead sign. For the ground-mounted sign reflectorization is achieved by using glass beads or other reflective elements that return part of the headlight beam back to the driver's eyes. Reflectivity enhances the color scheme of signs, but visibility is mainly achieved by contrast caused by the presence or absence of reflectorization. Because of the positioning of overhead mounted signs, reflectorized materials are not always effective, and therefore external illumination is necessary to provide contrast for nighttime visibility.

Illumination is used on many traffic facilities to illuminate the physical features of the roadway and to aid in the driving task. In order to provide satisfactory illumination, light sources must be placed at precise intervals along the roadway. However, the location of these light sources relative to signs may tend to decrease the contrast of sign legend and thus decrease legibility of the sign message. It is important that the effect of luminaire placement on sign visibility be investigated and necessary design criteria be established concerning the relative placement of signs and luminaires.

A luminaire is a complete lighting device that distributes light into patterns much as a garden-hose nozzle distributes water. Proper distribution of the light flux from luminaires is one of the essential factors in efficient roadway lighting. Light distributions are generally designed for a typical range of conditions, which include luminaire mounting height, overhang, longitudinal spacing, arrangement of luminaires, percentage of lamplight directed toward the pavement, maintained efficiency of the system, and the widths of roadway to be effectively lighted.

For effective design the many variables affecting roadway lighting are evaluated and the economic and aesthetic aspects considered. It is important that roadway lighting be planned on the basis of such traffic information as the night vehicular traffic, pedestrian volumes, and accident experience. In addition, some of the factors evaluated are the type of land use development abutting the roadway and distraction of background lighting and such geometric features as the number of traffic lanes, pavement surface, alignment, and channelization.

Spot illumination is usually provided at rural interchanges and channelized intersections. This is desirable, if only to obtain a reduction in the speed of vehicles approaching these types of facilities. Although continuous lighting for an entire interchange area cannot always be justified, it is important to illuminate the areas of geometric and traffic complexities such as the intersections and points of access and egress. In these cases, illumination should be extended beyond the central areas and graduated downward in level of intensity.

Too little attention has been given to the design of lighting systems for urban freeways. In most cases, the design has been merely an extension of the conventional mercury-vapor lighting systems used in the illumination of city streets. Due primarily to maintenance considerations, the 30-ft mounting height has been common practice in city street lighting. But, in order to produce an acceptable light pattern on urban freeways, conventional mercury-vapor luminaires mounted at 30-ft heights must be spaced longitudinally no greater than 160 ft. If this longitudinal spacing is exceeded, there is a significant decrease in uniformity of roadway brightness, which causes a noticeable reduction in visibility. The bright puddles of light under the luminaires and the apparent darkness between them create a "ladder" effect on the pavement.   This brightness pattern often conflicts with the alignment of the roadway and leaves areas of visual uncertainty between the luminaires.   Although required for adequate visibility, the close spacings make freeway lighting expensive and increase the potential hazard of collision with the lighting poles.

Even with the appropriate longitudinal spacing, the 30-ft mounting height has been found to be objectionable from the standpoint of visibility and visual comfort.   When it is used in conjunction with the conventional type 3 mercury-vapor luminaire, a considerable amount of glare is experienced.   Also, the level of adaptation brightness in such a system is much greater than the average roadway brightness, making objects on the roadway difficult to see.   In research conducted by the Texas Transportation Institute, it was found that a system of luminaires mounted at heights of 45 ft provided better visibility and visual comfort than a similar system mounted at 30 ft.[1]   There is a need for extended research in this area.

4.3   GEOMETRIC DESIGN

Geometric design deals with such roadway elements as cross section, gradient, curvature, sight distance, and clearances as well as with combinations of these elements.

There are many traffic factors which influence geometric design. Motor vehicles travel the highway under the control of individual operators, making it imperative that the abilities and limitations of the driver, vehicle, and the road, both individually and in combination, be considered.   However, it is of extreme importance that the need or desire to use the highway and the ability of various highway types to accommodate traffic be anticipated so that sufficient roadway is constructed.   Thus, traffic composition, volume, and speed as well as stopping and turning abilities are significant.   The geometry of the highway facility must be related to traffic performance and traffic demand in order to achieve safe, efficient, and economic traffic operations.

Concepts of geometric design, and other aspects of traffic engineering, emerge both from research and experience.   Two publications of the AASHO, namely, "A Policy on Geometric Design for Rural Highways"[2] and "A Policy on Arterial Highways in Urban Areas,"[2a] are the standards of accepted practice in the United States.   AASHO has emerged as probably the strongest force tending to standardize the differences existing in the various state highway departments.   This is mentioned because possibly the most important single rule in highway design is consistency.   Drivers are conditioned by the road they are traveling and by past experience to expect certain things and not others.   A sharp curve in an otherwise good alignment is the scene of more accidents than a curve of the same radius in poor alignment.   Only by making every element conform to the drivers' expectations and by avoiding abrupt changes in standards can smooth-flowing accident-free highways be conceived.

The physical-site characteristics, traffic data, capacity, and level of service serve to determine the type of facility required to serve the traffic needs, its precise location, and its geometric design.   The balancing of grades, drainage computations, and right-of-way considerations, though equally important, must never take precedence over geometric design. The characteristics of the vehicle serve to establish controls for many of the roadway elements.   The width of the vehicle affects the width of traffic lane; the height of the vehicle affects the clearance of grade separation structures.

A design vehicle is a motor vehicle whose weight, dimensions, and operating characteristics are used to establish highway design controls to accommodate vehicles of a designated type.   The American Association of State Highway Officials[2] has recommended that four design vehicles be used in geometric design.   These vehicles and their dimensions are given in Table 4.1.

Table 4.1   Design vehicle dimensions

Design vehicle			Dimensions, ft				
Type	Symbol	Wheelbase	Front overhang	Rear overhang	Overall length	Overall width	Height
Passenger car	P	11	3	5	19	7	
Single-unit truck	SU	20	4	6	30	8.5	13.5
Semitrailer combination, intermediate	WB-40	13 + 27 = 40	4	6	50	8.5	13.5
Semitrailer combination, large	WB-50	20 + 30 = 50	3	2	55	8.5	13.5

Source: "A Policy on Geometric Design of Rural Highways," p. 86, American Association of State Highway Officials, Washington, D.C., 1965.

The maximum safe speed that can be sustained over a specified section of highway is governed by geometric features of the highway when traffic and weather conditions are favorable and provides another criterion for design.   This assumed design speed should be a reasonable one, because once selected, all elements of the highway should be related to it to ensure balanced design. Sight distances used in both vertical and horizontal controls, as well as horizontal curvature and superelevation, are functions of the design speed.

The choice of design speed is based principally on what drivers will accept. The design speed should not be confused with the average speed, the 85 percentile speed, or the speed limit. Without enforcement, as many as 25 percent of the drivers may exceed a 70-mph speed limit. A low design speed cannot be arbitrarily assigned to all secondary roads, since many may be located in terrains which encourage high speeds. It should be remembered that selected design speeds represent minimum requirements; flatter horizontal curves and longer sight distances should be used if possible. Texas[3] recommends the design speeds shown in Table 4.2 for the main lanes of a controlled-access facility.

Table 4.2 Recommended design speeds for main lanes of a controlled-access facility

Design speed, mph	Desirable	Minimum
Rural		
Flat topography	70	60
Rolling topography	60	50
Mountainous topography	50	40
Urban		
Normal	60	50
Interchanges and downtown	50	40

Source: "Design Manual for Controlled Access Highways," Texas Highway Department.

4.4 THE CROSS SECTION

The complexity of highway cross sections varies directly with the importance of the facility. More through-traffic lanes are required on roads with higher design volumes. The need to maintain the integrity of through traffic on saturated facilities had led to the creation of separate left-turn lanes and the elimination of parking at the edge of through lanes, thus accounting for the increased importance of highway medians and shoulders. The cross section of the highway is composed of many elements, which for the purpose of classification fall into three broad groups: (1) the traveled portion of the road, (2) traffic separators, and (3) the road margins. Dimensions for each element are based on the evaluation of traffic characteristics and the interpretation of the design designation and level of service established for the proposed facility.

Pavement slope   Pavements are sloped from the center to each edge to prevent water from ponding on the pavement.   The effect on steering of slopes up to $\frac{1}{4}$ in./ft are barely perceptible.   Normally, high-type pavements, portland cement, and asphaltic concretes are sloped from $\frac{1}{8}$ to $\frac{1}{4}$ in./ft.   Lower types of pavements require greater slopes and generally range from $\frac{3}{8}$ to $\frac{1}{2}$ in./ft.

Shoulders   Shoulders should be provided continuously along all highways for all emergency stopping.   Experience has shown that paved shoulders are desirable for safety and for less structural failure along the outside of the pavement.   Shoulders of 8 to 10 ft in width should be provided on all except low-type roadways, where shoulders of 4 to 6 ft may be satisfactory. In selecting materials, the color and texture of the shoulders should provide adequate contrast between the shoulder and the adjacent pavement.

Lane widths   Although most design standards still permit, under special conditions, lane widths of less than 12 ft, experience has shown that the minimum width should be 12 ft.   The sacrifice in safety and comfort in providing lanes of less than 12 ft is not justified.

On high-speed rural two-lane roads, lane widths of 13 and even 14 ft have been found desirable.   Widths in excess of 14 ft are not desired because some drivers try to squeeze in another lane.   On multilane roads, 12-ft lanes have been found to be adequate and desirable.

Curbs   Curbs are used to control drainage, delineate proper vehicle paths, and deter vehicles from leaving the roadway or encroaching on certain areas.   Modern designs use curbs mainly in urban areas.

There are two general types of curbs, mountable and barrier.   Barrier curbs are intended to prevent or at least deter encroachments.   Mountable curbs should be designed so they may be crossed easily, without discomfort or undue hazard even at relatively high speeds.   Mountable curbs are generally low (less than 6 in.) and have a flat slope of about 2:1 or flatter.

Barrier curbs range from 6 to 20 in. in height, depending on the nature of encroachment they are to prevent.   They are generally vertical or battered not more than 1:3.   Barrier curbs are used on bridges and as protection around piers or along walls to prevent vehicles from striking them.   Where vehicles are expected to stop adjacent to the curb, its height should not exceed 6 in.

Continuous barrier curbs should be offset at least 1 ft from the edge of the through-traffic lane.   When curbs are introduced on sections of roadway otherwise having no curbs, they should be offset 2 or 3 ft from the through-traffic lane.

Medians   The portion of a divided highway separating the traveled ways for traffic in opposite directions is a median.   It serves to delineate the left extremity of the authorized path of vehicle travel, decreasing the amount of inadvertent vehicle encroachment and providing space for vehicles that run off the left edge of the pavement to regain control.   The exact function that a given median is expected to perform depends on the degree of control of access on the street or highway facility.   A median may serve to provide protection and control of left-turning or crossing vehicles;[4] refuge space for pedestrians, signs, and signals;[5] or refuge space for disabled vehicles.[6]

The importance of a median as space for drainage and snow storage depends on climate, topography, and the number of traffic lanes.[7]   Narrow medians in areas of heavy snowfall may require that most of the snow be pushed across several traffic lanes to the right for storage.   The importance of a median to reduce headlight glare varies with roadway alignment, driving speed, and the extent to which the roadway is lighted by other means.   To this end, landscaping in medians is often used to reduce headlight glare.[8]   The median concept offers flexibility to the planner and the designer by providing space for future expansion of the facility (addition of traffic lanes)[9] or the installation of other types of transportation facilities (rapid transit).[10]   Depending on terrain and land values, medians can provide grade-differential space for separate roadway profiles,[11] structural appurtenances, speed-change lanes, and access ramps.

Medians are sometimes classified as traversable, deterring, and barrier.   A traversable median—usually consisting of paint stripes, buttons, and an area of pavement of contrasting color or texture—does not present a physical barrier to traffic movement.   A deterring median is one which may incorporate any of the features of a traversable median plus a minor physical barrier such as a mountable curb or corrugations.   A barrier median consists of a guardrail, shrubbery, or some type of wall which traffic will not cross intentionally.   With traversable medians, the width should be great enough to prevent most vehicles which are out of control from reaching the opposing traffic lanes.   AASHO recommends 36 ft where topography, space, and right-of-way costs permit.   Using vehicle placement as a criterion of traffic behavior in comparing median designs, it was concluded[12] that a narrow (4 to 12 ft) median with a barrier fence had no significant effect on vehicle placement and was valuable in reducing the severity of accidents involving the median.

Fixed objects   All fixed objects near the road should be eliminated.   Trees should be surrounded by thick cushions of shrubbery and should be grouped to form a streamlined shape.   Guardrails at overpasses should be continuous with the bridge rails, effected through increased harmony between bridge and road designs.

Right-of-way width   In the past, right-of-way widths were selected depending upon the class of the highway to be built.   This arbitrary width then became a controlling factor in design considerations.   Although this practice still prevails to some degree, it is becoming more frequent for the requirements for right-of-way to be established based on the completed design.   Adequate attention must be paid to future requirements if widening or other changes can be anticipated.

In some cases, a lesser design may be determined to be adequate for a number of years; but future traffic is expected to require more lanes, grade separations, divided roadways, or other major changes.   In these cases, economy can be realized through stage construction.   The right-of-way should be procured initially to permit the ultimate development of the roadway.

4.5 VERTICAL ALIGNMENT

Vertical alignment consists of grades and vertical curves.   The grade line is established along a profile usually taken along the center of the pavement and is a series of straight lines connected by parabolic vertical curves. Grade is the number of feet of vertical rise or drop in each station (100-ft horizontal distance), usually expressed as a percent.

Driving practices with respect to grades vary greatly, but modern passenger vehicles are equipped with sufficient power to ascend grades up to 10 percent without an appreciable reduction in speed.[13]   Since grades in excess of 10 percent are rarely employed, the effect of steep grades on speed is essentially a study of the effect of these grades on truck speeds.   Table 4.3 summarizes maximum grade controls employed in terms of design speed for main rural highways and major streets and urban expressways as suggested by AASHO.[2]   Maximum grades for secondary highways may be in the range of 1.2 to 1.5 times those shown in Table 4.3.   However, grades in excess of 5 percent should be avoided, if possible, particularly in snow areas.

Table 4.3   Relation of maximum grades to design speed

Design speed, mph	30	40	50	60	70
Maximum grade, %	6–9	5–8	4–7	3–6	3–5

In addition to speed reduction, an ideal approach would be to balance the added annual cost of grade reduction against the added annual cost of vehicle operation without grade reduction. Although many recent studies[14,15] of vehicle characteristics deal with operating costs, as of this time, these data have not been translated into design warrants.

Although most grade problems in the past were worked out on a vehicle operating basis, today safety needs greater consideration. Easy downgrades are safer in icy weather, and easy upgrades reduce the chances of vehicles blocking the roadway in icy weather. Also, easy upgrades increase the speed of trucks and reduce the chances of rear-end collisions, which are one of the most serious types of accidents on a divided highway.

The gradual transition from one grade to another is accomplished by means of a vertical curve. If the point of intersection of the two tangents is above the road surface, the vertical curve is called a crest; if below, it is called a sag. Vertical curves should result in a design which is safe and comfortable in operation and adequate for drainage. Factors to be considered in the application of vertical curves are the sight distance along the curve, riding comfort, the economy of earthwork, and simplicity of calculations.

Types of curves which have been utilized in vertical-curve design are circular arcs, spirals, and parabolas. In this country, vertical curves are of the parabolic type. Parabolic curves are identified by their length L and the algebraic difference A of the intersecting grades G_1 and G_2. With this curve the vertical offsets from the tangent vary as the square of the horizontal distance from the curve end. Elevations along the curve are calculated simply as the proportions of the vertical offset at the vertical point of intersection (VPI), which is $AL/800$.

Safe vehicle operation demands that a clear line of sight of suitable length be provided. The sight distances to be provided for in the design of vertical curves are either *stopping-sight distance* or *passing-sight distance*. A safe stopping-sight distance is the minimum distance required for a driver to stop his vehicle while traveling at or near the design speed before reaching an object he sees in his path (see Sec. 2.6 and Table 4.4).[2] In addition, on two-lane highways the opportunity to pass slow-moving vehicles must be provided at intervals. The minimum distance ahead that must be clear to permit safe passing is called the *safe passing-sight distance* (see Fig. 4.1 and Table 4.5).[2]

Table 4.4   Minimum stopping-sight distance—wet pavements

Design speed, mph	Assumed speed condition, mph	Perception and reaction		Coefficient of friction	Braking distance on level, ft	Stopping-sight distance	
		Time, sec	Distance, ft			Computed, ft	Rounded for design, ft
V		T	D_r	f	D_b	$D_r + D_b$	
30	28	2.5	103	0.36	73	176	200
40	36	2.5	132	0.33	131	263	275
50	44	2.5	161	0.31	208	369	350
60	52	2.5	191	0.30	201	492	475
70	59	2.5	216	0.29	400	616	600
80	67	2.5	245	0.28	535	780	775
90	75	2.5	275	0.28	670	945	950
100	82	2.5	300	0.27	830	1,130	1,150

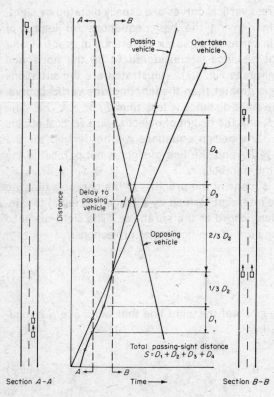

Fig. 4.1 Time-space relationship of passing-sight distance.

Table 4.5   Elements of safe passing-sight distance on two-lane highways

	30–40	40–50	50–60
Speed group, mph			
Average passing speed V, mph	34.9	43.8	52.6
Preliminary delay:			
$\quad a$ = average acceleration, mph/sec	1.40	1.43	1.47
$\quad T_1$ = time, sec	3.6	4.0	4.3
$\quad D_1$ = distance traveled, ft	145	215	290
Occupation of left lane:			
$\quad T_2$ = time, sec	9.3	10.0	10.7
$\quad D_2$ = distance traveled, ft	475	640	825
Clearance length:			
$\quad D_3$ = distance traveled, ft	100	180	250
Opposing vehicle:			
$\quad D_4$ = distance traveled, ft	315	425	550
Total distance:			
$\quad D_1 + D_2 + D_3 + D_4$, ft	1,035	1,460	1,915
Rounded for design	1,000	1,500	2,000

Minimum lengths of-crest vertical curves are usually dictated by sight-distance requirements. In deriving the basic equations for length of parabolic vertical curve L in terms of algebraic difference in grade A and sight distance S, the sight distance is considered to be the horizontal projection of the line of sight (see Fig. 4.2). Illustrated are the situations in which the sight distance is greater than the length of the vertical curve, or $S > L$, and in which the sight distance is less than L, or $S < L$. The height of the driver's eye H_1 and the height of object H_2 are vertical offsets to the tangent sight line. The design equations will be derived for the general case in which $H_1 \neq H_2$ and the line of sight is not parallel to the chord joining the ends of the parabola.

In case 1, $S > L$, use is made of the property of the parabola that the horizontal projection of the intercept formed by a tangent is equal to one-half the projection of the long chord of the parabola. The sight distance S may be expressed as the sum of the horizontal projections:

$$S = \frac{100H_1}{G_1} + \frac{L}{2} + \frac{100 I_2}{G_2} \qquad (4.5.1)$$

The problem is to find the slope of the sight line that will make S a minimum by setting $\partial S / \partial G = 0$:

$$\frac{\partial S}{\partial G} = \frac{-100H_1}{G_1{}^2} - \frac{100H_2}{G_2{}^2} = 0 \qquad (4.5.2)$$

Case 1. Sight distance greater than
length of vertical curve, $S>L$

$$L = 2S - \frac{200(\sqrt{H_1} + \sqrt{H_2})^2}{A}$$

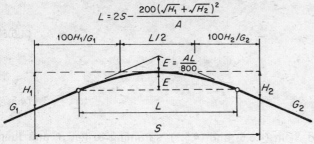

Case 2. Sight distance less than
length of vertical curve, $S<L$

$$L = \frac{AS^2}{100(\sqrt{2H_1} + \sqrt{2H_2})^2}$$

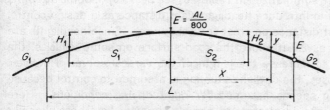

L = length of vertical curve
A = algebraic difference in grades, %
S = sight distance, ft
H_1 and H_2 = vertical height of eye and object, ft

Fig. 4.2 Sight distance on crest vertical curves.

Making use of $A = |G_1 - G_2|$, it can be shown that

$$G_1 = \frac{A\sqrt{H_1}}{\sqrt{H_1} + \sqrt{H_2}} \tag{4.5.3}$$

and

$$G_2 = \frac{-A\sqrt{H_2}}{\sqrt{H} + \sqrt{H_2}} \tag{4.5.4}$$

Substituting for G_1 and G_2 in (4.5.1) and solving for L gives the minimum length of vertical curve needed to provide the necessary sight distance S:

$$L = 2S - \frac{200(\sqrt{H_1} + \sqrt{H_2})^2}{A} \tag{4.5.5}$$

In case 2, $S < L$, use is made of the basic offset property of the parabolic curve giving

$$\frac{H_1}{AL/800} = \frac{S_1^2}{(L/2)^2} \tag{4.5.6}$$

and

$$\frac{H_2}{AL/800} = \frac{S_2^2}{(L/2)^2} \tag{4.5.7}$$

Solving for S_1 and S_2 in (4.5.6) and (4.5.7), summing to get S, and then solving for L is the following relationship:

$$L = \frac{AS^2}{100(\sqrt{2H_1} + \sqrt{2H_2})^2} \tag{4.5.8}$$

The criteria for height of eye and object used in the geometric design of crest curves are summarized in Table 4.6 (see Sec. 3.6). Some attempts have been made to introduce headlight-sight distance as a design control for stopping-sight distance on crest vertical curves. Since headlights are mounted as low as 2.0 ft above the road surface on some vehicles, the effect would be to increase the length of the vertical curve. However, for design purposes, the daylight situation is assumed to control because of lower running speeds assumed for night driving conditions. Here again, designers should attempt to provide the very best possible sight distance. Minimum sight distance should be used only where absolutely necessary.

Referring again to Table 4.6, the minimum length of vertical curve for passing is based on the distance S_p at which a driver whose eye is 3.75 ft above the pavement can see the top of an oncoming vehicle assumed to be 4.5 ft above the pavement. Curves based on the passing criteria

Table 4.6 Criteria in design of crest vertical curves

Parameter	Stopping-sight distance, ft	Passing-sight distance, ft
Height of eye H_1	3.75	3.75
Height of object H_2	0.5	4.5
Minimum length of curve L:		
$\quad S > L$	$2S_s - \dfrac{1,400}{A}$	$2S_p - \dfrac{3,290}{A}$
$\quad S < L$	$\dfrac{AS_s^2}{1,400}$	$\dfrac{AS_p^2}{3,290}$

must obviously be much longer than those based on the stopping criteria. For this reason, if a continuous passing opportunity is desired, a four-lane design with crest curves based on stopping-sight distances may be cheaper than a two-lane design based on passing-sight distances, assuming that the saving in earthwork offsets the cost of the two extra lanes.

There is no single widely used design criterion for establishing lengths of sag vertical curves. Four criteria have been used: (1) headlight-sight distance, (2) rider comfort, (3) drainage control, and (4) general appearance. Headlight-sight distance has been used by some authorities.[16] When a vehicle traverses a sag vertical curve at night, the portion of highway lighted ahead is dependent on the position of the headlight and the direction of the light beam. Figure 4.3 illustrates the conditions for deriving the minimum length of sag vertical curve L in terms of the distance

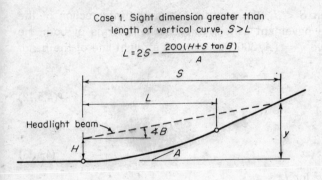

Case 1. Sight dimension greater than length of vertical curve, $S > L$

$$L = 2S - \frac{200(H + S \tan B)}{A}$$

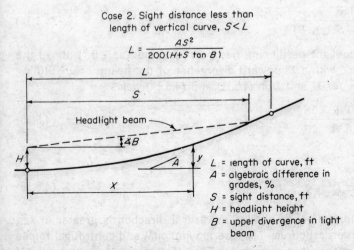

Case 2. Sight distance less than length of vertical curve, $S < L$

$$L = \frac{AS^2}{200(H + S \tan B)}$$

L = length of curve, ft
A = algebraic difference in grades, %
S = sight distance, ft
H = headlight height
B = upper divergence in light beam

Fig. 4.3 Headlight-sight distance on sag vertical curves.

between the vehicle and point where the light ray hits the pavement surface S, the height of the headlights above the pavement H, and the angle of the light ray above the horizontal B.

For case 1 (Fig. 4.3), when $S > L$, the coordinates of the intersection of the light beam with the pavement can be expressed in two ways:

$$y = H + S \tan B \tag{4.5.9}$$

and

$$y = \frac{AL}{200} + \frac{(S - L)A}{100} \tag{4.5.10}$$

Equating and solving for L:

$$L = 2S - \frac{200(H + S \tan B)}{A} \tag{4.5.11}$$

In case 2, when $S < L$, the coordinates of the intersection of the light beam with the pavement can be expressed in terms of both the parabola of the curve $y = (A/200L)x^2$ and the straight line of the headlight beam:

$$y = \frac{AS^2}{200L} \tag{4.5.12}$$

and

$$y = H + S \tan B \tag{4.5.13}$$

Equating and solving for L:

$$L = \frac{AS^2}{200(H + S \tan B)} \tag{4.5.14}$$

Since 1961, headlight positioning has been assumed as 2.0 ft above the highway surface with a 1° upward divergence of the beam. Substituting in these values for H and B in (4.5.11) and (4.5.14) yields

$$L = 2S - \frac{400 + 3.5S}{A} \qquad S > L \tag{4.5.15}$$

and

$$L = \frac{AS^2}{400 + 3.5S} \qquad S < L \tag{4.5.16}$$

The comfort effect of change in vertical direction is greater in sag than on crest vertical curves because gravitational and centrifugal forces are combined rather than canceled. The effect is not easily evaluated,[17] but attempts at its measurement have led to the general conclusion that riding on sag vertical curves is comfortable when the centripetal acceleration does not exceed 1 ft/sec^2. The general expression for lengths of sag vertical curves satisfying this comfort criterion is

$$L = \frac{AV^2}{46.5} \tag{4.5.17}$$

where V in the design speed in miles per hour. Of all the criteria for the minimum length of a sag vertical curve, the one based on the headlight-sight distance is the most rational.

When a sag vertical curve occurs at an underpass, the overhead structure may shorten the sight distance otherwise obtainable. In designing for proper sight distance and maintaining adequate clearance, the length of the vertical curve may be determined as shown in Fig. 4.4. The relation between sight distance S, the algebraic difference in grades A, the vertical clearance of the underpass C, and the length of the vertical curve may be obtained for the cases $S > L$ and $S < L$. In case 1, $S > L$, it is seen that

$$y:S/2 = AL/400:L/2 \tag{4.5.18}$$

where

$$y = C - \frac{H_1 + H_2}{2} + \frac{AL}{800} \tag{4.5.19}$$

Solving the ratio in (4.5.18) for y and equating to (4.5.19) gives

$$L = 2S - \frac{800}{A}\left(C - \frac{H_1 + H_2}{2}\right) \tag{4.5.20}$$

In case 2, $S < L$, the basic ratio takes the form

$$y':S/2 = AL/400:L/2 \tag{4.5.21}$$

where

$$y' = y - \left(\frac{x}{L/2}\right)^2 \frac{AL}{800} \tag{4.5.22}$$

$$y = C - \frac{H_1 + H_2}{2} + \frac{AL}{800} \tag{4.5.23}$$

and

$$x = \frac{L}{2} - \frac{S}{2} \tag{4.5.24}$$

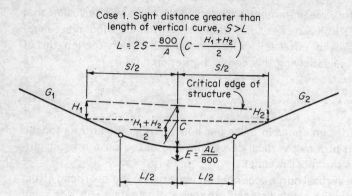

Case 1. Sight distance greater than
length of vertical curve, $S>L$

$$L = 2S - \frac{800}{A}\left(C - \frac{H_1 + H_2}{2}\right)$$

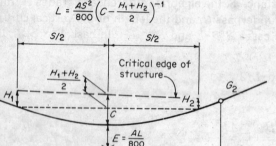

Case 2. Sight distance less than
length of vertical curve, $S<L$

$$L = \frac{AS^2}{800}\left(C - \frac{H_1 + H_2}{2}\right)^{-1}$$

L = length of vertical curve
A = algebraic difference in grades, %
S = sight distance, ft
C = vertical clearance of underpass, ft
H_1 and H_2 = vertical heights of eye and object, ft

Fig. 4.4   Sight distance at underpasses.

Substituting (4.5.23) and (4.5.24) into (4.5.22) and then equating (4.5.22) to (4.5.21) yields

$$L = \frac{AS^2}{800}\left(C - \frac{H_1 + H_2}{2}\right)^{-1} \tag{4.5.25}$$

4.6 HORIZONTAL ALIGNMENT

The alignment of a highway is a series of straight-section *tangents* joined by circular curves. The centrifugal force associated with a vehicle moving in a circular path requires that the roadways be *banked,* or superelevated, to overcome to a reasonable degree this centrifugal force. To facilitate a smooth change from a straight path to a curved path and to facilitate the introduction of the superelevation, transition or spiral curves should be used on all modern highways.

The horizontal alignment must be balanced to provide as nearly as possible continuous operation at the speed most likely to prevail under the general conditions of that section of road. For example, sharp curves should not be used after long straight tangents on which high-speed operation is likely. This is an important factor. Regardless of the "assumed design speed," the speed most likely to prevail must be given due consideration. Drivers can generally adjust to changes in conditions if these changes are obvious and reasonable. The element of surprise should, by all means, be avoided.

The designer should always attempt to use generally flat curves and should use maximum curvature only under the most critical conditions. Also, horizontal and vertical alignment must be considered together, not separately. Thus, a horizontal curve should be avoided on a crest to avoid the creation of a serious accident hazard.

In proper geometric design, it is necessary to establish the proper relationship between design speed, curvature, and superelevation. In the United States, horizontal curves are circular and are described by either their radius or *degree of curve.* In highway design the *arc definition* is used: The degree of curve is the central angle subtended by an arc of 100 ft.

The minimum radius, or maximum curvature, has a limiting value for a given design speed, as determined by the maximum rate of superelevation and the maximum side-friction factor (see Sec. 2.10):

$$R = \frac{V^2}{15(e + f)} \tag{4.6.1}$$

Use of sharper curvature for that design speed would necessitate a rate of superelevation beyond the limits considered practical or for operation with the tire friction beyond safe limits. Thus, the maximum curvature (minimum radius) is a significant parameter in alignment design.

The values of side-friction factors vary with speed and condition of the car tires and the roadway surface.[18,19] The values employed in curve design are governed by driver comfort: High centrifugal forces tend to cause the driver to slide across the seat of the automobile, and he instinctively slows the vehicle.[20] The recommended safe side-friction factors have a linear relationship with design speed, varying from 0.16 at 30 mph to 0.12 at 70 mph.[21] The designer should choose values which will encourage speeds on curves consistent with those speeds found on the tangent sections of the same highway.

Superelevation values also vary with design speed and usually range from 0.01 to 0.10 ft per feet of width.   However, since the design speed for a proposed facility is one of the design controls and is therefore fixed, superelevation rates are determined by curvature. [see Eq. (4.5.26)]. Other factors are ice problems, which limit the maximum superelevation, and drainage problems, which establish the lower limit of superelevation.

Table 4.7 Standards for curvature
(Maximum $e = 0.08$)

Design speed, mph	Maximum, f	Desirable maximum degree of curve	Absolute maximum degree of curve	Absolute minimum radius, ft
40	0.15	6°00′	12.4	465
50	0.14	4°00′	7.6	760
60	0.13	2°00′	5.0	1,145
70	0.12	1°30′	3.5	1,635

Source: "Design Manual for Controlled Access Highways," Texas Highway Department.

Table 4.7 illustrates values for degree of curve and radius of curvature for a superelevation of $e = 0.08$ and the side-friction factors shown. Table 4.7 is based on speed and does not consider sight distance.   An obstruction near the inside edge of the highway may interfere with the driver's view ahead as his vehicle travels around a horizontal curve.   Thus, the minimum radius of curvature may be determined by stopping-sight

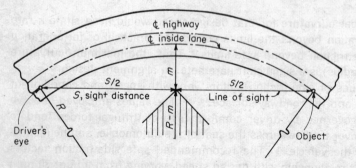

Fig. 4.5 Stopping-sight distance on horizontal curves. *(Source: modified AASHO chart.)*

distance.   The conditions assumed are that both the driver's eye and the object are in the center of the inside lane (see Fig. 4.5) and that the sight distance is less than the length of curve.   The relationship between minimum radius of curve R, stopping-sight distance S, and distance from the center line of the inside lane to the sight obstruction m, obtained by solving the right triangle whose sides are R, $R - m$, and $S/2$, is

$$R = \frac{m}{2} + \frac{S^2}{8m} \qquad (4.6.2)$$

A driver operating on a tangent section on which superelevation has been added is forced to steer opposite to the direction of the curvature ahead in order to stay on tangent.   When he reaches the beginning of the circular curve, he must immediately steer to the curved path.[20]   With high-speed operation this unnatural maneuver becomes quite critical.   At lower speeds, it has been found that the 12-ft width of lane provides enough clearance so that the "drift" toward the outside of the curve is not critical. As speeds become faster and faster, this compensation by the driver requires more and more distance and does become critical.

The use of compound circular or spiral curves is based on the principle of introducing the curvature by degrees, allowing for the corresponding superelevation to be introduced gradually.   This gradual change follows the path of the vehicle as it enters and leaves a circular curve.   The steering change and superelevation to offset centrifugal force should correspond at every point.

Superelevation may be rotated about:

1. The center line
2. The inside pavement edge
3. The inside crown line
4. The outside pavement edge
5. The outside crown line

Figure 4.6 illustrates the two cases where the pavement is revolved about the center line and about the inside edge.

Multilane roadways with a median may be rotated individually or as a unit about the master center line if the median is narrow or if there is a possibility of widening on the inside.

There are many advantages in flat curvature that are often overlooked. There is less danger of sliding sideways on icy pavement on a flat curve. Obviously, easier "all-weather" design is safer and superior to minimum "fair-weather" design.   Easy curves aid traffic volume; keeping horizontal and vertical curves far enough apart can increase traffic capacity significantly.

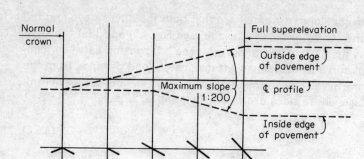

Pavement revolved about center line

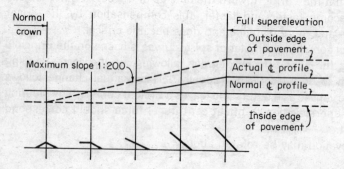

Pavement revolved about inside edge

Fig. 4.6 Diagrammatic profiles showing methods of attaining super-elevation. *(Source: Transition Curves for Highways, Public Road Administration.)*

Horizontal and vertical alignment must be coordinated not only on the main lanes of the roadway but also at ramps, intersections, and other locations where changes in direction are introduced.   In those cases where it is necessary to introduce horizontal curves that will reduce the safe operating speed, it is generally better to introduce them on an upgrade to take advantage of the improved deceleration conditions.   In all cases, sufficient sight distance and adequate panoramic view of the beginning of the curved roadway section are essential to safe operation.   When curves are introduced at the end of ramps, sufficient view of this condition and adequate deceleration distance are essential.

The visibility of the curved roadway is vital to safe operation.   If the vertical and horizontal alignments are not properly coordinated, the driver may get a completely inadequate perspective of the road ahead, which will affect his operation.

4.7 CONTROL DEVICES

Maximum efficiency and safety in the use of roads can be obtained only with the employment of the appropriate traffic control devices. Traffic control devices are all pavement markings, traffic signals, signs, and markers placed on, above, or adjacent to a roadway by authority of officials having the responsibility to regulate, warn, or guide traffic. All highway users—motorists, transit and truck operators, and pedestrians—must be informed, guided, and directed to ensure greatest convenience and safety.

The decision to use a particular control device at a particular location cannot be based on local decree. The need for high uniform engineering standards has been recognized in the profession for some time. To meet this need, a joint committee of the American Association of State Highway Officials and the National Conference on Street and Highway Safety published the "Manual on Uniform Traffic Control Devices" in 1935. The Institute of Traffic Engineers joined the committee in 1942, and in 1960 the committee was reorganized to include members from the National Association of County Officials and the American Municipal Association. The 1961 edition of the manual,[22] published by the U.S. Bureau of Public Roads, reflects both the official and professional thinking at various levels of government regarding the use of traffic control devices.

There are six elementary requirements for any traffic control device: (1) It should be capable of fulfilling an important need, (2) it should command attention, (3) it should convey a clear, simple meaning at a glance, (4) it should command the respect of road users, (5) it should be located to give adequate time for response, and (6) it should be sanctioned by law.

All traffic control devices are basically visual communication techniques. The traffic engineer uses signing as the principal means of visual communication with the driver of a vehicle. This signing must be designed to provide adequate visibility distance to enable the driver to respond and take positive action. To avoid hazardous situations, careful consideration must be given to sign design and to its placement with respect to the roadway. Design of the signing must assure that such features as size, contrast, colors, shades, composition, and lighting or reflectorization where needed are combined to draw attention to the device; that shape, size, colors, and simplicity of message combine to produce a clear meaning; that legibility and size combine with placement to permit adequate time for response; and that uniformity, reasonableness, size, and legibility combine to command respect. Placement of the signing must assure that it is within the zone of vision of the normal user so that it will command attention; that it is positioned with respect to the point, object, or situation to which it applies to aid in conveying the proper meaning; and that its location, combined with suitable legibility, is such that a driver traveling at normal speed has adequate time to make the proper response for daytime conditions — this presents a different problem to the designer for nighttime conditions.

Traffic engineers ensure that these requirements are met through the imaginative application of the standards and principles of good design, logical placement, responsible maintenance, and rigid uniformity. These four considerations are concessions to the capabilities of the vehicle and limitations of the driver.

4.8 PROJECT SKYHOOK

In 1963, the Texas Transportation Institute, in cooperation with the Texas Highway Department and the U.S. Bureau of Public Roads, initiated a research project devoted to improving the design of roadside sign supports to reduce the injury and damage effects of motor vehicle accidents involving these signs. Considerable priority was given this research effort from the beginning because of the increasing number and the severity of accidents involving roadside sign structures in the state of Texas and in the nation. Accident statistics pointed strongly toward interstate highways and other high-speed access-controlled facilities which require large roadside signs to adequately guide the high-speed high-density traffic.

A *reverse tow* procedure was developed whereby riderless vehicles can be sped along a guide rail and hurled into full-size sign supports, under controlled conditions (see Fig. 4.7). Next, engineers of the Bridge Division of the Texas Highway Department developed several designs for reducing the structural rigidity or stiffness of the supports and the condition of fixity of the base.

One design that is proving very successful in the crash tests utilizes a shear plate in the base just below vehicle bumper level. This plate is held in place by bolts in open-end slots to disconnect the lower end of the post on impact and to allow it to bend upward to clear the careening vehicle. The bending occurs readily because at a point above automobile roof level the post, which is usually a steel beam, is already seven-eighths severed from the front side and has a readily breakable cast-iron strip, or *fuse,* bolted across the saw and cut for stability of the sign in high winds. The bending in the beam is a condition that is known structurally as a *plastic hinge.*

The results of this research are being applied in the modifications of existing sign supports as well as in new installations (see Fig. 4.7). Several

Fig. 4.7   Highway sign-support research immediately saves lives.

of the districts of the Texas Highway Department have instituted practices of modifying sign supports in the field, especially the EXIT signs, with their own maintenance forces.   There have been at least four accidents involving these modified signs without a reported personal injury.   The following account of one of the accidents was taken from the October, 1965, issue of the *Texas Transportation Researcher.*

> Recently a driver lost control of his vehicle and skidded into an "EXIT" sign on IH 10 near Beaumont.   Less than twenty-four hours before this incident, the T.H.D. district maintenance force had modified the sign supports at that location, incorporating the slip base and hinge features of the experimental breakaway structures.   The driver and passenger were uninjured, and the major portion of the $400 vehicle damage was attributed to the vehicle side-swiping a smaller fixed base sign after colliding with the "EXIT" sign.
>
> Eleven months earlier a driver alone in his car was killed when he apparently fell asleep and struck the "EXIT" sign at the same location.   At that time the sign supports were of the fixed-base type.

4.9   THE ROAD AS A SYSTEM

The highway transportation engineer is concerned with the planning, design, construction, operation, maintenance, control, and administration of highway facilities.   In this chapter much has been said regarding geometric design and some reference has been made to highway operation and maintenance.   It should be pointed out that the word *design* as used in *geometric design* is a misnomer—perhaps more properly designated as *proportioning.*   Design includes all the above functional aspects from planning to administration, including proportioning.

Quite recently there has evolved a systems approach to engineering design which is receiving increasing emphasis in teaching and practice.   In this approach, operating the total project is emphasized rather than considering the efficiency of the various component parts independently. This, in itself, is nothing new; it has always been the object of good engineering.   What is new, however, is that the systems approach is a rational, rather than an intuitive, approach, utilizing such formalized techniques as game theory, queueing theory, linear programming, dynamic programming, control theory, critical path methods, network theory, and various optimization techniques.

A fundamental principle of system design as related to the design of classical systems is simply to maximize performance for a given cost or to minimize cost for a given performance. In highway design the latter half of this axiom should govern. This is the essence of the level-of-service concept discussed in the next chapter.

Two additional principles of classical system design are the principle of events of low probability and the principle of suboptimization. The former states that the fundamental objective of the system should not be compromised to accommodate events of low probability. One should not design a culvert for a 100-year rainfall nor a grade separation structure for 150-mph wind loads; one should not try to PERT a construction schedule with 100 percent assurance of completing the project on time nor provide enough freeway lanes so that there is *never* congestion.

Suboptimization, though easy to identify even in highway design (see Fig. 4.8), is quite difficult to overcome. In the organization of system design—whether it be a highway or missile system—one must have a rational way of setting the boundaries between subsystems. Two obvious ways are (1) through geographic considerations and (2) by technical specialty. The first, though undesirable, is certainly necessary at some level. An interstate route cannot be designed as an entity from coast to coast. In practice, each state assumes the responsibility for the portion within its boundaries and in turn may delegate its responsibility to "districts" within its highway department. The Texas Highway Department has 25 districts, each with a district engineer, who in turn might place portions of a facility like an interstate freeway under the direction of his resident engineers.

At some point in this geographic or political hierarchy, the system becomes sectionalized on the basis of technical specialties. In the design of a highway system, these specialties are essentially the traditional disciplinary divisions of civil engineering: (1) materials engineering, (2) structural engineering, (3) soil mechanics, (4) construction engineering, (5) hydraulics and fluid mechanics, (6) planning, (7) surveying, and (8) traffic engineering. If each of these disciplines is treated as a subsystem of the highway system design (as they often are), the number of paths between subsystems becomes quite large. For the eight subsystems cited, the number of combinations is $\binom{8}{2} = 28$.

Herein lies the key to sectionalization: boundaries set between subsystems so as to minimize the number of inputs and outputs for each subsystem. Three logical subsystems in the highway system design problem are (1) geometric design, (2) structural design, and (3) pavement design. The number of paths between subsystems has been reduced from twenty-eight to three. Although each of the three subsystems suggested is a combination of the eight technical specialties listed above, interdisciplinary

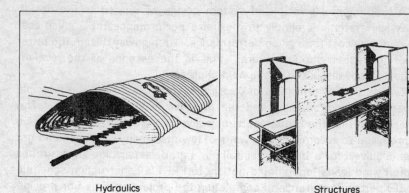

Hydraulics

Structures

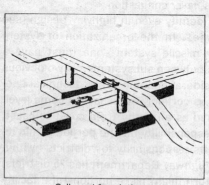

Soils and foundations

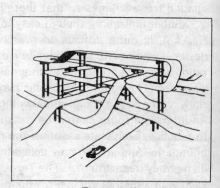

Transportation

Materials

Fig. 4.8 View of a highway by different specialists.

teams have actually been formed with, it is hoped, each member of the team acting as more than just a narrow specialist in a phase of the design.

In the past, the complexity of highway engineering problems and the large amounts of data often forced an engineer to unnaturally decompose a problem into noninteractive tasks, with many of the feedback aspects of the problem ignored.   The highway transportation engineer did not have the necessary tools to coordinate the data and perform the design with all the interactions considered, let alone weigh alternate designs.   Computers now offer engineers these capabilities.

The Civil Engineering Systems Laboratory of M.I.T., supported by the IBM Corporation, has provided a computer system called ICES (Integrated Civil Engineering System), specifically designed for use by civil engineers.[25] A feature of the system is effective computer utilization by an engineer who maintains his basic decision-making role.   This is accomplished by engineering-oriented input and output, which eliminates the "cult of the programmer."

ICES consists of subsystems corresponding to the various engineering discipline areas.   The availability and integration of these subsystems allow engineers to achieve complete problem solutions where all factors and data interactions are properly considered.   Each subsystem has a series of problem-oriented language commands associated with it, which the engineer uses to communicate his problem-solving requests to the computer. It is therefore unique in that it has been designed to satisfy the specifications of the programmers developing it and the engineers who will use it.

4.10 CHALLENGES

There is an intimate connection between the safety of a highway and the complex interrelations between curvature, grade, and sight distance.   For low- to moderate-speed operation the empirical methods currently used are perhaps adequate.   However, for the high-speed operation now approaching, a closer look at the assumptions and criteria for geometric designs is mandatory.

If past trends are to be taken seriously, the next generation of highways will be based on design speeds of 100 mph or more.   This will have a profound effect on geometric design and even structural design.   Flatter curves and slopes will necessitate more and longer spans on elevated structures.   There will be a demand for 200-ft prestressed concrete beams and perhaps prefabricated structures.   The entire right-of-way must be made free of such obstructions as trees and utility poles.   Signs should be of the harmless breakaway type.   Guardrails and barriers are collision objects, themselves, so that rational warrants should be developed for their installation.

PROBLEMS

4.1. Sketch typical cross sections for a design speed of 70 mph for highway facilities of the following types:
  a. A two-lane highway
  b. A rural freeway
  c. An urban expressway
  d. A city street

4.2. Discuss horizontal and vertical control standards for each of the above.

4.3. How might such information as traffic volumes, percentage of trucks, and turning movements alter the designs suggested in Probs. 4.1 and 4.2?

4.4. Define safe stopping-sight distance and safe passing-sight distance, explaining when each controls in geometric design.

4.5. Under what conditions might a four-lane highway design be cheaper than a two-lane highway design?

4.6. Determine the value for each element of the passing-sight distance depicted in Fig. 4.1 for a design speed of 80 mph.

4.7. Determine the stopping-sight distance and maximum safe speed over a crest formed by grades of +4 percent and −2 percent if the vertical curve is 800 ft long.

4.8. What is the minimum length of a crest vertical curve formed by grades of +6 percent and +1 percent on a two-lane rural highway with a design speed of 80 mph?

4.9. For the vertical curve designed in Prob. 4.8, find the maximum safe speed if climbing lanes are added to make a four-lane cross section. (Reconsider Prob. 4.5.)

4.10. A +4 percent grade meets a −2 percent grade at station 20 + 00, elevation 100.00 ft. A culvert crosses the roadway at station 19 + 00 with the top of culvert elevation at 90.00 ft. If at least a 3-ft cover is needed over the culvert, will this curve meet the minimum design requirements for a design speed of 60 mph?

4.11. In an effort to discourage drivers from exceeding the design speed of a road, superelevations sometimes are set at values that develop the maximum permissible side friction. Determine the superelevation that should achieve this on a 3° curve at a design speed of 70 mph.

4.12. For a highway having a design speed of 40 mph in a state where the maximum rate of superelevation is 0.10 ft², determine the minimum horizontal radius of curvature and make a sketch showing the method of attaining the superelevation.

4.13. For a freeway exit ramp curving to the right located on a curve to the left, illustrate by cross sections and the appropriate profiles the method of attaining the superelevation of the ramp. How could the exit ramp design be altered to eliminate the crossover crown effect without moving the ramp location?

4.14. A corner of an existing building is 25 ft from the center line on a curved portion of a level two-lane highway. If the lane widths are 12 ft, what is the safe operating speed on a 12° curve?

4.15. Sketch an entrance ramp and an exit ramp and briefly describe their elements and desired operation. Explain why traffic might not use an adjacent parallel speed-change lane in each case.

4.16. Sketch a wye intersection and a tee intersection. Assuming the intersections are in an urban area with high pedestrian volumes, sketch the required channelization element.

4.17. An exit ramp is planned from the right roadway of a divided expressway to the right frontage road which is parallel to the expressway lanes. If the ramp is to be in the form of a reverse curve, show that the properties of the curve can be given by the relationship

$$R_1 + R_2 = \frac{x^2 + y^2}{2y}$$

where R_1, R_2 = curve radii

$\qquad y$ = distance between expressway roadway and frontage road

$\qquad x$ = distance along expressway center line between point of curvature on expressway and point of tangency on frontage road

4.18. A single rectangular bridge span of length L and width W is used on a horizontal curve R. If the roadway is 24 ft wide and a minimum clearance of 3 ft is desired between the edge of the pavement and the bridge rail, show that the minimum radius of curvature R is given by the following expression:

$$R = \frac{L^2}{8\,(W - 30)} + \frac{W - 30}{2}$$

REFERENCES

1. Rowan, N. J., and P. T. McCoy: A Study of Roadway Lighting Systems, *Texas Transportation Inst. Res. Rept.* 75-1, Texas A&M University, College Station, August, 1966.
2. "A Policy on Geometric Design of Rural Highways," American Association of State Highway Officials, Washington, D.C., 1965.
2a. "Highways in Urban Areas," American Association of State Highway Officials, Washington, D.C., 1957.
3. "Design Manual for Controlled Access Highways," The Texas Highway Department, Austin, Tex., 1960.
4. Planning Highway Systems and Development of Standards, *Am. Road Builders' Assoc. Bull.* 142, pp. 34–36, 1948.
5. Johnson, A. E.: Highway Signs for the Interstate System, *Civil Eng.*, vol. 27, p. 251, 1957.
6. Webb, George M., and Karl Moskowitz: Freeway Traffic Flow, *Calif. Highways Public Works*, vol. 35, no. 7-8, p. 5, July–August, 1956.
7. Simonson, Wilbur H.: The Sectional Layout of Multiple Lane Highways, *Highway Res. Board Proc.*, vol. 18, pp. 62–64, 1938.
8. Landscaping: Full Partner in Turnpike Design, *Eng. News-Record*, vol. 158, no. 28, pp. 62–64, June 13, 1957.
9. Build to the Middle, *Eng. News-Record*, vol. 154, no. 16, p. 17, Apr. 21, 1955.
10. Gilliss, C. M.: Los Angeles MTA Selecting Tomorrow's Super Rapid Transit System, in "Going Places," pp. 10–11, General Electric Company, January–February, 1960.
11. Lang, C. H.: Geometric Design of Modern Highways, *Civil Eng.*, vol. 30, no. 1, p. 63, 1960.
12. Keese, C. J., and C. Pinnell: "The Effect of Freeway Medians on Traffic Behavior," Texas Transportation Institute, Texas A&M University, College Station, 1959.
13. Saal, C. C.: Hill Climbing Ability of Motor Trucks, *Public Roads*, vol. 23, no. 3, p. 33, May, 1942.
14. Claffey, P. J.: Time and Fuel Consumption for Highway User Benefit Studies, *Highway Res. Board Bull.* 276, 1960.
15. Sawhill, R. B., and J. C. Firey: Motor Transport Fuel Consumption Rates and Travel Time, *Highway Res. Board Bull.* 276, 1960.

16. Thompson, D.: Sight Distance on Sag Vertical Curves, *Civil Eng.*, vol. 14, no. 1, January, 1944.

17. McConnell, W. A.: Human Sensitivity to Motion as a Design Criterion for Highway Curves, *Highway Res. Board Bull.* 149, 1957.

18. Barnett, J.: Safe Side Friction Factors and Superelevation Design, *Highway Res. Board Proc.*, vol. 35, 1936.

19. Moyer, R. A., and D. S. Berry: Marking Highway Curves with Safe Speed Indications, *Highway Res. Board Proc.*, vol. 39, 1940.

20. Taragin, A.: Driver Performance on Horizontal Curves, *Highway Res. Board Proc.*, vol. 33, 1954.

21. Stonex, K. A., and C. M. Noble: Curve Design and Tests on the Pennsylvania Turnpike, *Highway Res. Board Proc.*, 1940.

22. "Manual on Uniform Traffic Control Devices," U.S. Bureau of Public Roads, Washington, D.C., June, 1961.

23. Hawkins, D. L., N. J. Rowan, and R. M. Olson: "Project Skyhook: Safer Highway Sign Supports," Texas Transportation Institute, Texas A&M University, College Station, May, 1966.

24. Keese, C. J., and D. R. Drew: Geometric Design, *Traffic Eng.* (To be published by the Automotive Safety Foundation, Washington, D.C.)

25. Roos, Daniel: "ICES System Design," The M.I.T. Press, Cambridge, Mass., 1966.

CHAPTER FIVE
THE CIRCULATION SYSTEM

Expatiate free o'er all this scene of man;
A mighty maze! but not without a plan.

Alexander Pope

5.1 INTRODUCTION

In the past, the geometrics of an individual road were developed primarily in relation to safety, right-of-way, physical controls, and economic feasibility. First a design and ruling grade were established after determining such factors as the road's importance, the terrain, and the availability of funds. The design speed and ruling grade in turn provide the basis for setting minimum standards for vertical and horizontal alignment. Then, the designer fitted these standards, or higher standards, to the terrain to produce a plan and profile for the highway.

However, because the construction of new facilities vitally affects traffic operations, the design of fixed facilities has come to be of special concern to traffic engineers, and more and more design engineers have begun to consider traffic operations. Therefore, *geometric design* might more properly be referred to as *traffic design*. The principles of location and design must be developed with reference to the needs of through and local traffic and their effect on the community and other traffic facilities in the system.

The distinguishing feature of any system is that although its performance depends on all its components, it transcends that of any one such that its performance cannot be determined by the analysis of the individual components alone. The principal factors to be considered in designating traffic systems are the travel desires of road users, the access needs of adjacent land development, the network pattern of existing streets, and the existing and proposed land uses. Requisite to the designation of traffic systems is a thorough understanding of the function of each component in the highway or street network.

5.2 FUNCTIONAL CLASSIFICATION

Geometric features of highways should be designed to provide mobility, convenience, economy of operation, and safety. Highways of high standards tend to fulfill these goals, whereas highways of low standards may compromise them. A high design standard is warranted where there are sufficient benefits to road users to justify the additional cost above that of a low design standard. The development of highway classification of varying standards is consistent with the need of administrative engineers to apportion available funds for improvements while adequately maintaining and operating all roads.

Functional highway classification is defined as the grouping of roads and streets into classes, or systems, according to the character of service they will be expected to provide. In general, highways and streets are grouped into interstate primary- and secondary-road classes in rural areas and into freeway, expressway, major-arterial, collector-street, and local-street classes in urban areas. Considerations for highway classification are the travel desires of the public, access requirements, and the continuity of the system.

In urban areas, street design, though necessarily influenced by topography and construction costs, should be determined by population density, land development, the character and composition of traffic movements, and present and future traffic volumes. Based on these criteria, the street network is divided into subsystems with each subsystem accommodating either movement or access to a varying degree. The function of each street system governs the selection of the geometries and the choice of control devices, in order that the desired service of each be maintained. These, briefly, are the characteristics and functions of the four basic categories of urban streets.[1]

Local-street subsystem Local streets are divided according to the area they serve into residential, commercial, and industrial. In all three cases

their main purpose is to provide access.   Service to through-traffic movement on local streets not only is deemphasized but in some instances may
be deliberately discouraged.   This control should be effected at the design
level through alignment, street width, and division into short sections by
collector streets.

Collector-street subsystem   Collector streets serve a dual purpose by
providing a means for local through-traffic movement within an area and
direct access to abutting property.   In order to distribute traffic effectively
and provide sufficient capacity, the collector system should form a network
of streets spaced from $\frac{1}{2}$ to 1 mile apart.

Major-arterial subsystem   This subsystem, together with expressways,
must serve as the principal network for through-traffic movement.   It
provides direct connections to the principal traffic generators—the central
business district, major employment centers, goods distribution and transfer centers, and transportation terminals in all portions of the urban area.
Major arterials also provide extensions of rural arterials.

Expressway subsystem   The function of the urban expressway is to expe-
dite traffic movement.   By handling very high traffic volumes efficiently,
it helps to remove through traffic from other parts of the overall street
system, thus enabling all streets to fulfill their roles more effectively.   In
order to ensure safety for heavy volumes, points of access should be
controlled.   Standards for the above street classification are summarized
in Table 5.1.

   Although city streets support sanitation, admit light and air, provide
walkways for pedestrians, allow fire and police protection, and are the
right-of-way for underground utilities, the primary function of a city's
circulation system is to provide for the movement of persons and goods.
This movement ranges from pedestrians to autos, buses, and trucks—all
for a variety of purposes, including work, shopping, entertainment, education, and business.

   Unfortunately, the circulation system, itself, is a holdover from the
premachine era.   Many towns were laid out in gridiron sections with rectangular blocks divided into narrow rectangular lots.   Other cities utilized
the radial-circumferential design to break the monotony; but as with the
conventional grid, only the widths of the streets were varied to distinguish
locals, collectors, and arterials.[2]

   The practice of varying the width of a street to establish its integrity
can lead to narrow local streets with excessive weaving about parked cars.
A more rational approach seems to lie in the reduction of the total length of

Table 5.1 Street classification and standards

Type of facility	Function and design features	Spacing	Right-of-way, ft	Pavement	Desirable maximum grades, %	Speed, mph	Other features
			Widths				
Freeways	Limited access; no grade crossings; no traffic stops	Variable; related to regional pattern	200–300	12 ft per lane; 8–10 ft for shoulders; 8–60 ft for median strip	2	70	Preferably depressed through urban areas. Service roads desirable
Expressways	Limited access; channelized grade crossings; parking prohibited	Variable; generally radial or circumferential	150–250	12 ft per lane; 8–10 ft for shoulders; 8–30 ft for median strip	3	50	Generally at grade. Service roads desirable
Major roads (major arterials)	Usually form boundaries for neighborhoods; parking generally prohibited	1½–2 miles	120–150	84 ft maximum for four lanes, parking, and median strip	4	35–40	Require 5-ft-wide sidewalks with planting strips 5–10 ft wide or more
Secondary roads (minor arterials)	Main feeder streets. Signals where needed; stop signs on side streets	¾–1 mile	80	60 ft	5	35–40	Require 5-ft-wide sidewalks with planting strips 5–10 ft wide or more
Collector streets	Main interior streets. Stop signs on side streets	¼–½ mile	64	44 ft (2–12 ft for traffic lanes; 2–10 ft for parking lanes)	5	30	Require at least 4-ft wide sidewalks; vertical curbs
Local streets	Local service streets. Non-conductive to through traffic	300–800 ft	50	36 ft where street parking permitted	6	25	Sidewalks at least 4 ft in width

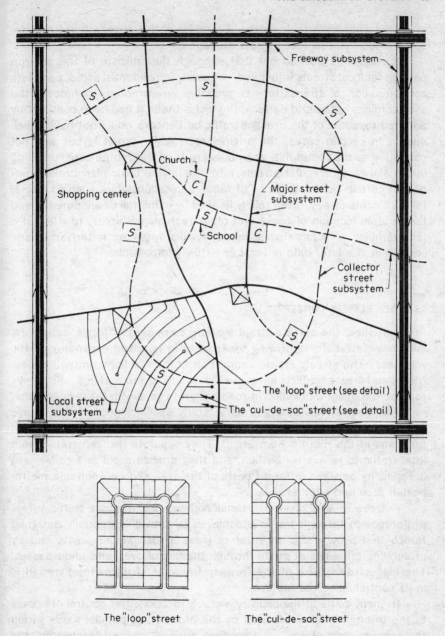

Fig. 5.1 Diagrammatic layout illustrating functions of street systems.

streets, rather than their width.   Examples of this approach are culs-de-sac and loop streets (see Fig. 5.1 on page 83).

It should be pointed out that although the criticism of the gridiron pattern for local streets is justified, especially in residential areas, complete condemnation of the pattern is probably unwarranted. Whatever the shortcomings of the grid pattern, the fact is that if it had been consistently adhered to, some of the present traffic bottlenecks would not have developed. In certain cases, the original rectangular street layout was not continued as the community grew outward from the central core, resulting in a mélange of irregular streets. In others, the basic plan broke down only at certain points, because of topographical obstacles such as rivers; railroad rights-of-way, often solidly flanked by industrial developments; and the random location of parks and other recreational areas, to cite but a few examples.   Factors such as these have disrupted the pattern and made correction at a later date difficult or virtually impossible.

5.3   THE FREEWAY CONCEPT

In most cities, the surface-street system, composed of locals, collectors, and major arterials, is proving inadequate in handling expanding traffic volumes. The streets create congestion at intersections, imperil pedestrians, and create conflicts among vehicles moving and parking. Freeways represent a major innovation in road design to cope with these conditions.

The freeway has emerged as a facility specifically created to move large volumes of traffic at high speeds safely. They perform two main functions in the metropolitan area. They separate through traffic from local traffic to relieve congestion, and they provide rapid and convenient accessibility between different parts of the area and between one metropolitan area and another.

A freeway is a divided arterial highway for through traffic. This preference to through traffic is achieved by providing specially designed ramps that allow traffic to enter or leave at designated points and by prohibiting crossings at grade through the use of over- and underpasses. The biggest departure of the freeway from the surface-street design is in its controlled-access feature.

In many parts of the country, notably in Texas, this control of access to the through lanes is ensured by the provision of parallel roads within the freeway right-of-way. A local-road auxiliary to, and located on, the side of a main highway for service to abutting property and for control of access is called a frontage road. Its function is to control access to the traveled way for through traffic, to provide access to the property adjoining the highway, and to maintain circulation of traffic on the street system on each side of the main highway.   Access from frontage roads to the main traveled way is permitted only at a specially designated location.   In urban areas, frontage roads should be operated as one-way pairs.   In rural areas, it is often necessary to operate two-way frontage roads because of extreme distances between grade separations with the main highway.

The frontage-road system can add tremendous flexibility to the operation of a freeway when utilized as an auxiliary facility.   It provides a means of utilizing a development program by stages for the construction of urban freeways as well as a means of handling traffic flow during the construction of the main freeway lanes.   During the design life of a freeway facility, a continuous frontage-road system provides maximum land service to properties abutting the freeway.   It greatly increases the flexibility of the interchanges and becomes an integral part of the overall street system.   During periods of saturated flow on urban freeways, frontage roads provide the operational flexibility often required to operate freeway surveillance and control systems.

In many areas the term *expressway* is used to denote freeways.   However, this term normally implies that there may be some intersections at grade along the route.   Similarly, parkways tend to fit the freeway definition, although only noncommercial traffic is normally permitted.   Actually, the freeway should be defined not by its physical characteristics alone but by the level of service it provides.   The freeway concept represents the most vigorous attempt to date to eliminate traffic congestion.   It is a superhighway, eliminating vehicle-to-pedestrian conflicts and reducing vehicle-to-vehicle conflicts.

In major urban areas the freeway system takes many forms, depending on the age and development of the city, topographical features, historical landmarks, and traditional developments.   The most common type of freeway system configuration is the radial-loop type.   Loops are usually a downtown loop (inner loop) and an outer loop.   The radials primarily serve commuter movements to and from the central area.

5.4 MEASURING TRAFFIC VOLUMES

The distinguishing feature of a traffic system lies in the assurance that the traffic flow can be optimized with respect to some criteria.   Although the selection of the proper criteria for evaluating traffic-system performance is the subject of much theoretical research, three basic tests of efficiency of a street system are (1) volume handled, (2) overall speed, and (3) overall accident rate.

Volume is defined as the number of vehicles moving in a specified direction or directions on a given lane or roadway that pass a given point during a specified period of time.   A determination of traffic volume is basic to the evaluation of traffic movement.   Because it furnishes a basic scale of comparison, it shows the relative importance of different facilities in highway planning and design and in the estimation of highway revenues. For example, traffic engineers utilize volumes in order to establish priorities for the construction of new facilities as well as improvements in existing facilities.   Some of the parameters employed in the study of volume char-acteristics are the magnitudes, composition, and time and directional dis-tribution of vehicular flows.   Practical aspects of the problem are volume studies, including the collection, presentation, and application of the data.

Traffic volumes vary from hour to hour, day to day, and month to month; but it is the traffic at certain peak hours (peak-hour volume) and the average daily traffic throughout the year which are needed —the former as the basis for evaluating efficiency and the latter as the basis for meas-uring safety.   The process of determining efficiency in the service afforded by a street system has two broad aspects.   One is consideration of the main characteristics of individual streets.   The second is concerned with the efficiency of the street system as a whole.

Statistics on vehicular volume offer a quantitative yardstick of the traffic system.   To obtain complete accuracy in measuring average daily traffic, it would be necessary to count vehicles continuously on every stretch of road within the city.   Since this is cut of the question, some plan based on sampling must be formulated to provide volume figures to the necessary degree of accuracy at a reasonable expenditure in time and manpower.

Since the inception of the first counting programs, which were oriented to rural systems, the highway problem has shifted from our rural highways to our city streets.   The procedure manual "Measuring Traffic Volumes," prepared by the National Committee on Urban Transportation, describes a series of systematic studies designed to provide the city with the minimum necessary volume data on the traffic which uses the existing street plan.[3] The recommended studies include control counts, coverage counts of both major- and minor-street systems, cordon counts around the CBD, and counts of selected screen lines.

Control counts are used to provide information on annual trends and seasonal, daily, and hourly variation patterns.   One major control station should be located on each major street.   The purpose of this major con-trol count is to obtain a pattern of hourly and directional variations in traffic volume which is typical of the entire route and free of capacity restrictions. Of particular importance is tne determination of expansion factors to be applied to the short sample counts of the coverage-count operation.   The minimum recommended duration and frequency of major control counting is a 24-hr directional machine count every second year.

Minor control stations should be located to fulfill the primary aim of sampling typical streets of a class.   The selection of minor control stations requires a good knowledge of the land use characteristics of the local-street system and the application of ingenuity in determining their locations. The number of minor control stations selected will vary with the size of the city and with the importance attached to the determination of local-street volumes.   The procedure manual suggests a minimum of nine minor control stations on the local-street system in a very small city, with the recommended duration and frequency of control counting for the minor control stations being a 24-hr nondirectional machine count performed twice a year. The purpose of this type of count is to determine an expansion factor to be applied to the moving-vehicle method of obtaining traffic volumes on the minor-street system (see Sec. 12.6).

In order to obtain an index of daily and seasonal variations in traffic volume for each of the various classes of streets, key counts are utilized. Since the purpose is to determine daily and monthly expansion factors, the counting durations and frequencies for key-counting stations are a nondirectional 7-day count performed annually to obtain the daily expansion factors and a 24-hr nondirectional (weekday) count performed monthly to obtain the monthly and seasonal expansion factors.   A key station should be designated on one expressway, one freeway, one major arterial, one collector street, one residential street, one commercial street, and one industrial street to represent all streets with these specific classes of streets.

It is recommended that one nondirectional 24-hr weekday coverage count be taken within each control section on the major-street system. The 24-hr total is the only figure required, since this value can be expanded to an estimate of average daily traffic by application of the daily and monthly factors.   By repeating this expansion for all the control sections, vehicle volumes in both directions as well as the total vehicles per mile on the major-street system are ascertained.   The frequency of coverage counts on the major-street system should be taken at 4-year intervals or after a 10 percent growth in volume, whichever comes first.

In addition to the systematic traffic counts employed by the states, there are several special volume studies employed by state and city traffic personnel.   Street counts are used in determining trends, in preparing traffic-flow maps, and in developing design volumes.   These counts consider the total volume without regard to direction.   For planning improvements and capacity analyses, such as timing signals, directional counts are used.   Because intersections are the most critical points in traffic control, intersection counts (including both turning movement and pedestrians) are used to design channelization, justify turn prohibitions, determine phase lengths, and make accident analyses.   The distribution of total volume into the various types of vehicles is accomplished by classification counts.   These are of importance in establishing structural and geometric design criteria, estimating highway-user revenues, adjusting highway capacity, and correcting machine counts.

Three types of volume studies usually associated with origin-destination surveys are occupancy counts, cordon counts, and screen-line counts. Occupancy counts are made to determine the distribution of passengers per vehicle, the accumulation of persons within an area, and the proportion of persons utilizing transit facilities and in general to contribute to the understanding of trip characteristics. Cordon counts are made at the perimeter of an enclosed area such as the CBD, a shopping center, or an industrial area. The information from the cordon count is used to determine the mode of travel of persons entering and leaving and the accumulation of persons and vehicles in the cordon area. The screen-line study is the principal source of study on the annual growth of traffic movements and is used to check the accuracy of projections made from past studies. The screen line is an imaginary line bisecting an area, and its procedure calls for classified counts to be taken on all streets crossing this line.

5.5 VOLUME SURVEY DEVICES

Volume counts may be made by hand tally, manual counters, and machine counters. To make volume counts by hand tally, the observer is equipped with a clipboard, watch, pencils, and a supply of the appropriate recording forms. Manually operated counters obviate the need for detail counting sheets, since only the totals from the counters need be recorded. Manual counts are used in turning-movement counts, classification counts, occupancy counts, pedestrian counts, and multilane freeway counts. Machine counters may be either permanent or portable and recording or non-recording. Permanent counters are installed to obtain control counts, such as used to provide factors for adjusting short counts to average daily traffic (ADT); types of detectors utilized are photoelectric cells, pressure, magnetic, radar, sonic, and infrared. Portable counters are used to obtain temporary or short-term counts. They consist of an electrically operated counter actuated by air impulses from a road tube stretched across the highway. A correction factor obtained from classification counts is used to minimize the error introduced by multiaxle trucks.

Automatic volume survey devices may best be described in terms of their components (detection, transmission of data, recording and measuring). In a comprehensive report prepared by the Institute of Traffic Engineers Technical Committee 7-G,[4] devices available for volume surveys and their principles of operation, operating features, and advantages and limitations are shown.

The most common type of detector is the pneumatic detector, or roadtube. The device consists of a flexible tubing fastened to the pavement at right angles to the direction of travel.   One end of the tubing is sealed, and the other end is attached to a pressure-actuated switch.   The passage of a vehicle axle over the tubing displaces a volume of air, creating a detectable pressure at the switch.   The accuracy of vehicle detection by means of a pneumatic tube is high, and the probability of error from simultaneous wheel passings can probably be neglected.   The pneumatic roadtube is, however, vulnerable to many hazards, including tire chains, street cleaners, snowplows, and vandals.   Many of these limitations can be overcome by use of a permanent treadle-type electric contact detector.

The detecting of objects by photoelectric equipment is accomplished when an object passes between a source of light and a photocell.   Appropriate types of electric traffic counters may then be connected to the photocell and activated by its circuits.   Overcounts occasioned by such things as upright stakes on truck beds can be minimized by using two independent beams spaced some 30 in. apart.   However, one rather serious limitation is the inability to count individual lanes of traffic.

The sonic, radar, and infrared detectors are all similar in that they direct energy at the oncoming vehicle from above or from the side.   The sonic detector transmits a signal toward the roads every $\frac{1}{15}$ sec.   After transmission, the detector listens for the echo.   If a certain predetermined time elapses, the pulse of energy has had time to hit the pavement and return, indicating no vehicle.   However, if the elapsed reflection time is shorter, the presence of a vehicle is indicated.   Since a very short pulse would indicate the presence of a tall vehicle such as a truck, sonic detection could be utilized for some classification counts.   The reflection of these short-time pulses causes a relay to be closed, indicating the presence of a vehicle.   The pulse generator and reflection pickup are contained in a head which is mounted over the roadway on a bridge, sign structure, or mast arm, whereas the unit containing the electronic circuitry may be mounted in a more accessible location and connected to the head by cable.

The radar detector beams high-frequency energy toward the oncoming vehicle.   The doppler phase shift of the reflected energy is detected and causes a relay to be closed.   The phase shift of the reflected signal is directly proportional to the velocity of the vehicle.   Since the vehicle serves as a signal reflector, a radar detector is a self-contained unit.   The radar unit is normally counted above the center of the lane or lanes for which detection is desired.   The form of detection requires a high initial cost but little maintenance.

5.6 VOLUME CRITERIA

The high cost of right-of-way and construction dictates the need for thorough engineering analyses in designing a facility to accommodate traffic.   In rural areas, traffic data may include traffic volumes for an annual average day of the year; volumes by hour of the day at strategic points on the system; and the distribution of vehicles by type, weight, and performance. In urban areas, the high concentration of traffic and restricted space make the need for current and comprehensive data all the more imperative.

Volume data are expressed in relation to time, with the particular time parameter being determined by the type of information desired and the application in which it is to be used. Thus, traffic estimates for a particular urban route are usually ADT volumes obtained from origin and destination studies. Current ADT is the total traffic for the year divided by 365. Although the ADT volume is important in determining classification of a highway or street, programming capital improvements, and designing the structural elements, the direct use of ADT in geometric design is not practical because it does not indicate the cyclical traffic patterns. In most cases, the period used as the basis for design is 1 hr. Periods of less than 1 hr are now being considered in urban design problems.

Because traffic assignments are made in terms of two-way daily traffic whereas one-way hourly volumes are used as the basis of design, the two most important traffic patterns are the daily time pattern and the directional distribution. The hourly variations on most urban facilities depict two distinct daytime peaks for the five weekdays. This is due to the repetitive travel to and from the central business district or some other major traffic generator. Generally, rural freeways some distance from metropolitan areas exhibit only one peak, which occurs in the late afternoon.

The traffic volume representing traffic expected to use a facility at some designated future date, used as the basis for design, is called the *design hourly volume* (DHV). Analysis of the relation between peak hourly flows and annual average daily traffic on rural highways suggests the thirtieth-highest hour (30 HV) as a practical criterion of needed capacity. It has been shown that at the thirtieth-highest hour, the ratio of benefit to expenditure is near the maximum. Although a facility designed to accommodate the thirtieth-highest hour will be, theoretically, inadequate for 29 hr during the design year, it is not economically feasible to design for all the hours of greatest traffic demand.

In urban areas the highest of the afternoon peaks for each week may be averaged for the year to represent the hourly volume for design. Assuming that all morning peak-hour volumes are less than afternoon peaks, the average of the 52 weekly afternoon peak hourly volumes would be exceeded 26 times, thus approximating the twenty-sixth-highest hourly volume. Therefore, the thirtieth-highest hourly volume can also be accepted for use in urban design.

Traffic is usually estimated for a particular urban route as ADT volumes derived on an area basis from comprehensive transportation studies, as

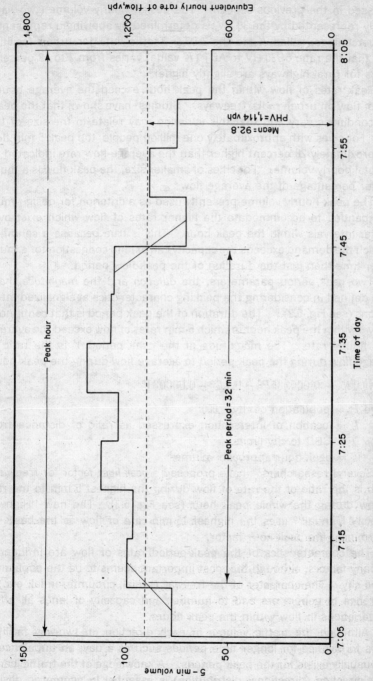

Fig. 5.2 Peak-hour 5-min traffic demands. Heights Street approach at Sixth Street, Houston, Dec. 20, 1961.

discussed in the previous section. The design hourly volume (two-way), usually represented by the 30 HV, is determined by applying a representative percentage, called K, to the ADT. Traffic data for urban facilities show that the ratio of 30 HV to ADT (K value) varies from 7 to 18 percent. Values for rural highways are slightly higher.

Peak rates of flow within the peak hour exceed the average hourly rate of flow on urban radial freeways. Studies[5] have shown that the peak-flow conditions vary and that this variation may relate to the size of the city. For cities with approximately one million people, the peak 5-min flow is approximately 20 percent higher than the average-flow rate indicated by the total hourly volume. For cities of smaller size, the peak flow is a much greater percentage of the average flow.

The peak hourly volume presently used as a criterion for design must be expanded to accommodate the higher rates of flow which exist over shorter intervals within the peak hour. This is true because a condition in which the demand exceeds the capacity can extend congestion for a much longer time than just the duration of the peak-flow period.

Two peak-period parameters, the *duration* and the *magnitude,* have been defined in considering the peaking characteristics at signalized intersections (see Fig. 5.2).[6] The duration of the peak period is that continuous interval within the peak hour in which 5-min rates of flow exceed the average peak hourly rate. The magnitude of the peak period Y is the ratio of average flow during the peak period to average flow during the peak hour:

$$Y = 1.225 - 0.000135P - 0.1L + 0.00003V$$

where P = population of city/1,000

$\quad\quad L$ = location of intersection expressed as ratio of distance from CBD to city limits

$\quad\quad V$ = peak-hour approach volume

Several researchers[7,8] have proposed a *peak-hour factor* for freeways, which is the ratio of the rate of flow during the highest 5 min to the rate of flow during the whole peak hour (see Fig. 5.3). The new "Highway Capacity Manual"[9] uses the highest 15-min rate of flow as the basis for determining the peak-hour factor.

The characteristics of the peak-period rates of flow are influenced by many factors, although the most important seems to be the population of the city. The character of the freeway (radial, circumferential, etc.) in reference to generators and to number and capacity of lanes all affect the variations in flow within the peak hours.

Although the traffic volume in each direction on two-way facilities tends to balance for longer time periods such as a day, an unbalance of flow usually exists for the peak periods. A knowledge of the traffic load in each direction (directional distribution) is essential in geometric design.

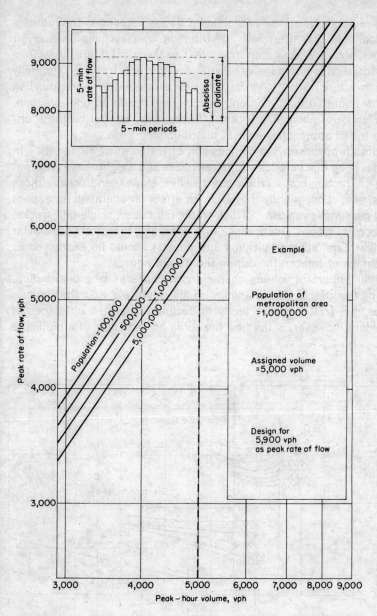

Fig. 5.3 Determination of the rate of flow for the highest 5-min interval from the rate of flow for the whole peak hour.

On urban radial facilities, the predominant flow is inbound in the morning and outbound in the afternoon.   Distribution by directions on a given facility does not vary, as a rule from year to year; therefore, the directional distribution percentage (D value) for current traffic is generally applicable to future traffic.   Typical values for rural highways and suburban areas show about 60 to 80 percent of peak-hour traffic in one direction, with an average of about 67 percent.   In and near central business districts, the traffic approaches a 50:50 distribution.

For design purposes, the composition of traffic must be known.   In Chap. 2, it is seen that vehicles of different sizes and weights have different operating characteristics.   Trucks are heavier, slower, and occupy more roadway space; consequently they have an effect on operation equivalent to several passenger vehicles.   Truck traffic (all buses, single-unit trucks, and truck combinations except those with operation characteristics similar to passenger cars, such as pickups and panels) should be expressed as the percentage of total traffic during the design hour T.

Another important volume characteristic concerns lane distribution. Because of the design and markings of the highways and the driver's experience, the traffic stream tends to become separated into lanes of flow.   Seldom, however, is the flow the same on each lane of a multilane facility.

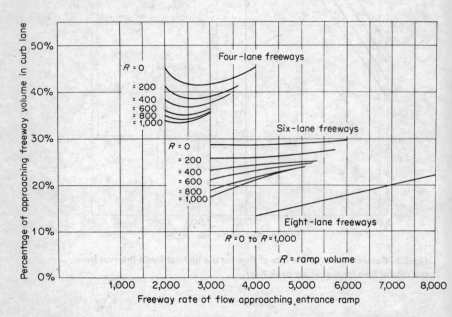

Fig. 5.4 Relationship between percentage of total freeway volume in the outside lane and freeway and ramp volumes.

It has been illustrated in Fig. 5.3 how the rate of flow for the highest 5-min interval can be determined from the rate of flow for the whole peak hour.   These flows represent the demand on all lanes and as such are inadequate for making a capacity-demand analysis at critical sections of the freeway.   Such critical sections often exist adjacent to ramps, and if a certain level of service is to be assured the motorists, it is necessary to give close attention to such areas.   Because the merging problem directly involves traffic in the outside lane and the entering ramp traffic, the percent of total freeway traffic using the outside lane is a desirable parameter.

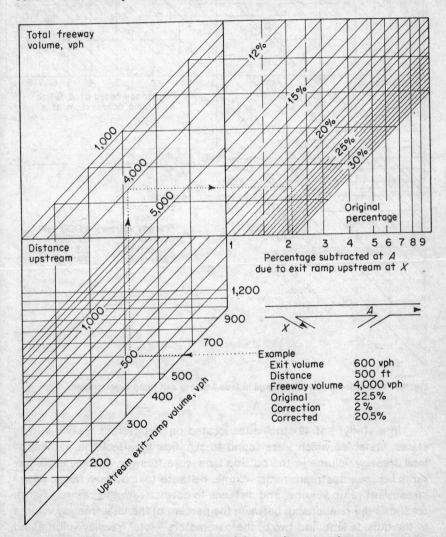

Fig. 5.5 Correction to the percentage in lane 1 due to exit ramp upstream.

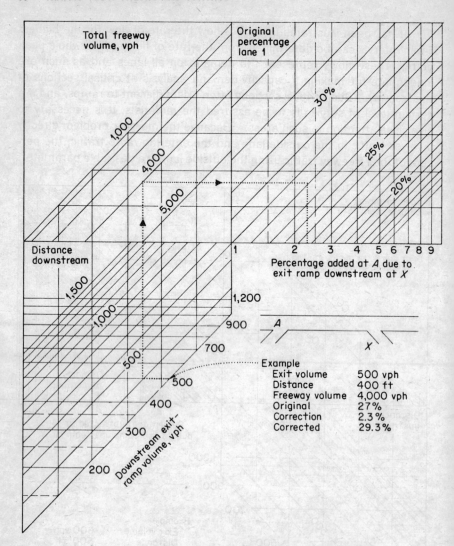

Fig. 5.6 Correction to the percentage in lane 1 due to exit ramp downstream.

In a study[7,8] of 49 study sites located on 14 different freeways in 10 states, variables which were found to significantly affect the percent of total freeway volume in the outside lane were freeway volume, entrance-ramp volume, upstream-ramp volume, distance to upstream ramp, downstream exit-ramp volume, and distance to downstream exit ramp. Figure 5.4 shows the relationship between the percent of the total freeway volume in the outside lane and two of the parameters—total freeway volume and entrance-ramp volume. The nomographs shown in Figs. 5.5 to 5.7 represent the relationships between the percent of the total traffic in the outside lane and the volume "on," and distance to, an upstream exit ramp; the volume "off," and distance to, a downstream exit ramp; and the volume "on," and distance to, an upstream entrance ramp, respectively. Distances are referenced to the ramp nose in each case. A downstream entrance ramp was considered to have no effect on the percent of traffic in the outside lane.

Intuitively, one would think that a driver's choice of lane would be based on more than ramp configuration and relative volumes. Drivers

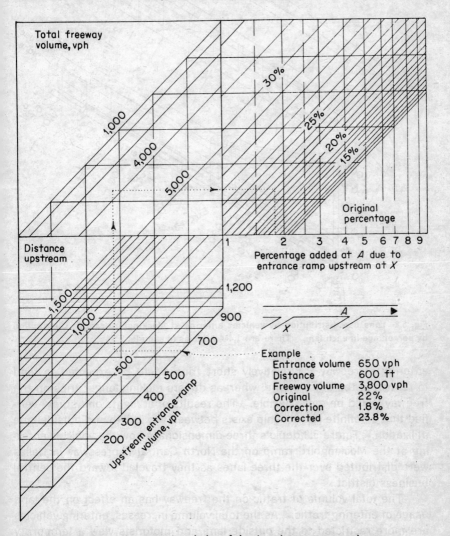

Fig. 5.7 Correction to the percentage in lane 1 due to entrance ramp upstream.

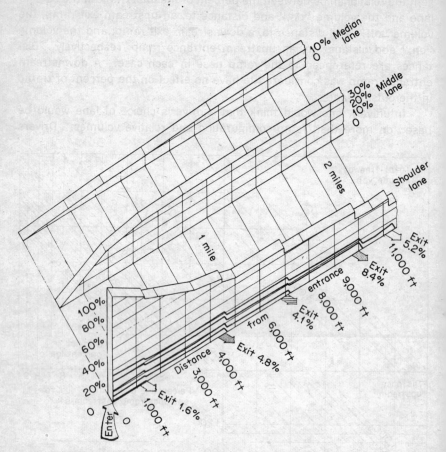

Fig. 5.8 Lane use distribution by vehicles entering at Mockingbird ramp. Distribution by percentage in each lane. There are 1,444 vehicles in sample.

entering a freeway for relatively short trips would be expected to tend to remain in the outside lane, whereas drivers making longer trips on the freeway would be more flexible.   The results of a "lights on" study[7] verified that a definite relationship exists between trip length and outside-lane utilization.   Figure 5.8 depicts three-dimensionally how the vehicles entering at the Mockingbird ramp on the North Central Expressway in Dallas were distributed over the three lanes as they traveled toward the central business district.

The total volume of traffic on the freeway has an effect on the lane usage of entering traffic.   As the total volume increases, entering vehicles are more restricted to the outside lane and motorists view a temporary lane change less worthwhile in view of the fact that another lane-change opportunity must be found to return to the outside lane prior to exiting. Figure 5.9 shows the relationship between the trip length and the percent of the entering traffic in the outside lane.   The upper portion of Fig. 5.9 pertains to only light total freeway-traffic volumes of less than 3,000 vehicles per hour (vph), one way.   The lower portion of Fig. 5.9 corresponds to conditions of moderate to heavy total freeway traffic volumes of over 3,000 vph, one way.   Although the observation points are widely spaced, the restrictive effect of the higher volumes is noticeable.   It is also apparent that vehicles traveling less than 3 miles cannot be expected to reach that *steady-state* lane distribution which is characteristic of "through" vehicles.

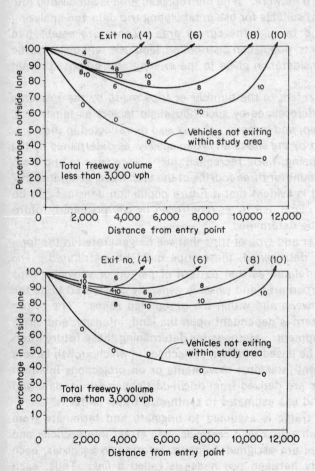

Fig. 5.9 Outside-lane use relation with trip lengths.

5.7 FUTURE TRAFFIC

It has been shown how the traffic engineer measures present traffic in order to evaluate existing operation. The design of highway facilities necessitates a knowledge of future traffic demand. Many relatively new highways are now overcrowded because they were designed for traffic volumes far lower than those that actually developed. In fact, there probably never has been an overdesigned highway.

Today's methods for estimating future traffic generally utilize a three-step procedure: (1) trip generation forecasting, (2) distribution of these trips between zones in the urban area, and (3) the assignment of the interzonal trips to a future network. The metropolitan area is subdivided into smaller areas (zones) suitable for use in tabulating trip data and analyzing the travel within and through the survey area. Zones are established based on census tracts, population distribution, land use, and major traffic generators, with consideration given to the existing and proposed street system.

Trip generation refers to the number of trips made by the residents of a household, as determined by such household factors as family size, automobile ownership, and income. It may also be reflected in the number of trips attracted by the site of an urban activity, as determined by the type of activity, shopping, work, recreation, etc. Since trips to and from residential areas account for three-fourths of the total travel generated in a metropolitan area, it is evident that if future population densities can be estimated for known land use zones (controlled by city planning), future trip generation can be determined.

After the number and type of trips that will be generated in the forecast year has been determined, these trips must be distributed. *Trip distribution* may be defined as that pattern of person or vehicular trips which exists in an urban area and which is generally considered to be the transfer of trips between and within the designated zones. The future trip distribution pattern is dependent upon the kind, intensity, and direction of urban development. Methods for determining these future interzonal transfers may be based either on projections in which growth factors are applied to present interzonal movements or on projections in which travel characteristics are derived from origin-destination survey data and applied to future land use estimates to synthesize travel patterns.

All distributed traffic is assumed to originate and terminate from central points in each zone called *centroids*. All street intersections and freeway interchanges are assigned numbers referred to as *nodes;* each segment of a facility between two nodes is called a *link*. Thus, each centroid is connected to every other centroid in the system by a series of links called a *tree*. A computer program is utilized to assign traffic

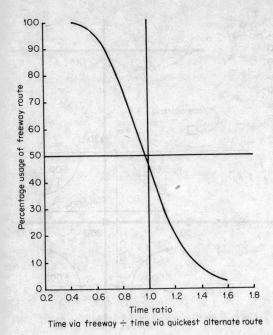

Fig. 5.10 Traffic diversion curve using time ratio.

volumes to each link based on some criterion. The simplest method, the all or nothing, assigns all trips to the shortest freeway route or to the shortest alternate arterial route, based on a comparison of the travel times. Another method, diversion, is illustrated in Fig. 5.10.

Initial assignments are usually for nondirectional 24-hr volumes. Usually two systems are built: one to portray the existing major network and another to include the proposed arterials and freeways in the future network. A typical future assignment network is shown in Fig. 5.11.

It should be pointed out that several attempts have been made to develop an assignment model which will make allowance for overassignments to a given route or link. These are referred to as *capacity restraint models* and include (1) models in which the travel-impedance parameter on a link is varied with volume and the assignment of interzonal trips to the system is done in an iterative procedure until the change in the parameter is less than some specified value and (2) models in which the minimum path is determined, the next best path is determined, and a diversion curve is used to proportion the trips to alternate routes.

From the point of view of the designer, there is a need for "programmed assignments"—or 5-, 10-, and 15-year projections as well as 20-year. In this manner, the transportation plan would be continually updated. Assignments should be based on the peak hours since this is

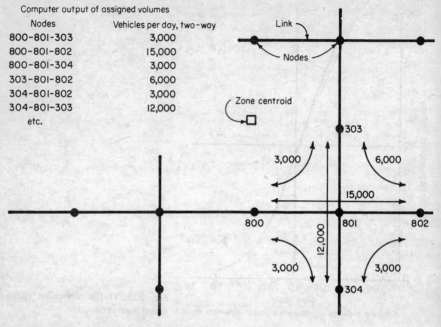

Computer output of assigned volumes

Nodes	Vehicles per day, two-way
800-801-303	3,000
800-801-802	15,000
800-801-304	3,000
303-801-802	6,000
304-801-802	3,000
304-801-303	12,000
etc.	

Fig. 5.11   Illustration of assignment network.

the period in which the driver measures the efficiency of a facility.   The two most important trips are to work and to his home.

5.8   CAPACITY

Capacity is the maximum number of vehicles that can pass a point on a lane or roadway during a given period of time under prevailing roadway and traffic conditions.   Capacity is a particular rate of flow—the maximum rate.   Since the usual time modifier is the hour, capacity is usually expressed in so many vehicles per hour per lane or roadway, that is, 2,000 vph per freeway lane.

Capacity is dependent on the physical features of the roadway, the composition of traffic, and such ambient conditions as weather and daylight.   The maximum observed traffic volumes, as obtained in the counting procedures discussed in Sec. 5.5, have been used as guides in selecting numerical values for the capacity of different types of facilities under various prevailing conditions.   This tends to limit highway capacity to the amount of traffic demand and is similar to estimating the capacity of a steel rod by the load on the rod.   The only way to be sure that one has actually measured capacity is to load either the highway facility or steel rod to failure. This is not easily discernible in the case of a highway since failure is not as dramatic as it is in the case of a steel rod.

There is evidence that studies and application of the capacity concept date back to the 1930s. On the basis of 5-min traffic counts carried out as a cooperative effort among the U.S. Bureau of Public Roads, the State Roads Commission of Maryland, and the University of Maryland, Johnson[10] determined some "working capacities" at 51 stations in the New England area. Capacity was marked as the rate of flow at which congestion becomes evident.

A year later, in 1931, Johannesson[11] based his roadway capacity on the average minimum vehicle spacing, center to center, of the vehicles. He writes, "The traffic capacity of a roadway is reached when any further increase in traffic volume, all other factors remaining constant, results in a marked decrease in traffic speed."

In 1933, Greenshields[12] used a time-lapse photographic technique to measure the speeds and distances between individual vehicles. He plotted the speeds against the spacings and fitted the straight line to the points:

$$S = 21 + 1.1V$$

where V is speed in mph and S is distance in feet. It can be seen that if V is converted to feet per second, the coefficient of V becomes 0.75. The units of this coefficient are in time, and its significance is the reaction time of the average driver. A year later, Greenshields[13] used the reciprocal of the spacing values for the abscissa and, calling it density, changed traffic engineering from an art to a science.

The vehicular capacities of major-street intersections are an index of the capacity of the arterial-street system as a whole. Intersection capacity is dependent upon its geometric characteristics, such as surface width of streets, and upon such variables as composition of traffic, percentage of vehicles making turns, and the kinds of traffic and parking controls in operation.

The intersection capacity per lane for any green period is measured by the number of vehicles which can enter the intersection during this period. The average minimum headways which develop when a queue of vehicles is released at the stoplight are shown in Table 5.2. As the speeds between successive vehicles as they cross the stop line become constant and equal to the vehicles' desired speed for that facility, the headways tend to become a constant of about 2.0 sec, suggesting a lane capacity approaching 1,800 vph of green.

The determination of highway capacity is as fundamental to traffic engineering as the study of strength of materials is to structural engineer-ing. We have seen that there is a significant body of literature dating

Table 5.2a   Typical reduction of time intervals between successive vehicles into departure headways and starting delays

Vehicles	Interval I	Headway D	Delay K_1
0–1	2.8	0.0	2.8
1–2	2.6	2.0	0.6
2–3	2.1	2.0	0.1
3–4	2.1	2.0	0.1
4–5	2.0	2.0	0.0
5–6	2.0	2.0	0.0
Summations: 6	13.6	10.0	3.6

D (departure headway) = 2.0 sec
K_1 (starting delay) = $\Sigma(I - D)$ = 3.6 sec

Table 5.2b   Summary of departure headways and starting delays for through and turning movements

Type of movement	D	K_1
Through	2.0	4.0
Left turn	2.0	3.9
Right turn	2.0	4.1
Side-by-side turns:		
Inside lane	2.2	4.7
Outside lane	2.4	5.3

back more than 30 years, illustrating that early practitioners appreciated the fact that capacity and quality of service should be considered. However, it did not begin to be generally accepted as a useful tool in highway design, operation and control, and regulation and management until after World War II. More recently, capacity has been the subject of many theoretical studies and the source of many important traffic applications, as shall be discussed in some of the later chapters in this text.

5.9   LEVEL OF SERVICE

The formation of the Highway Capacity Committee in the Highway Research Board about 1944 brought together under the chairmanship of the late O. K. Normann some of the individuals who were involved in developing the early capacity concepts. This early work provided the basis for the method of measuring capacity reported in the 1950 edition of the "Highway Capacity Manual." This contribution of the Capacity Committee of the Highway Research Board has served vitally in modern highway design.

For the most part, the manual was based on two capacity levels — possible and practical. *Possible* represented the maximum hourly volume under prevailing roadway and traffic conditions and *practical*, "the maximum number of vehicles that can pass a given point on a roadway or in a designated lane during one hour without the traffic density being so great as to cause unreasonable delay, hazard, or restriction to the driver's freedom to maneuver under prevailing roadway and traffic conditions."[14] Under proper application, this concept served most satisfactorily for a number of years.

For rural conditions, the basic concepts of the 1950 manual are probably still quite applicable. The desire for higher speeds and better conditions, however, makes it desirable to review the concept even for strictly rural conditions. In urban conditions, the desire for a more dependable and higher quality of service has brought about a change in the application of "capacity" in the design and operation of streets and highways. This new application has been termed the *level-of-service concept.*

Level of service, as applied to the traffic operation on a particular roadway, refers to the quality of the driving conditions afforded a motorist by a particular facility. Factors which are involved in the level of service are (1) speed and travel time, (2) traffic interruption, (3) freedom to maneuver, (4) safety, (5) driving comfort and convenience, and (6) vehicular operational costs.

Each of the foregoing factors is somewhat related to all the others. The volume of traffic using a facility affects all the factors, and in general, the greater the volume, the more adverse are the effects. As the ratio of the volume of traffic on a facility to the volume of traffic the facility can accommodate approaches unity, congestion increases. *Congestion* is a qualitative term, long used by the general public as well as traffic engineers, which refers to what can quantitatively be defined as *vehicular density.* The result of an oversupply of vehicles is the formation of a queue of stopped (or "crawling") vehicles at bottleneck locations (a "breakdown" of the operation) such that volumes momentarily drop to zero, leaving only congestion on the facility until a "clearout" can be effected.

Traffic volumes are known to be continuously variable; even at very low hourly volumes there will be infrequent short-term occasions when a relatively large number of vehicles will pass a given point. There also are regions on a facility which, because of the geometry, inherently will tend to accommodate fewer vehicles. This implies that bottlenecks do exist, and thus the level of service on a given facility may vary even with a "constant" hourly volume. Bottlenecks may be fixed in space because of the aforementioned geometrical considerations of the facility and may be studied at the particular location.   Such geometrical aspects as entrance and exit ramps have been studied and evaluated as bottlenecks.   It is possible, however, that the random "bunching" of vehicles at any point in space may result in "bottlenecking," because of the statistically variable nature of streams of vehicles.   In this case, the designers should be able to predict such peaking characteristics in order to assure acceptable levels of service.

Basically, congestion will be the direct result of the nature of the "supply and demand" on a facility.   The supply, in terms of traffic engineering, has been referred to as capacity.   The demand placed on the facility is, as it implies, the number of motorists who seek to use the facility; it can be estimated by origin and destination surveys if the times of the desired trips are obtained.   It is often futile to measure the flow of traffic on an existing facility with the objective of determining the demand on that facility.   About the only relationship between existing volumes and actual demand which can be determined from such measurements is whether the demand is, in fact, as large as the capacity during any significant length of time during a day.   In borderline instances, peaking occurring within a peak hour might show where capacity is exceeded by demand for intervals of time less than 1 hr.   Even this feature cannot be exhibited by a traffic system which is so inadequate that it limits (or "meters") the input of vehicles such that the volumes are less than the capacity of the particular facility being studied.

If it is possible, in a given system, for more vehicles to enter than the facility can handle, congestion will result whenever the demand exceeds capacity; and the accompanying inefficiency results in fewer numbers of vehicles being accommodated by the facility in a given period of time.   It is theoretically true that there is some maximum number of vehicles which can use a facility.   This *possible capacity* is the volume of traffic during the peak rate of flow that cannot be exceeded without changing one or more of the conditions that prevail.   From this value more restrictive conditions of roadway and traffic conditions are imposed to describe the measure of level of service that a given lane or roadway should provide.   If the conditions are associated with highways or streets to be constructed at a future date, it is defined as *design level of service*.   If the conditions express prevailing traffic flow conditions, it is designated as *operational level of service*.

Various volume levels can cause various levels of operating conditions, or levels of service.   For any volume of vehicles using a particular facility, there is an associated level of service afforded these vehicles.   It is possible that the input of vehicles into a particular facility will be regulated such that traffic volumes will not exceed a predetermined suitable level of service volume. Such operational control procedures are being investigated and seem to offer considerable promise.

Although the use of a design-level-of-service volume has considerable appeal in that it conforms to traditional engineering practice, the determination of such a volume, relative to various levels of service, is complex. There are regions on a freeway which are subject to more restrictive vehicular operation, such as in the vicinity of an entrance ramp or exit ramp. Such regions should be considered when determining the design volume of a facility, and a knowledge of the operational characteristics and traffic requirements at such locations is necessary for proper planning, in order to avoid future bottlenecks.

In effect, the 1966 "Highway Capacity Manual" is not a major departure from the 1950 manual as far as "capacity" is concerned. It does, however, attempt to provide those responsible for the design and operation of traffic facilities with a range of service volumes (rather than one "practical" capacity) which will be related to various conditions of operation, or levels of service. The highest quality, or level, of service can then be provided, based on economics and other management and engineering controls.

The new manual takes into account several characteristics for which data were not available in the preparation of the previous manual. For one thing, the variation of flow or peak flows within the hour has been considered. The short-period peak rate of flow (5- to 15-min period) is accounted for through the application of a peaking factor, as discussed in Sec. 5.6 of this chapter. The consideration of these short-peak flows will much more accurately relate traffic flow (volume) to traffic operation or conditions experienced by the motorists.

Both capacity and level of service are functions of the physical features of the highway facility and the interaction of vehicles in the traffic stream. The distinction is this: A given lane or roadway may provide a wide range of levels of service but only one possible capacity. The various levels for any specific roadway are a function of the volume and composition of traffic. A given lane or roadway designed for a given level of service at a specified volume will operate at many different levels of service as the flow varies during an hour and as the volume varies during different hours of the day, days of the week, periods of the year, and during different years with traffic growth. In other words, fluctuations in demand do not cause fluctuations in capacity but do effect changes in the quality of operation afforded the motorist. In a very general way then, highway planning, design, and operational problems become a case of whether a certain roadway (capacity) can handle the projected or measured demand (volume) at an acceptable level of service (speed, etc.). Because of both observed and theoretical speed-volume relationships on freeway facilities, which shall be considered later, it is possible to anticipate to some degree just what level of service can be expected for a given demand-capacity ratio. The obvious weakness lies in the fact that most of the qualitative factors affecting level of service have not yet been related directly to traffic volume.

Greater dependency on motor vehicle transportation has brought about a need for greater efficiency in traffic facilities. The motorist is no longer satisfied to be "out of mud." In fact, fewer and fewer people remember the days of unpaved roads. The freeway is an outgrowth of the demand for highways which provide higher levels of service. The place that motor vehicle transportation plays in our society demands that dependable service be provided by traffic facilities, and the attraction to the freeway illustrates this point. It is very important that the engineer clearly understands the factors affecting efficiency, or level of service, of our highways and streets.[14]

An individual street or highway is a part of a network, or system, of facilities that provides access and movement of motor vehicle transportation. Various types of streets and highways in the system serve different functions and are generally grouped according to the two major functions of access and traffic movement. Those designed primarily for land access are generally referred to as local streets; collector facilities serve both access and traffic movement more or less equally. Major arterials must accommodate greater volumes of traffic for longer distances, and therefore movement is primary; but the land-access function is also served. The freeway-type facility is designed primarily for traffic movement, and access is not part of its function.

The higher types of facilities, major arterials and freeways, generally require careful analysis of capacity and level of service. The intersection is the major capacity problem on at-grade facilities such as major streets. The freeway epitomizes the level-of-service concept since it is designed to provide a high quality of traffic movement.

5.10 CHALLENGES

At the present time, expediting traffic movement on existing street systems is attacked with standard traffic engineering procedures on a section-by-section basis. There is a need to describe the problem in terms of network theory and optimize traffic movement on that basis. Part of the problem stems from the inadequate understanding of the nature of interaction between flows on these complex networks. A necessary requisite to its solution is an analytic description of this flow process.

PROBLEMS

5.1. Select a criterion for the evaluation of a freeway system for a large city, serving commuters from the suburbs to the CBD, shoppers destined for centers located along the freeway, and highway traffic passing through the city.

5.2. Describe the cyclic variations normal to traffic volumes.

5.3. Identify each of the following traffic counts.
  a. Classification counts
  b. Cordon counts
  c. Screen-line counts
  d. Control counts
  e. Occupancy counts

5.4. Define each of the following and explain its relation to design.
  a. Thirtieth-highest hourly volume
  b. Peak hourly volume
  c. Peak-period rate of flow

5.5. Describe the procedure you would follow in setting up a volume count program for a city, discussing survey equipment requirements.

5.6. List the four types of streets and describe the function of each. How would these functions affect their design in a large new subdivision?

5.7. Sketch several types of interchanges and explain how you would go about determining the number of lanes for each segment of the freeway system including the interchange intersections.

5.8. How are the traffic assignments and other necessary traffic data determined for the design of a proposed freeway?

5.9. Find the volume of traffic in the outside lane of a six-lane freeway at the nose of an entrance ramp if the total freeway volume on the through lanes is 5,000 vph and the entrance ramp volume is 400 vph.

5.10. Repeat Prob. 5.9 for the following cases:
  a. The entrance ramp is 800 ft downstream from an exit ramp with a volume of 400 vph.
  b. The entrance ramp is 500 ft downstream from an entrance ramp with a volume of 400 vph.

5.11. Using appropriate studies investigate the applicability of short-count techniques to estimate hourly and peak-period traffic volumes.

5.12. Select an approach to a signalized intersection and count the traffic demand from 7 to 8 A.M., grouping the volumes by 5-min intervals. Calculate the peak-magnitude factor and compare it with the regression equation in Sec. 5.6.

5.13. Selecting appropriate signalized intersections and using a stopwatch, determine the departure headways and starting delays for through and turning movements and compare them with the values in Table 5.2.

REFERENCES

1. "Better Transportation for Your City," National Committee on Urban Transportation, Library of Congress Catalog Number 57-14978.
2. Gallion, A. B., and S. Eisner: "The Urban Pattern, D. Van Nostrand Company, Inc., Princeton, N.J., 1963.
3. "Measuring Traffic Volumes," National Committee on Urban Transportation, Library of Congress Catalog Number 57-14978.

4. Institute of Traffic Engineers Technical Committee 7-G: Volume Survey Devices, *Traffic Eng.*, vol. 31, no. 6, pp. 44–51, March, 1961.

5. "Design Manual for Controlled Access Highways," The Texas Highway Department, Austin, 1960.

6. Drew, D. R., and C. Pinnell: A Study of Peaking Characteristics of Signalized Urban Intersections as Related to Capacity and Design, *Highway Res. Board Bull.* 352, 1962.

7. Drew, D. R., and C. J. Keese: Freeway Level of Service as Influenced by Volume and Capacity Characteristics, *Highway Res. Board Record* 99, 1965.

8. Lipscomb, J. L.: "The Effect of Ramps in the Lateral Distribution of Vehicles on a Six-lane Freeway," thesis, Texas A&M University, 1962.

9. "Highway Capacity Manual," *Highway Res. Board Spec. Publ.*, 1966.

10. Johnson, A. N.: Traffic Capacity, *Highway Res. Board Proc.*, vol. 10, 1930.

11. Johannesson, S.: Highway Economics, *Highway Res. Board Proc.*, vol. 8, 1928.

12. Greenshields, B. D.: The Photographic Method of Studying Traffic Behavior, *Highway Res. Board Proc.*, vol. 13, 1933.

13. Greenshields, B. D.: A Study of Traffic Capacity, *Highway Res. Board Proc.*, vol. 14, 1934.

14. "Highway Capacity Manual," U.S. Bureau of Public Roads, Washington, D.C., 1950.

CHAPTER SIX
TRAFFIC RESEARCH

Research is not only getting the facts, not only experimentation,
not only building equipment—although these may be
indispensable aids; research is the conception of a new
relationship between the variables entering into the problem.

Unknown

6.1 INTRODUCTION

There has been a long-standing need for a systematic and thorough attack on the major problems facing highway transportation. Gradually, integrated and balanced programs of research and development are being developed by such agencies as the U.S. Bureau of Public Roads, the Highway Research Board, and several highway departments, with the problems of highway safety and efficient highway transportation emerging as being of the highest priority.

Traffic efficiency and safety depend significantly upon the nature of the interrelations in streams of vehicles. The basic mechanics underlying this flow process are, however, incompletely understood, with the consequence that there is limited capacity to control flow or the accidents that arise therefrom. A necessary prerequisite to the practical solution of this problem is a definition of the nature of traffic flow on real-world facilities.

The empirical approach represents the traditional approach to the explanation of traffic movement. It has provided most of the present knowledge in the field of traffic engineering. As of today, the knowledge which traffic engineers draw upon to solve problems and to design roads is largely empirical knowledge built up through observations, measurements, and statistical analyses. This approach has not been able to answer satisfactorily many of the important questions about road traffic: Why build 12-ft traffic lanes if a 9-ft traffic lane has the same capacity? What causes traffic to move in an accordion-like fashion? What is the reduction in ramp capacity of a short acceleration lane?

6.2 THE SYSTEMS APPROACH

In the previous chapter, one observes a liberal, and perhaps vague, use of the term *system*. A wholly satisfactory definition of the word system does not seem to exist. According to Hall,[1] "A system is a set of objects with relationships between the objects and between their attributes." "Objects" here might refer to the traffic elements discussed in Chaps. 3 to 5—the vehicle, the driver, and the road. Attributes are the properties of these objects or elements. Attributes, therefore, include driver vision and reaction time; vehicle speed, accelerating capability, and headway; street width and control.

A system is a collection of interacting diverse human and machine elements integrated to achieve a common desired objective by manipulation and control of materials, information, energy, and humans. Here, the traffic elements—the driver and the vehicle—are mentioned explicitly; the objective is the driver's motivation for safety and time-distance economy. On the other hand, the objective might refer to the engineer's objective in providing efficiency through the use of information and control (signs and signals). Relationships describe how the traffic units interact with each other and with the environment, where the environment is the road, street, or highway or the circulation system in general.

Traffic problems result from many specific factors and from the interaction of various combinations of these factors. The traffic phenomenon is so complex that single problem situations can seldom be corrected by changes in any one factor or by uncoordinated changes in several factors.

Systems engineering is the comprehensive examination of a total problem situation. Important steps in this examination are problem definition (including objectives, synthesis, analysis), selection of the optimum system from the alternatives based on some criterion, and evaluation. This approach often involves the formulation of a mathematical model, the analysis of the sensitivity of the system with respect to its elements, the analysis of the compatibility of the various components and subsystems, and the determination of the stability of the system when subjected to various inputs.

The definition of the objectives of a highway system is one of the most difficult tasks facing the transportation planner, the geometric designer, and the traffic engineer. This definition is important because in the systems approach it provides the means of optimization and the basis for choosing among alternates. A good example lies in freeway design. Should interchanges be spaced at great distances to minimize turbulence, or should they be closely spaced to minimize indirection in the total trip? Even more basic is the description of optimum freeway operation, which characteristically degenerates to some intuitive value system such as good, bad, congested. On the other hand, many objectives are rendered ineffective because of their lofty approach. Consider the announced objective of transportation planning "to develop an integrated system of transportation to provide an improved quality of service consistent with anticipated travel demands within the economic capabilities of the area and compatible with the requirements of the ultimate development of the area."

In the systems engineering concept, synthesis and analysis require the construction of a model which relates the topological properties of the system to its inputs and outputs. Synthesis involves the combination of parts to achieve a whole, usually accomplished by extrapolating or interpolating existing techniques and results to achieve objectives. These in turn are subject to analysis. Analysis is a separation of the system into components in order that all consequences in terms of objectives can be determined. Synthesis and analysis in practice cannot be separated but go hand in hand. Various techniques, many popularized in operations research, such as linear programming, queueing theory, simulation, and critical-path methods, are employed in this synthesis-and-analysis phase.

6.3 THE MATHEMATICAL MODEL

One of the most difficult tasks facing the traffic researcher is the translation of a real traffic problem situation, comprising drivers, vehicles, control devices, and the highway, into a set of mathematical symbols and relationships that reproduces their behavior. The fundamental conceptual device, which enables one to regard this interaction as a whole, is a model.

A model is an idealized representation of reality. It must be constructed in such a way as to reproduce the behavior of the real world with acceptable accuracy, recognizing that no abstraction can be identical to reality. This attempt to establish a correspondence between the problem and rational thought may be realized by forming a physical model or a theoretical model.   Physical models may be scalar or they may be analogs of the prototype, such as a wind tunnel for testing aircraft and an hourglass for measuring time.   Kirchhoff's dynamical analogy illustrates that the critical load on an axially loaded structural column may be determined by studying the oscillations of a pendulum of equal length.   Numerically oriented analog models may be even more subtle.   The slide rule and nomograph are numerical analogs; the odometer of a vehicle gives us the area under a vehicular speed-vs.-time curve and as such is an integrator.

A theoretical model is essentially a hypothesis.   For example, Newton's three laws of motion provide a theoretical model of our physical world. Most theoretical models are mathematically oriented for obvious reasons. Unless the hypothesis and the situation it describes are very simple, the only practical method of studying the many manifestations of a complex system is with the aid of mathematics.   The trick, in representing a traffic situation, is to define the relevant parts and arrive at a set of relationships between them which, though simple, will result in meaningful predictions.

The traffic engineer has only a limited influence on the traffic variables. True, he can add traffic lanes so as to reduce the rate of flow per lane, he can set speed limits to discourage high speeds, and he can erect traffic signals to alternate the right-of-way between conflicting streams.   However, traffic demand, vehicular capabilities, and driver performance are a few of the variables over which he has little influence.

As in the case of other disciplines, many problems can be reduced to finding the maximum or minimum of some function.   Thus, the determination of the number of traffic lanes is contingent upon finding the maximum rate of flow (capacity), whereas traffic-signal cycle apportionment may be based on minimizing delay.   If the mathematical model chosen enables precise computation of what will happen to one variable if a specified value is chosen for another variable, the model is said to be *deterministic*.   This may be contrasted with *stochastic* models, in which allowance is made for the probability of a variable assuming various values.

The deterministic aspects of vehicular traffic may be explored by studying a single unit of traffic or the traffic stream as a whole.   The effect of the motion or the headway of one vehicle on another vehicle is referred to as the local, or *microscopic*, properties of traffic.   On the other hand, the relation between traffic flow over an extended period of time to traffic concentration over a portion of the roadway defines the overall, or *macroscopic*, properties of traffic.

In very general terms, any sequence of experiments that can be subjected to a probability analysis is called a stochastic, or random, process. Such a process can be either *independent* or *Markov*.   An independent process occurs when experiments are performed in such a way that the outcome of any one experiment does not influence the outcome of any other experiment.   If the outcome of any particular trial or experiment depends on the outcome of the immediately preceding trial, it is called a Markov process.

The organization and relationship of many of the traffic models that will be discussed in the remaining chapters are illustrated in Fig. 6.1.   It will be shown that the theoretical phases of many investigations can utilize either the stochastic or the deterministic approaches or both.   Consider the freeway merging process, for example.   Merging is a probabilistic phenomenon involving a distribution of ramp arrivals, distribution of freeway gaps, distribution of waiting times for ramp vehicles in a position to merge, and even distribution of driver reaction times and vehicular accelerating capabilities.   After the merge, a certain percentage of the ramp vehicles will move to other lanes, during successive incremental sections of the freeway.   This appears to be of the nature of a Markov process, which is subject to fluctuations before a stationary process, or steady state, is reached.

Continuing with the merging example, in the vicinity of the merging area, shock waves are produced in the outside lane, spreading to the other lanes.   The quantitative description of this effect is best described by the utilization of the deterministic approach as manifest in the energy-momentum concept, since this offers a rational description of both capacity and level of service.

6.4 EXPERIMENT AND EVALUATION

Inherent in the research process are theory formulation, experimentation,

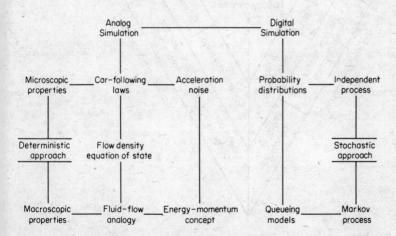

Fig. 6.1 Classification and organization of traffic models.

and evaluation. Theory formulation includes establishing the criteria for system optimization and formulating a mathematical model. Experiment can be of either a controlled or uncontrolled nature. Controlled experiments, in traffic research, may be either in the laboratory or in the field. Examples of the former are simulation, both analog and digital; examples of the latter are test tracks, such as the one operated by General Motors and to a lesser degree the various tunnel and freeway control projects.

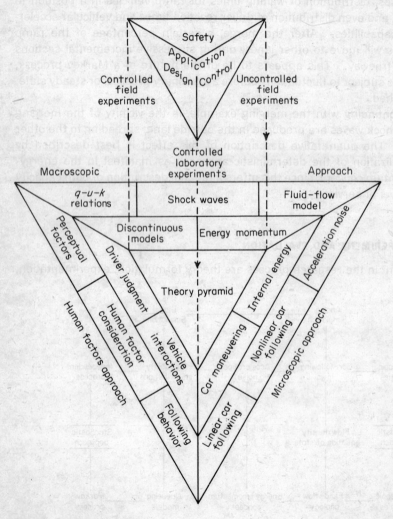

Fig. 6.2 Research pyramid.

The steps in the traffic research process are diagramed in Fig. 16.1. The implication is that the controlled laboratory experiments are an attempt at solution where other models fail.   The controlled field experiments are useful in evaluating theories, but actual validation must be based on real-world traffic, the uncontrolled field test.   The last step, and probably most important, is presentation.

This contemporary approach to understanding the mechanics of traffic flow is illustrated in the *theory pyramid* of Fig. 6.2.   The macroscopic approach is based on such physical analogies as kinematic waves, compressible-fluid flow, heat flow, and energy-momentum concepts.   The microscopic approach treats the driver-vehicle combination as a closed-loop system and then investigates the tracking characteristics using control-system concepts.   The human-factors approach seeks to define the processes by which a driver locates himself with reference to other vehicles and to the highway, in order to describe the information-handling and decision-making processes involved in the driving task.   Each of these three scientific approaches is appropriately represented as a surface of the deterministic theory pyramid since all three are needed to rationally explain traffic movement.

The microscopic approach best describes stability in vehicular interactions, so important in highway safety.   The macroscopic approach is particularly appropriate for the steady-state phenomenon of flow and hence best describes operational efficiency.   The human-factors approach strives to understand the unique sensory and motor capabilities and limitations of the driver, thus minimizing situations where he is unable to handle adequately the functions of steering control, absolute velocity control, or relative velocity control.

Figure 6.2 shows the three basic experimental techniques used in traffic research—the controlled laboratory, the controlled field, the uncontrolled field.

This research approach and all present research are significant for the simple reason that time is running out.   In a half-dozen years the present 41,000-mile Interstate System will be virtually complete.   After this what do we build?

The chances are that 500-mile stretches of wide clear "runways" will be proposed, to be built for travel at up to 150 mph between our metropolitan centers.   Double-decker aquatic freeways will arc around large seaports far enough underwater to allow ship traffic and will surface like a whale in places to provide a breakwater a few hundred feet offshore.   If these dreams are to become reality, this research must help find a source of communication "from the ivory tower to the marketplace."

At the beginning of the twentieth century there was a time lag of 25 to 50 years between the discovery of a new scientific fact and its practical application. This time lag has been reduced steadily, especially when a sense of urgency or national security has been involved. The facts that established the possibility of a nuclear explosion were discovered in 1939, and in 1945 the first nuclear explosion took place—a total of 6 years from basic discovery to practical application.

This reduction in time lag has resulted in new relationships in research and application. The application of new scientific theory to modern complex systems creates the need for more exact theory. The first task facing the traffic researcher is the translation of a real traffic problem situation into a set of mathematical symbols and relationships that reproduces their behavior. The second task facing the traffic researcher is the application of this knowledge. As the sophistication of the application increases, more scientific theory must be made available.

Research workers of varying interests have provided a wealth of theory toward the description of traffic flow. However, a gap exists between theory formulation and application. Although the theorist has limited his efforts almost exclusively to formulation, the practitioner has been forced to seek out the solutions to his day-to-day problems. The organization, testing, validation, and evaluation of existing traffic theory must be undertaken in order to close the gap between formulation and application.

PROBLEMS

6.1. Select an intersection in your city which is signalized and frequently encounters operational problems. Sketch the intersection. Discuss the aspects of its operation that you would study, the various elements you would consider, and where you would draw the boundary of the environment in conducting a *systems analysis* of the intersections.

6.2. A driver went through a light barricade which had been erected to prevent vehicles from using a structure that still lacked the other approach. When the police arrived at the scene, they measured the horizontal distance from the abutment to the center of the vehicle and found it to be 120 ft. If the structure is 20 ft high, what was the speed of the vehicle when it left the structure? List the assumptions necessary to make it possible to solve the problem according to the model you selected to describe the phenomenon. Indicate which assumptions you consider significant.

6.3. Find five examples of the application of empirical research results in "A Policy on Geometric Design of Rural Highways, 1965" by AASHO.

6.4. You are asked to develop a model to describe the operation of an airport in a city of 300,000 people at present which will be applicable to the operation of the facility as the city grows to a population size of 1 million persons. Outline your study considerations in attempting to develop such a model.

6.5. As the newly appointed chairman of a state commission on traffic safety with a $1.5 million annual budget which includes a major portion allocated to traffic safety research and education, in what approaches to traffic research would you be most interested? Define problems on which you would invite traffic research groups to propose research.

REFERENCES

1. Hall, A. D.: "A Methodology for Systems Engineering," D. Van Nostrand Company, Inc., Princeton, N.J., 1962.
2. Rekoff, M. G.: "Systems Engineering," unpublished.

CHAPTER SEVEN
TRAFFIC EVENTS

There are three kinds of lies: lies, damned lies, and statistics.

Disraeli

7.1 INTRODUCTION

In order to evaluate the chance that a certain traffic event (an arrival at a stoplight, an accident on an overpass, an available parking space) will occur, one may adopt either of two procedures. One may rely on past experience with similar situations and assume that future results will be approximately the same, or one may analyze the various ways in which the event can happen or fail to happen and thereby compute a theoretical probability. The former method is that of statistical probability, which has considerable value and utility in connection with statistics; the latter gives rise to mathematical probability.

Probability and statistics have been likened to two people approaching the same house from opposite ends of the street. In probability the contributing factors are known, but a result can never be predicted with absolute certainty. In statistics the end product is known, but the causes are in doubt.

The collective behavior of the millions of Americans driving in random fashion over an infinite variety of roads is thoroughly predictable, although it is impossible to foresee that driver *A* will crash into driver *B* on Route 66. The theory of probability does not preclude the possibility that driver *B* will suffer a streak of bad luck and become involved in several accidents. The laws of chance begin to act as laws only when many instances, in this case many collisions, are involved.   Driver *B* has only a remote chance of being consistently unlucky, that is, consistently doing worse than a probability prediction would warrant.   On the other hand, a long run of good luck does not decrease the chance that an individual will again be lucky on a given occasion.   A salesman who drives thousands of miles a year does not incur a greater risk of an accident every time he takes his car out of the garage.

Whereas the individual driver is interested in probability, his insurance company applies statistics.   An insurance actuary knows the amount of damage that many thousands of drivers will do collectively.   Based on these statistics, he accepts driver *B*'s $100-a-year bet that his car will not cause $20,000 worth of damage.   Driver *B*'s premium ends up being only slightly more than his share of all the damage that he and many thousands of drivers like him will do collectively.

One difficulty in applying the laws of probability lies in determining all the possible ways in which an event can take place.   In most games of chance, a gambler is essentially drawing chances out of a grab bag in which the kinds of payoffs and their relative proportions are known in advance.   However, in most practical traffic situations, it is seldom possible to enumerate all the possible outcomes.   Again, one must be impressed with the inseparability of probability and statistics since the usual approach is to take a well-judged experimental sampling from the grab bag and then to assess to what degree of confidence it represents the entire contents of the bag.

7.2   THE BINOMIAL DISTRIBUTION

Early theories of traffic flow were, as might be expected, statistical.   They included methods for measuring the means and standard deviations for such traffic characteristics as speed and volume.   It was learned, however, that the traffic phenomenon could not be completely described by single measures of central tendency and dispersion but only by functions called *frequency,* or *probability, distributions.*

Everyone has had occasion to deal with positive integral powers of a binomial expression, that is, with expressions of the form $(p + q)^n$, where

n is a positive integer. By repeated multiplication of $(p + q)$ by itself, one obtains

$$(p + q)^n = \sum_{x=0}^{n} \frac{n!}{x!(n - x)!} p^x q^{n-x} \qquad (7.2.1)$$

The statement of this result, given in (7.2.1), is known as the *binomial theorem*. The coefficients in the expansion are called *binomial coefficients*, often expressed symbolically as $\binom{n}{x}$ where

$$\binom{n}{x} = \frac{n!}{x!(n - x)!} \qquad (7.2.2)$$

Equation (7.2.2) gives us the combination of n things taken x at a time. If we assume that all vehicles are to be classified as compacts, standard passenger cars, pickups, trucks, and buses, the combinations of two vehicles would be

$$\frac{5!}{2!3!} = 10$$

Although n has been restricted to positive integral values, no restrictions have been placed on the quantities p and q, as yet. We next consider the question of successive independent trials of an event. Let p denote the probability that a certain event will occur in a given trial, so that $q = 1 - p$ is the probability that the event will fail to occur in a single trial. Clearly since $p + q = 1$, the sum of the $n + 1$ terms of the series in the right side of (7.2.1) must also equal unity. Therefore, each term of the binomial expansion represents a probability; and all the terms represent a probability distribution, appropriately called the *binomial distribution*. The binomial distribution is expressed as

$$P(x) = \binom{n}{x} p^x q^{n-x} \qquad (7.2.3)$$

where p = probability of success on any given trial
$\qquad q$ = probability of failure on any given trial
$\qquad n$ = number of independent trials
$\qquad x$ = number of successes
$\qquad P(x)$ = probability of x successes in n trials

To illustrate consider an intersection approach in which studies have shown 25 percent right turns and no left turns.   The probability of one out of the next three vehicles turning right is, from (7.2.3),

$$P(1) = \frac{3!}{1!2!} (0.25)^1 (0.75)^2 = 0.422$$

7.3 MOMENTS OF A DISCRETE DISTRIBUTION

In elementary mechanics moments are associated with the physical properties of bodies of mass. The first moment about the origin is related to the
center of gravity, and the second moment about the center of gravity is
known as the moment of inertia. Similarly, for probability distributions,
the concept of moments is extremely important. The analogy between
bodies of unit mass and probability distributions lends to the usefulness
of the analogy between the moment properties. Just as some bodies are
completely characterized by their moments, some probability distributions
are completely characterized by their moments.

The kth moment about the origin of a discrete distribution $P(x)$, such
as the binomial distribution, is defined to be μ_k where

$$\mu_k = \sum_{x=0}^{\infty} x^k P(x) \tag{7.3.1}$$

The first moment μ_1, or just μ, is called the mean. The second moment μ_2
is useful in finding the variance σ^2, since

$$\sigma^2 = \mu_2 - \mu^2 \tag{7.3.2}$$

The first moment of the binomial distribution is found by substituting
(7.2.3) in (7.3.1) and rearranging terms to give

$$\mu = np \tag{7.3.3}$$

The second moment is calculated in a similar manner by using the identity
$x^2 = x(x - 1) + x$, eventually reducing to

$$\mu_2 = n(n - 1)p^2 + np \tag{7.3.4}$$

If (7.3.3) and (7.3.4) are placed in (7.3.2), the variance for the binomial
distribution becomes

$$\sigma^2 = npq \tag{7.3.5}$$

7.4 THE POISSON DISTRIBUTION

We know that the probability of getting some number on the roll of a die
may be expressed as the ratio of the number of successful possibilities
to the total number of possibilities, or 1:6. In this simple application,
as in the illustration of the binomial distribution in the previous section,
both the number of times the event can occur and the number of times
the event cannot occur are known. However, there is another type of
problem, in which the number of failures of an event to occur is impertinent

Moroney[1] suggests the example of a man counting the number of lightning flashes in a storm without having any idea how many times the lightning did not flash. The same is true in considering vehicle arrivals at an intersection—one has no way of knowing how many arrivals did not arrive. In situations of this type, the Poisson distribution must be used.

The form of the Poisson distribution utilizes the famous mathematical constant e, which has the value

$$e = \frac{1}{0!} + \frac{1}{1!} + \frac{1}{2!} + \frac{1}{3!} + \cdots = 2.7183 \tag{7.4.1}$$

This number is the basis of the natural logarithms as opposed to common logarithms, which utilize 10 as a basis. If e is raised to some power m, then the result may also be expressed in the form of an infinite series

$$e^m = 1 + m + \frac{m^2}{2!} + \frac{m^3}{3!} + \cdots \tag{7.4.2}$$

Now, in order for any distribution to qualify as a probability distribution, the sum of all its terms must equal unity. In the case of the binomial distribution, the value of $p + q = 1$, so that $(p + q)^n$ in (7.2.1) must also equal unity no matter what value is given to n. In order to utilize the expansion of e^m as a probability distribution, one has only to divide both sides of (7.4.2) by e^m:

$$1 = e^{-m}\left(1 + m + \frac{m^2}{2!} + \frac{m^3}{3!} + \cdots\right) \tag{7.4.3}$$

Equation (7.4.3) gives the sum of the terms of the Poisson distribution; therefore the form of the distribution is given by the general term

$$P(x) = \frac{e^{-m}m^x}{x!} \qquad x = 0, 1, 2, \ldots \tag{7.4.4}$$

It has been found that this distribution describes very satisfactorily the occurrence of isolated events x in terms of the mean number of occurrence of these events μ.

In order to evaluate the mean, one has only to substitute (7.4.4) into (7.3.1),

$$\mu = 0 + me^{-m} + \frac{2m^2 e^{-m}}{2!} + \frac{3m^3 e^{-m}}{3!} + \cdots$$

$$\mu = me^{-m}\left(1 + m + \frac{m^2}{2!} + \cdots\right) \tag{7.4.5}$$

and note that the term in the parentheses is e^m.   Thus, the parameter m in (7.4.4) is the mean of the Poisson distribution.   Applying (7.3.1) to get the second moment and substituting in (7.3.2), we see that the variance of the Poisson distribution is equal to the mean.

7.5   STATISTICAL TESTS FOR RANDOMNESS

The first step in the fitting of the Poisson distribution to experimental data such as arrivals at an intersection is the classification of the arrivals by frequency.   For the study period the number of 1-min intervals in which 0, 1, 2, 3, etc., vehicles arrive is tabulated.   These constitute the observed distributions, and the inference is then made that the postulated theoretical distribution is in fact the true population.   The use of the chi-square test as an index of the goodness of fit of observed and expected frequencies of occurrences is well established in testing such a hypothesis.

The value for the chi-square is simply the summation of all the deviations between the observed f and the theoretical frequencies F, squared and divided by the theoretical frequencies:

$$\chi^2 = \sum \frac{(f - F)^2}{F} \qquad (7.5.1)$$

The chi-square value obtained is used to enter a chi-square table to determine the probability of a larger value occurring purely by chance.   If this probability is 0.05 or less, the test is usually assumed to be significant, and the hypothesis that the observed distribution and the theoretical distribution are the same is rejected.

So far in the analysis, only the frequency has been considered.   In the event the chi-square test verifies a Poisson distribution for a period, it is well to check the independence of arrivals for successive intervals since this is imperative for true randomness.   Thus, if the average number of arrivals during the study period is 10 vehicles per minute, the probability of 10 or more arrivals is 0.54 [obtained through successive application of (7.4.4) or from tables of the Poisson distribution].[2]   Similarly, during any consecutive pair of 1-min intervals, the probability of 10 or more arrivals for both intervals is 0.54×0.54.   The probability of less than 10 arrivals for both intervals is 0.46×0.46.   Two additional possibilities remain: (1) 10 or more in the first interval and less than 10 arrivals in the second and (2) the reverse.   The probability of either of these combinations is 0.54×0.46.   The point is that a run of consecutive 1-min arrivals above the mean, followed by a similar run of arrivals below the mean, might not yield a significant chi-square value in the analysis of frequency of arrivals and yet could obviate true randomness.

7.6 THE PROBABILITY–GENERATING FUNCTION

In the case of a discrete distribution, it is frequently of great convenience to work with what is called the *probability-generating function* (pgf) of the distribution. The pgf is defined as

$$Z_x(\theta) = E(\theta^x) = \sum_{x=0}^{\infty} \theta^x P(x) \tag{7.6.1}$$

The convenience referred to arises in finding the moments of discrete distributions from the following operation:

$$\frac{d^k Z_x(\theta)}{d\theta^k} \bigg|_{\theta=1}$$

Thus, from the definition,

$$Z_x'(\theta) = E(x\theta^{x-1}) = \sum_{x=0}^{\infty} x\theta^{x-1} P(x)$$

and setting $\theta = 1$ gives the first moment

$$Z_x'(1) = E(x) = \sum_{x=0}^{\infty} x P(x) = \mu \tag{7.6.2}$$

Accordingly,

$$Z_x''(1) = \mu_2 - \mu \tag{7.6.3}$$

and

$$\sigma^2 = Z_x''(1) + \mu - \mu^2 \tag{7.6.4}$$

The pgf for the binomial distribution is

$$Z_x(\theta) = (q + \theta p)^n \tag{7.6.5}$$

The derivatives of the pgf evaluated at $\theta = 1$ are

$$Z_x'(1) = np$$

and

$$Z_x''(1) = n(n-1)p^2$$

Thus, the mean and variance are np and npq, respectively, as before.
For the Poisson distribution,

$$Z_x(\theta) = e^{-m(1-\theta)} \tag{7.6.6}$$

and the mean and variance may be verified by (7.6.2) and (7.6.4).

7.7 THE NEGATIVE BINOMIAL DISTRIBUTION

Consider, again, the binomial distribution

$$P(k) = \binom{n}{k} p^k q^{n-k} \qquad (7.7.1)$$

which gives the probability of k successes in n trials.   Assume that we are interested in finding the probability $P(k,n)$ of the kth success on the nth trial.   This occurs if and only if among the first $n - 1$ trials there are exactly $k - 1$ successes

$$P(k - 1) = \binom{n-1}{k-1} p^{k-1} q^{n-k} \qquad (7.7.2)$$

and the next trial results in a success p, giving

$$P(k,n) = \binom{n-1}{k-1} p^k q^{n-k} \qquad (7.7.3)$$

Equation (7.7.3) defines the negative binomial probability distribution.

As will be seen in the next section in addition to the classical interpretations expressed by Eqs. (7.2.3) and (7.7.3), the binomial and negative binomial distributions may be used to describe the distribution of events. To use the negative binomial in this manner, one sets the number of events $x = n - k$ so that the probability of occurrence of x events is

$$P(x) = \binom{x+k-1}{k-1} p^k q^x \qquad (7.7.4)$$

The importance that the moments of a density function play in estimating the parameters of the function is illustrated repeatedly in the text. Since the moments of a distribution are so important, it would be useful if a function could be found that would provide a representation of all the moments.   Such a function is called a *generating function*.   The pgf for the negative binomial distribution is denoted by

$$Z_x(\theta) = p^k (1 - q\theta)^{-k} \qquad (7.7.5)$$

Taking the first and second derivatives of (7.7.5) gives

$$Z'_x(\theta) = kqp^k (1 - q\theta)^{-(k+1)} \qquad (7.7.6)$$

and

$$Z''_x(\theta) = (k + 1)kq^2 p^k (1 - q\theta)^{-(k+2)} \qquad (7.7.7)$$

Setting $\theta = 1$ in (7.7.6) gives the first moment about the origin, or mean, of the negative binomial distribution

$$\mu = \frac{kq}{p} \qquad (7.7.8)$$

The variance of the distribution is obtained from (7.7.7) using

$$\sigma^2 = Z_x''(1) + \mu - \mu^2 = \frac{kq}{p^2} \qquad (7.7.9)$$

7.8 ANALYSIS OF ARRIVALS

The three discrete distributions discussed in this chapter are listed in Table 7.1 along with a few of their properties.

Table 7.1   Moments of common discrete distributions

Name	Form	Mean	Variance
Binomial	$P(x) = \binom{n}{x} p^x q^{n-x}$	np	npq
Poisson	$P(x) = \dfrac{e^{-u}\mu^x}{x!}$	μ	μ
Negative binomial	$P(x) = \binom{x+k-1}{k-1} p^k q^x$	$\dfrac{kq}{p}$	$\dfrac{kq}{p^2}$

Gerlough's classic paper discusses in detail the application of the Poisson distribution in highway traffic. Some of the uses open to the traffic engineer are shown to include the analysis of arrival rates at a given point, determination of the probability of finding a vacant parking space, studies of certain accident locations, and the design of left-turn pockets.

Inherent in the Poisson law, however, is the assumption that the probability of an event remains constant, which in highway traffic practice is seldom true. Any variation in the expectation of an event—in particular, any tendency for one event to increase or decrease the probability of another—will increase or decrease the variance of the distribution relative to the mean; and the Poisson distribution no longer describes the data. It is well known that the presence of traffic control devices can alter arrivals and preclude randomness. Finally, some traffic phenomena may be Poisson when observed for an interval of one length but nonrandom when observed for an interval of a different length.

One way in which the traffic engineer may test his data is to assume it follows some distribution and then test this assumption by the chi-square test.   The example in Table 7.2 is taken from Gerlough's [3] article in "Poisson and Traffic," page 17, and is based on the number of arrivals in each of 360 ten-sec intervals.   Here the failure of the test (nonacceptability of fit) indicates that the distribution of arrivals *does not conform to a Poisson distribution*.   The example is repeated in Table 7.3 where the negative binomial distribution is used with the same mean and variance and a

Table 7.2   Test of Poisson distribution using chi-square test

Cars per interval	Observed frequency	Theoretical frequency	Observed frequency/ theoretical frequency
x	f	F	f^2/F
0	139	129.6	149.1
1	128	132.4	123.7
2	55	67.7	44.7
3	25	23.1	27.1
4	13	7.2	23.5
	360	360.0	368.1

Mean = 1.022 $\chi^2 = \sum \dfrac{f^2}{F} - n = 368.1 - 360.0 = 8.1$

Variance = 1.200 $\chi^2_{0.05} = 7.81 < 8.1$ $\therefore$ reject Poisson distribution

Source: "Poisson and Traffic," p. 17.

Table 7.3   Test of negative binomial distribution using chi-square test

Cars per interval	Observed frequency	Theoretical frequency	Observed frequency/ theoretical frequency
x	f	F	f^2/F
0	139	138.0	140.0
1	128	121.5	134.9
2	55	63.1	47.9
3	25	25.2	24.8
4	13	12.2	13.8
	360	360.0	361.4

Mean = 1.022 $\chi^2 = \sum \dfrac{f^2}{F} - n = 361.4 - 360.0 = 1.4$

Variance = 1.200 $\chi^2_{0.05} = 5.99 > 1.4$ $\therefore$ accept negative binomial remarkably close fit is obtained.   The explanation of this will become apparent in the following discussion.

Table 7.1 indicates that the mean of the negative binomial distribution is kq/p and the variance is kq/p^2.   Thus as p is necessarily less than 1, the variance is always greater than the mean.   Any variation in the Poisson parameter—e.g., any tendency for one event to increase the probability of another event—will increase the variance of the distribution relative to the mean, and the negative binomial distribution will invariably fit the data better.

The procedure for fitting a discrete distribution to observed data is to first compute the mean $\bar{x}$ and variance s^2 of the observed data.   If the mean and variance are approximately equal, the Poisson distribution can be used to compute the theoretical probabilities.   However, if the variance is appreciably greater than the mean, the negative binomial can be used; if the variance is appreciably less than the mean, the binomial distribution can be used.   In the last two cases the mean and variance of the observed data are equated to the first two moments of the desired distribution in order to estimate the parameters of the distribution.   This procedure is summarized in Table 7.4.   Once the parameters in the table have been

Table 7.4 Summary of procedure for fitting discrete distributions to data with mean $\bar{x}$ and variance s^2

Distribution	Binomial	Poisson	Negative binomial
Frequency function, $P(x)$	$\binom{n}{x} p^x(1-p)^{n-x}$	$\dfrac{\mu^x}{x!}e^{-\mu}$	$\binom{x+k-1}{k-1}p^k(1-p)^x$
Mean	np	μ	$\dfrac{k(1-p)}{p}$
Variance	$np(1-p)$	μ	$\dfrac{k(1-p)}{p^2}$
Mean/variance	$(1-p)^{-1} > 1$	1	$p < 1$
Estimation of parameters	$p = \dfrac{(\bar{x}-s^2)}{\bar{x}}$	$\mu = \bar{x}$	$p = \dfrac{\bar{x}}{s^2}$
	$n = \dfrac{\bar{x}^2}{(\bar{x}-s^2)}$		$k = \dfrac{\bar{x}^2}{(s^2-\bar{x})}$

determined, the probability of x events occurring on any trial (during any time interval) is computed from the frequency function $P(x)$.   Sources of tables for the binomial,[4] Poisson,[2] and negative binomial[5] distributions, which facilitate computations, are listed in the bibliography for this chapter.

7.9 THE GEOMETRIC DISTRIBUTION

Another type of sequence that has considerable importance in traffic queueing is constructed from the geometric progression

$$p + pq + pq^2 + pq^3 + \cdots$$

where q is called the common ratio since it is the ratio of any term to the one preceding it in the progression.   The sum S_{x+1} of the first $x + 1$ terms is

$$S_{x+1} = p + pq + pq^2 + \cdots + pq^x \qquad (7.9.1)$$

If we multiply both members of (7.9.1) by q, we get

$$qS_{x+1} = pq + pq^2 + \cdots + pq^x + pq^{x+1} \qquad (7.9.2)$$

Now subtracting corresponding members of (7.9.2) from (7.9.1), we find

$$(1 - q)S_{x+1} = p(1 - q^{x+1})$$

or

$$S_{x+1} = \frac{p(1 - q^{x+1})}{1 - q} \qquad (7.9.3)$$

It is important to note that, as yet, no restrictions have been placed on p, q, and x.   However if q is chosen as less than unity and x is allowed to approach infinity, the sum of the $x + 1$ terms may be written as follows:

$$\lim_{x \to \infty} S_{x+1} = \frac{p}{1 - q} \qquad (7.9.4)$$

Now, if p and q are further restricted such that $p + q = 1$, the sum of all terms must equal unity

$$p + pq + pq^2 + \cdots = \sum_{x=0}^{\infty} pq^x = \frac{p}{1 - q} = 1 \qquad (7.9.5)$$

and the probabilities described by these terms

$$P(x) = pq^x \qquad (7.9.6)$$

form a distribution called the *geometric distribution*.   Equation (7.9.6) gives the probability that the first success occurs after x failures,

$$x = 0, 1, 2, \ldots$$

It is apparent that the geometric distribution is the special case of the negative binomial distribution [Eq. (7.7.4)] for $k = 1$.

To show how the distribution is important in queueing, it is necessary to give another interpretation to the variables.   Referring to (7.9.6), we have the waiting time for the first success or the probability of x vehicles waiting in a line, where p and q are functions of the rate of arrivals and departures.   This will be discussed in more detail in a later chapter.

The pgf for the geometric distribution is

$$Z_x(\theta) = p(1 - \theta q)^{-1} \qquad\qquad (7.9.7)$$

It will be recalled from (7.7.5) that (7.9.7) is the special case of the negative binomial distribution pgf for $k = 1$.   The negative binomial distribution is thus referred to as the "k-fold convolution of the geometric distribution with itself."[6]

7.10   POISSON APPROXIMATION TO THE BINOMIAL

If the number of trials n in (7.2.3) is large, the computations involved become quite lengthy; therefore a convenient approximation to the binomial distribution would be very useful.   It turns out that for a large n and small p, the Poisson distribution gives a good approximation.

In order to verify that (7.4.4) may be used for (7.2.3) under the stipulated conditions, consider what happens to the binomial distribution when n becomes infinite and p approaches zero in such a manner that the $\mu = np$ remains fixed.

First rewrite (7.2.3) as follows:

$$P(x) = \frac{n(n - 1) \, \cdots \, (n - x + 1)}{x!} \, p^x(1 - p)^{n-x}$$

If numerator and denominator are multiplied by n^x and the indicated algebraic manipulations are performed,

$$P(x) = \frac{n(n - 1) \, \cdots \, (n - x + 1)}{n^x x!} \, (np)^x(1 - p)^{n-x}$$

$$= \frac{n(n - 1) \, \cdots \, (n - x + 1)}{n \cdot n \, \cdots \, n} \frac{\mu^x}{x!} (1 - p)^{n-x}$$

$$= \left(1 - \frac{1}{n}\right)\left(1 - \frac{2}{n}\right) \cdots \left(1 - \frac{x - 1}{n}\right) \frac{\mu^x}{x!} \frac{(1 - p)^n}{(1 - p)^x} \qquad (7.10.1)$$

Next, express $(1 - p)^n$ in the form

$$(1 - p)^n = [(1 - p)^{-1/p}]^{-\mu} \qquad\qquad (7.10.2)$$

Now use the definition of e

$$\lim_{p \to 0} [(1 - p)^{-1/p}]^{-\mu} = e^{-\mu} \tag{7.10.3}$$

Letting $n \to \infty$ in (7.10.1) and applying the results of (7.10.2) and (7.10.3) in (7.10.1), it will be seen that

$$\lim_{n \to \infty} P(x) = \frac{e^{-\mu}\mu^x}{x!}$$

Although the Poisson distribution has been presented here as an approximation of the binomial distribution, we have seen in Sec. 7.4 that it is a well-known and useful distribution in its own right. Since the Poisson distribution can be derived from the binomial distribution by letting n increase without limit and p approach zero—so that the mean can be held fixed at np—it is not surprising that the Poisson distribution is obtained as a limiting form of the negative binomial distribution.

Solving Eq. (7.7.8) for p and then q yields

$$p = \left(1 + \frac{\mu}{k}\right)^{-1} \tag{7.10.4}$$

$$q = \frac{\mu}{k + \mu} \tag{7.10.5}$$

Substituting (7.10.4) and (7.10.5) in (7.7.4), we obtain

$$P(x) = \frac{(x + k - 1)!}{x!(k - 1)!} \left(1 + \frac{\mu}{k}\right)^{-k} \frac{\mu^x}{(k + \mu)^x} \tag{7.10.6}$$

Rearranging terms,

$$P(x) = \left(\frac{x + k - 1}{k + \mu}\right)\left(\frac{x + k - 2}{k + \mu}\right) \cdots \left(\frac{k}{k + \mu}\right)\left(1 + \frac{\mu}{k}\right)^{-k} \frac{\mu^x}{x!} \tag{7.10.7}$$

Consider first the limit of (7.10.7) as $k \to \infty$, and the quantity in the brackets approaches 1. Next, express the second term in the form

$$\left(1 + \frac{\mu}{k}\right)^{-k} = \left[\left(1 + \frac{\mu}{k}\right)^{k/\mu}\right]^{-\mu} \tag{7.10.8}$$

Now from the definition of e

$$\lim_{y \to 0} (1 + y)^{1/y} = e \tag{7.10.9}$$

Hence letting $y = \mu/k$

$$\lim_{k \to \infty} \left[\left(1 + \frac{\mu}{k}\right)^{k/\mu}\right]^{-\mu} = e^{-\mu} \tag{7.10.10}$$

because $k \to \infty$ as $y \to 0$.   By applying these two results to the right side of (7.10.7), it will be seen that

$$\lim_{k \to \infty} P(x) = e^{-\mu}\frac{\mu^x}{x!} \tag{7.10.11}$$

7.11 APPLICATION TO SIGNALIZATION

In traffic, both capacity and demand are expressed as volumes, or vehicles per interval of time.   The *capacity* of a highway facility is dependent upon its geometrics and the characteristics of the traffic using the facility. *Demand* on a highway facility, like runoff to a culvert or a wind load on a structure, is variable and can only be estimated in terms of a probability of occurrence.

Nowhere is this capacity-demand relationship more vividly expressed than at a high-type signalized facility.   This may be a single at-grade intersection or a system of at-grade intersections in the form of an interchange. If conflicting movements are separated through the signal phasing, lane controls, parking restriction, and high-type geometric design, the intersection or interchange is referred to as a *high-type facility*.   In general, where high-volume expressways cross other high-volume expressways, urban highways, or major arterials, a signalized interchange can be utilized to reduce the number of costly grade separations inherent in fully directional interchanges or to reduce the amount of right-of-way typical of the cloverleaf.   The high-type intersection may be considered as the special case of the signalized interchange in which the roadways are not separated.

In designing a facility to handle the movements between two intersecting high-volume roadways, the designer must, on the basis of traffic demand, decide on the number of approach lanes needed and then apportion the phases of the signal system.   In most cases, whether it be the reconstruction of an existing intersection or planning a new freeway through a city, there are few alternatives concerning the number of approach lanes. The cross sections of existing major arterials on the route of the proposed freeway are well defined, and there are practical limitations on the number of lanes on ramps and frontage roads in the freeway system.   Therefore, the problem becomes primarily one of signal time apportionment for an assumed geometric condition.

There are several methods available for apportioning green time to the various phases of a signal cycle.   The relative precision of these solutions is proportional to the degree of realism achieved in the hypotheses regarding the arrival and departure rates.   The simplest procedure is based on the assumptions that the arrival rate is constant from cycle to cycle throughout the design hour and that the departure rate, and hence the departure headways, are constant throughout the green interval. Thus, the ratio of the duration of a given phase to the total cycle length is equal to the demand on the given phase divided by the demand on all phases. This has been referred to as the G/C method, where G and C refer to the durations of the green phase and the cycle length.

We have seen in a previous chapter how the minimum average departure headways, which result when a queue of vehicles is released by a light, are gradually reduced until about the fifth or sixth vehicle in line, when a constant headway is developed. If arrivals are still assumed to be uniform, the computation of cycle length and apportioning of phases is still rational. The maximum capacity for a given phase may be obtained by a direct analysis of headways with allowances for time lost starting and stopping the queue, as suggested later in this section.

Most existing procedures have been based on the assumption of a constant or average demand. However, traffic tends toward a random arrangement; the number of vehicles arriving at a given point in any interval of time can vary appreciably from the mean. The Poisson distribution is well established in predicting vehicle arrivals at intersections. The Poisson equation expresses the probability of a given number of vehicle arrivals per cycle based on the average number of arrivals per cycle. Since it is obvious that for any reasonable cycle length some cycle failures must be expected, the number of tolerable failures may be used as a criterion for the cycle length determinations. The assumptions of uniform arrivals or Poisson arrivals as models for traffic demand have limitations. The distribution of traffic over a 24-hr period is obviously neither uniform nor random; most daily traffic patterns show approximately 70 percent of total daily traffic movement occurring between 7 A.M. and 7 P.M. Now that the A.M. and P.M. peak hours have arbitrarily replaced the day (ADT) as the basic traffic design intervals, it is just as important to establish the time pattern of traffic flow within the peak hours.

Currently, one of two assumptions is employed: (1) Poisson arrivals throughout the peak hour or (2) uniform arrivals throughout the peak hour. Figure 7.1 shows the relationships between arrivals predicted by these respective assumptions and the observed arrivals at various approaches studied.[7] Superimposed on the graphs are the Poisson and uniform arrival curves for the peak period within the peak hour. The curves for uniform arrivals are straight lines, representing the average arrivals for the peak hour and the peak period, respectively. It is apparent that the best estimator of the observed demand is the assumption of a Poisson distribution during the peak period, and the least reliable is the assumption of the uniform arrivals throughout the peak hour.

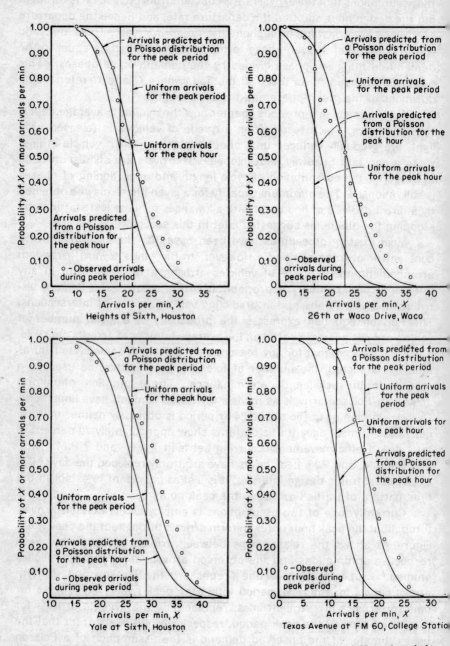

Fig. 7.1 Relationship between observed and predicted arrivals during A.M. peak periods.

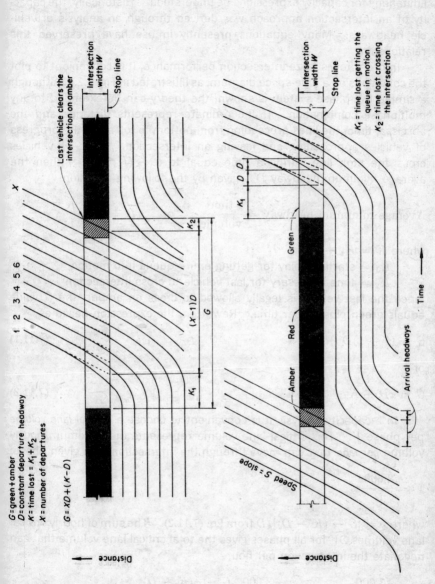

G = green + amber
D = constant departure headway
K = time lost = $K_1 + K_2$
X = number of departures

$G = XD + (K - D)$

K_1 = time lost getting the queue in motion
K_2 = time lost crossing the intersection

Last vehicle clears the intersection on amber

Intersection width W

Stop line

Intersection width W

Stop line

Distance

Distance

Speed S = slope

Amber

Red

Green

Arrival headways

Time

Fig. 7.2 Time-space relationships for two-phase system.

Since any design procedure is actually a systematic attempt to resolve the capacity-demand relationship, it is important that the development and limitations for capacity expressions be understood.   Historically, the capacity of an intersection approach was derived through an analysis of vehicle headways.   Many equations presently in use have preserved this relationship.

In order to visualize intersection performance, it is convenient to plot the conditions on a time-space diagram, as illustrated in Fig. 7.2.   Although a simple two-phase system is shown, the theory can be extended to any multiphase combination.   If the ordinate represents distance and the abscissa, time, the lines proceeding from bottom to top show the progress of vehicles approaching and leaving an intersection.   If $x - 1$ vehicles cross the stop line during a time equal to $G - (K_1 + K_2)$, then the average minimum headway D is given by the following equation:

$$\text{Average minimum headway} = \frac{\text{time}}{\text{volume}} \quad \text{or} \quad D = \frac{G - K}{x - 1}$$

where $K = K_1 + K_2$
K_1 = starting delay for getting entire queue into motion
K_2 = time necessary for last vehicle to cross intersection
Since the last vehicle is legally allowed to cross on amber, the G value equals green plus amber time.   Rewritten, the expression becomes

$$G = (x - 1)D + K \tag{7.11.1}$$

or

$$G = xD + K - D \tag{7.11.2}$$

In a capacity analysis it is convenient to choose a *critical lane volume per phase* V.   This critical lane volume represents the maximum hourly volume *per lane* that can move through the intersection on a given phase:

$$V = \frac{3{,}600}{C} x$$

where $x = [G - (K - D)]/D$ from Eq. (7.11.2).   The sum of hourly critical lane volumes ΣV for all phases gives the total critical lane volume that can negotiate the intersection per hour.

$$\Sigma V = \frac{3{,}600}{C} \Sigma x \qquad \Sigma V = \frac{3{,}600}{C} \frac{\Sigma G - \phi(K - D)}{D}$$

where $C = \Sigma G$
ϕ = number of phases

$$\therefore \Sigma V = \frac{3,600}{C} \frac{C - \phi(K - D)}{D}$$

$$\Sigma V = \frac{3,600}{D} - \frac{3,600\phi(K - D)}{CD} \tag{7.11.3}$$

Substituting $K = 6.0$ sec and $D = 2.0$ sec, we can obtain an expression for ΣV in terms of cycle length C for both three- and four-phase intersections.

$$\Sigma V(\phi = 3) = 1,800 - \frac{21,600}{C} \tag{7.11.4}$$

$$\Sigma V(\phi = 4) = 1,800 - \frac{28,800}{C} \tag{7.11.5}$$

It is seen that as C approaches infinity, ΣV approaches 1,800 vph per lane. Last, Eq. (7.11.3) may be solved for cycle length C:

$$C = \frac{3,600\phi(K - D)}{3,600 - D\Sigma V} \tag{7.11.6}$$

It should be remembered that Eqs. (7.11.1) through (7.11.6) are based on the assumption of uniform arrivals for every cycle over a 60-min period. This, of course, does not occur. However, if the hourly approach volumes are increased by some peak-magnitude factor, the capacity equations become applicable for uniform arrivals during the peak period.

Since it has been established that during the peak period the assumption of Poisson arrivals provides the best estimate of actual demand, it is well to consider its application in capacity determinations. Equation (7.11.7) is the cumulative Poisson expression for determining the probability of $x + 1$ arrivals or more per cycle during the peak period, based on an average of m arrivals per cycle.

$$P(x + 1) = \sum_{x+1}^{\infty} \frac{m^{x+1}e^{-m}}{(x + 1)!} \tag{7.11.7}$$

where $m = V/(3,600/C)$
 $x = [G - (K - D)]/D$ from Eq. (7.11.2)
 $C = \Sigma G$

These four equations can be reduced by successive approximations. This reduction is greatly facilitated through the use of a graph of cumulative Poisson curves (Fig. 7.3). The philosophy is to provide the designer with a figure of merit in the form of P the percentage of *cycle failures*, with a cycle failure defined as any cycle during which approach arrivals exceed the capacity for departures.

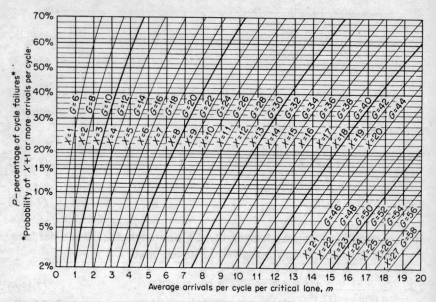

Fig. 7.3   Green requirements for nonconflicting movements.

7.12   ILLUSTRATION OF CAPACITY–DESIGN PROCEDURE

Figures 7.4 to 7.6 illustrate seven steps to be followed in the design and signalization of a future high-type intersection.   These steps are as follows:

Step 1.   Determine the three conditions (population, location, and volumes) which affect the magnitude of peak period.

Step 2.   Since the volumes are given in terms of ADT, they must be converted to peak hourly volumes.   In an actual problem the A.M. peak would also be checked.

Step 3.   The peak-magnitude factor Y' for each approach is calculated.

Step 4.   The peak-magnitude factors Y' are applied to the peak hourly approach volumes to arrive at an hourly rate of flow equivalent to the arrivals during the peak period.

Step 5.   Consistent with the assumption of a high-type facility, all conflicting movements must be separated by the signal phasing.

Step 6.   Design combinations are assumed by varying the number of approach lanes on the two streets.   Volumes are assigned to each lane, assuming equal lane distribution during the peak period. The maximum lane volume required to move on a given phase is called the critical lane volume.   The sum of these critical lane volumes for all phases provides the basis for calculating the minimum cycle length by use of Eq. (7.11.6).   In the example chosen, design alternative A yielded an unreasonable cycle length, and therefore only alternatives B and C merited further consideration.

Step 7. The average arrivals per cycle m are calculated from the critical lane volumes. These values are used to enter the graph of Poisson curves (Fig. 7.6), and the phase lengths G are tabulated for various probabilities of failure P. Any combination of $G_A + G_B + G_C + G_D$ that equals the assumed cycle length C is acceptable.

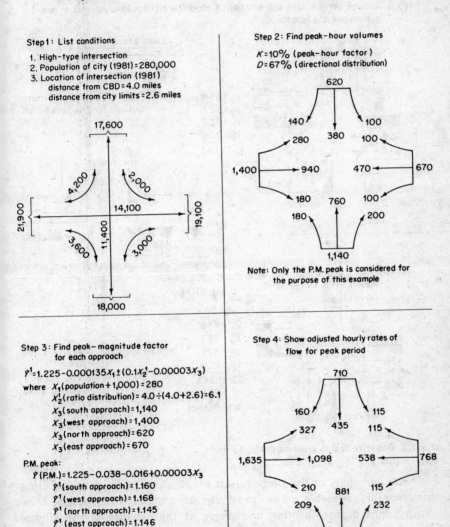

Step 1: List conditions

1. High-type intersection
2. Population of city (1981) = 280,000
3. Location of intersection (1981)
   distance from CBD = 4.0 miles
   distance from city limits = 2.6 miles

Step 2: Find peak-hour volumes

$K = 10\%$ (peak-hour factor)
$D = 67\%$ (directional distribution)

Note: Only the P.M. peak is considered for the purpose of this example

Step 3: Find peak-magnitude factor for each approach

$$\hat{Y}^1 = 1.225 - 0.000135X_1 \pm (0.1X_2^1 - 0.00003X_3)$$

where X_1(population + 1,000) = 280
X_2^1(ratio distribution) = $4.0 \div (4.0 + 2.6) = 6.1$
X_3(south approach) = 1,140
X_3(west approach) = 1,400
X_3(north approach) = 620
X_3(east approach) = 670

P.M. peak:
$\hat{Y}$(P.M.) = $1.225 - 0.038 - 0.016 + 0.00003X_3$
$\hat{Y}^1$(south approach) = 1.160
$\hat{Y}^1$(west approach) = 1.168
$\hat{Y}^1$(north approach) = 1.145
$\hat{Y}^1$(east approach) = 1.146

Step 4: Show adjusted hourly rates of flow for peak period

Fig. 7.4 Capacity-design procedure, steps 1 to 4.

Step 5 : Assume phasing

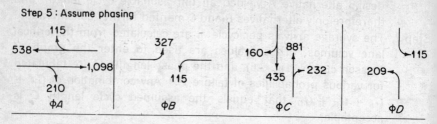

Step 6 : Assume various lane combinations. Determine critical lane volumes V and minimum cycle lengths C.

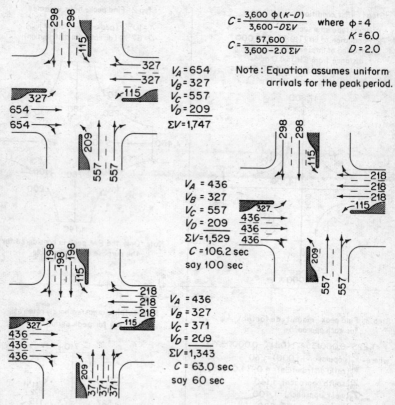

$$C = \frac{3,600 \; \phi(K-D)}{3,600-D \Sigma V} \qquad \text{where} \quad \phi = 4$$

$$C = \frac{57,600}{3,600-2.0 \; \Sigma V} \qquad\qquad K = 6.0$$

$$D = 2.0$$

Note : Equation assumes uniform arrivals for the peak period.

$V_A = 654$
$V_B = 327$
$V_C = 557$
$V_D = 209$
$\Sigma V = 1,747$

$V_A = 436$
$V_B = 327$
$V_C = 557$
$V_D = 209$
$\Sigma V = 1,529$
$C = 106.2$ sec
say 100 sec

$V_A = 436$
$V_B = 327$
$V_C = 371$
$V_D = 209$
$\Sigma V = 1,343$
$C = 63.0$ sec
say 60 sec

Fig. 7.5   Capacity-design procedure, steps 5 and 6.

The versatility of the procedure is emphasized in the many phasing combinations available. The proximity of another intersection or a ramp might dictate favoring one phase at the expense of the others. Therefore, it would be possible to prevent excessive queues leading to interference on adjacent facilities and perhaps progressive failures.

Design alternative B (assuming cycle length equals 100 sec)

Phase ϕ	Average arrivals per cycle m^*	Phase length G for various percentages of failure P†								
		2	5	10	20	30	40	50	60	70
A	11.9	42	39	36	33	30	28	26	25	23
B	9.1	34	32	29	26	24	22	21	19	18
C	15.5	52	48	44	40	38	36	34	32	30
D	5.8	25	23	21	19	17	16	14	13	12

$$* \ m = \frac{V}{3,600/C} \qquad C = \Sigma G = \text{between } 102 \text{ and } 95$$

where m = average arrivals per cycle per critical lane
  V = hourly rate of arrivals per peak period
$3,600/C$ = number of cycles per hour
Interpolating between 102 and 95 for $C = 100$, it is seen that $P = 43$ percent. However, since P need not be the same for all phases, there can be an infinite combination of phase lengths as long as their summation equals the assumed cycle length.
† Using the design chart for Poisson arrivals during the peak period (Fig. 7.3).

Design alternative C (assuming cycle length equals 60 sec)

Phase ϕ	Average arrivals per cycle m^*	Phase length G for various percentages of failure P†								
		2	5	10	20	30	40	50	60	70
A	7.2	29	26	24	21	20	18	17	15	14
B	5.5	25	22	20	18	16	15	14	13	11
C	6.2	26	24	22	19	18	16	15	14	13
D	3.5	19	17	15	13	12	11	10	9	8

$$* \ m = \frac{V}{3,600/C} \qquad C = \Sigma G = 60$$

where m = average arrivals per cycle per critical lane
  V = hourly rate of arrivals per peak period
$3,600/C$ = number of cycles per hour
† Using the design chart for Poisson arrivals during the peak period (Fig. 7.3).

Fig. 7.6 Capacity-design procedure, step 7.

It is at this point in the procedure that the engineer's judgment must be utilized.

It should be remembered that the percentage of failures P is based on the number of cycles during the peak period, not the number during the peak hour.   If a duration of 25 min for the P.M. peak period is assumed, the number of failures and total failure time may be calculated for designs B and C (Table 7.5).

Table 7.5

Length of cycle, min	100	60
No. of cycles in peak period	15	25
Failures, %	43	40
No. of cycle failures	6.5	10
Total "failure time," min	10.7	10.0

Although additional research is needed in deciding just what percentage of failures may reasonably be allowed, it seems that a level of 30 to 35 percent during the peak period represents a practical design level (remembering that this would be only about 10 to 15 percent of the peak hour).   Step 7 for designs B and C could be repeated assuming longer cycle lengths (say 120 and 80 sec, respectively) to obtain a lower level of failure.   The excessive cycle length of 120 sec required for design B might preclude its use.   However, since the conditions utilized in the calculations are for projected volume data, design alternative B might possibly offer 15 years of desirable operation and thus still merit consideration.

PROBLEMS

7.1. Street A forms a tee intersection with street B so that all traffic approaching the intersection on A must turn.   Random arrivals on A are observed to turn left 70 percent of the time and right 30 percent.   What is the probability of exactly three of the next ten vehicles turning right?   What is the probability of two or more of four consecutive vehicles turning left?

7.2. A stop sign is studied, and it is observed that no cars are waiting at the stop sign twice as often as one car, one car is waiting to enter the intersection twice as often as two cars are waiting, and so on.   What is the probability of zero cars waiting to enter the intersection?   What is the average number of cars expected to be waiting?

7.3. If the probability of being killed on an airplane flight from Los Angeles to New York is 0.001, what is the probability of being killed if a commuter makes 1,000 flights from Los Angeles to New York?

7.4. Select a location on a street and make a peak-period directional count.   Attempt to include 30 min of counts either before the peak flow builds up of after it has diminished. Count 1-min volumes and test the Poisson distribution as a description of arrivals for the total count period, the off-peak, and the peak period using the chi-square test.

7.5. While counting random arrivals on an approach to a signalized intersection, you notice that the probability of $x + 2$ arrivals per cycle is twice the probability of $x + 1$ arrivals per cycle and that the probability of $x - 1$ arrivals per cycle is one-third the probability of x arrivals per cycle. After 1 hr, having counted a total of 480 vehicles, calculate the cycle length in seconds.

7.6. Calculate the sample mean and variance of the data in Prob. 7.4. On the basis of these two statistics, would you recommend testing the Poisson, negative binomial, or binomial distribution as an arrival distribution for the off-peak, the peak, and the total count period?

7.7. You are counting two-way traffic on University Drive and find the directional distribu- is 60:40. What is the probability that three of the next four vehicles are going in the lighter flow direction?

7.8. A driver is waiting at an intersection to cross a one-way street which has Poisson traffic flowing on it. As soon as a gap between vehicles greater than T sec appears, he will cross. The probability p of a gap being greater than T is e^{-qT}. The proba- bility of waiting while n gaps pass the intersection is $p_n = (1 - p)^n p$. What is the average number of gaps a driver will let pass before crossing?

7.9. An intersection formed by a one-way street crossing a two-way street has the following critical lane volumes and corresponding phase lengths:

$$V_a = 500 \text{ vph} \qquad G_a = 34 \text{ sec}$$
$$V_b = 200 \text{ vph} \qquad G_b = 16 \text{ sec}$$
$$V_c = 300 \text{ vph} \qquad G_c = 22 \text{ sec}$$

What must be the values of K and D if the demand is equal to the capacity on each approach?

7.10. Determine the number of lanes on all approaches and phase lengths needed for the hourly volumes shown. Assume $K = 5$ sec and $D = 2$ sec. Indicate the *probability of failure* on each approach lane. The approaches are numbered clockwise.

Approach	Left turns, vph	Through, vph	Right turns, vph
1	150	350	150
2	150	200	100
3	200	200	100
4	300	300	300

7.11. The following are the number of vehicles that arrived each minute to turn left at an intersection: 0, 3, 0, 1, 2, 1, 0, 0, 2, 1, 1, 0, 2, 1, 2, 0, 1, 1, 0, 2. Test this sample to determine whether it may have come from a population described by the Poisson law

$$P(x) = \frac{e^{-m}m^x}{x!}$$

A standard median alteration would provide a storage bay for five vehicles. What is the probability of getting more than five vehicles arriving in 1 min (assuming the Poisson law fits)?

7.12. If the following are the number of times the given queue length was observed at the beginning of a green phase, develop a general expression for the probability of a queue being x vehicles long:

Queue length	0	1	2	3	4	5	≥ 6
Number of queues	25	24	21	16	9	0	0

7.13. A certain police department has estimated that only 10 percent of the traffic citations are prosecuted.
  a. What is the probability that a motorist who commits three violations will be apprehended exactly once?
  b. The chief of police is more interested in the probability of no apprehensions out of three violations. What is it?

7.14. Planes arrive randomly at an airport at a rate of one per 2 min. What is the probability that more than three planes arrive in a 1-min interval?

7.15. Telephone calls during the busy hour arrive at a rate of three per minute. What is the probability that none arrive in a period of (a) $\frac{1}{2}$ min; (b) 1 min; (c) 3 min?

REFERENCES

1. Moroney, M. J.: "Facts from Figures," Penguin Books, Inc., Baltimore, 1956, 474 pp.
2. Molina, E. C.: "Poisson's Exponential Binomial Limit," D. Van Nostrand Company, Inc., Princeton, N.J., 1942.
3. Gerlough, D. L.: Use of Poisson Distrbution in Highway Traffic, in "Poisson and Traffic," The Eno Foundation for Highway Traffic Control, Saugatuck, Conn., 1955.
4. Weintraub, Sol: "Tables of the Cumulative Binomial Probability Distribution for Small Values of p, Collier-Macmillan Limited, London, 1963.
5. Williamson, Eric, and M. H. Bretherton: "Tables of the Negative Binomial Probability Distribution," John Wiley & Sons, Inc., New York, 1963.
6. Feller, William: "An Introduction to Probability Theory and Its Applications," John Wiley & Sons, Inc., New York, 1964, 461 pp.
7. Drew, D. R., and C. Pinnell: A Study of Peaking Characteristics of Signalized Urban Intersections as Related to Capacity and Design, *Highway Res. Board Bull.* 352, pp. 1–54, 1962.

CHAPTER EIGHT
CONTINUOUS TRAFFIC VARIABLES

The figure 2.2 children per adult female was felt to be in some respects absurd, and a Royal Commission suggested that the middle classes be paid money to increase the average to a rounder and more convenient number.

Punch

8.1 INTRODUCTION

Until now only distributions of variables which take on discrete values have been considered. In this chapter continuous random variables, which range over a continuum of values, shall be considered. Such general measurements as length and time are continuous variables; specific traffic examples are space gaps, time headways, and vehicular speeds.

In the case of a discrete random variable, one can plot the probabilities $P(x)$ against the discrete values x. There are no values between the discrete points, and it is incorrect to draw a smooth curve through the points. In the case of a continuous random variable, an infinity of points with infinitely small separations between them exist. Instead of plotting probabilities $P(x)$, one plots a probability function $f(x)$, alternately referred to as a *continuous frequency distribution*. Whereas the sum of the probabilities over all values of x for a discrete frequency distribution is written

$$\sum_{x=0}^{\infty} P(x) = 1$$

for the continuous distribution the area under the curve over the entire range of x is

$$\int_0^{\infty} f(x) = 1$$

Although the distribution of x is determined by the probability density function $f(x)$, for many purposes it is more convenient to work with other functions equivalent to $f(x)$. Two such are the cumulative distribution functions

$$P(x < X) = \int_0^X f(x)\, dx \tag{8.1.1}$$

and

$$P(x > X) = \int_X^{\infty} f(x)\, dx \tag{8.1.2}$$

The kth moment of a continuous distribution with frequency function $f(x)$ is defined by

$$\mu_k = \int_0^{\infty} x^k f(x)\, dx \tag{8.1.3}$$

It is often desirable to calculate moments of a function of x, say $g(x)$, rather than x itself. A general definition in terms of an arbitrary function $g(x)$ is the following:

$$\mu_k = \int_0^{\infty} g^k(x) f(x)\, dx \tag{8.1.4}$$

8.2 THE MOMENT-GENERATING FUNCTION

Even though the direct computation of theoretical moments from the definition may not be difficult in some cases, it is convenient for many applications to be able to calculate such moments indirectly using an auxiliary function. One such function, useful when working with discrete distributions, is the probability-generating function. Another is the moment-generating function.

As the name implies, the *moment-generating function* is a function that generates moments. It is defined symbolically as follows:

$$M_x(\theta) = E(e^{\theta x}) = \sum_{x=0}^{\infty} e^{\theta x} f(x) \tag{8.2.1}$$

In order to see how $M_x(\theta)$ produces moments, one simply expands $e^{\theta x}$ in a power series

$$M_x(\theta) = \sum_{x=0}^{\infty} \left(1 + \theta x + \frac{(\theta x)^2}{2!} + \cdots \right) f(x)$$

$$= 1 + \theta\mu + \frac{\theta^2}{2!}\mu_2 + \cdots \qquad (8.2.2)$$

The kth moment about the origin is the coefficient of $\theta^k/k!$ in this expansion. Therefore, if a particular moment is desired, one has only to evaluate the proper derivative at $\theta = 0$

$$\mu_k = \frac{d^k M_x(\theta)}{d\theta^k} \bigg|_{\theta=0} \qquad (8.2.3)$$

The mgf's for some of the discrete distributions discussed in the previous chapter are:

Binomial distribution: $\qquad M_x(\theta) = (q + e^\theta p)^n$ $\qquad\qquad (8.2.4)$

Poisson distribution: $\qquad M_x(\theta) = e^{-m(1-e^\theta)}$ $\qquad\qquad (8.2.5)$

Geometric distribution: $\qquad M_x(\theta) = p(1 - e^\theta q)^{-1}$ $\qquad\quad (8.2.6)$

By analogy, the moment-generating function for a continuous distribution is

$$M_x(\theta) = \int_0^\infty e^{\theta x} f(x)\, dx \qquad (8.2.7)$$

Since expansion of $e^{\theta x}$ in (8.2.7) yields the same form as (8.2.2), Eq. (8.2.3) holds for both discrete and continuous distributions.

8.3 THE RECTANGULAR DISTRIBUTION

Particularly useful in theoretical statistics because it is convenient to deal with mathematically is the *rectangular distribution*, also called the *uniform distribution*. It is constant over some interval (a,b) and is 0 elsewhere:

$$f(x) = (b - a)^{-1} \qquad a < x < b \qquad (8.3.1)$$

The probability that an observation will fall in any interval $c < x < d$ within (a,b) is equal to $(b - a)^{-1}$ times the length of interval or

$$P(c < x < d) = (b - a)^{-1} \int_c^d dx$$

$$= \frac{d - c}{b - a} \qquad (8.3.2)$$

The rectangular distribution is the simplest continuous frequency function on which to illustrate general formulas. In traffic, it has been used to describe the distribution of delay in starting headways and as a gap acceptance function.

8.4 THE NORMAL DISTRIBUTION

A great many of the techniques used in applied statistics are based upon the normal distribution. The density function is given by

$$f(x) = \frac{1}{\sigma \sqrt{2\pi}} e^{-(x-\mu)^2/2\sigma^2} \tag{8.4.1}$$

The function actually represents a two-parameter family of distributions, the parameters being the mean μ and variance δ^2. The mgf is given by

$$M_x(\theta) = e^{\mu\theta + (\sigma\theta)^2/2} \tag{8.4.2}$$

Although the mean is the most important measure of location of the center of a distribution, other measures such as the median and the mode are sometimes useful. The median x_m of a probability distribution of a continuous variable is defined as

$$\int_0^{x_m} f(x)\,dx = \int_{x_m}^{\infty} f(x)\,dx = 0.5 \tag{8.4.3}$$

The mode is simply the most probable value obtained by setting the derivative of the probability density function equal to zero, that is,

$$\frac{df(x)}{dx} = 0 \tag{8.4.4}$$

The two points of inflection occur at points one standard deviation on either side of the mean. The function $f(x)$ in (8.4.1) is difficult to integrate; however the areas under the curve may be found in statistical tables.

The normal distribution is widely used in traffic engineering practice as a model for describing the distribution of spot speeds. When spot-speed data for a given condition are collected and arranged in order of magnitudes as shown in Table 8.1, they form a frequency distribution. The mean speed of vehicles is computed by multiplying the frequency for each class by the mean speed of that class. In the example[1] furnished by Table 8.1, the mean speed is 30.8 mph. The cumulative frequency curve is most useful in determining the median and the percent of vehicles traveling at or below a given speed. Figure 8.1 is a typical cumulative frequency curve, based on the data in Table 8.1. The graph of Fig. 8.1 utilizes a *probability scale* in the ordinate, which has the effect of transforming the normal ogive into a straight line. The 85th and 15th percentiles

Table 8.1 Illustrative spot-speed frequency distribution

Speed class, mph	Frequency	Cumulative, %
13.6–16.5	1	1
16.6–19.5	2	3
19.6–22.5	6	9
22.6–25.5	12	21
25.6–28.5	13	34
28.6–31.5	20	54
31.6–34.5	18	72
34.6–37.5	17	89
37.6–40.5	4	93
40.6–43.5	5	98
43.6–46.5	1	99
46.6–49.5	1	100

are shown on the figure; the values are located at approximately one standard deviation on either side of the mean. These percentiles are used in establishing maximum and minimum speed limits.

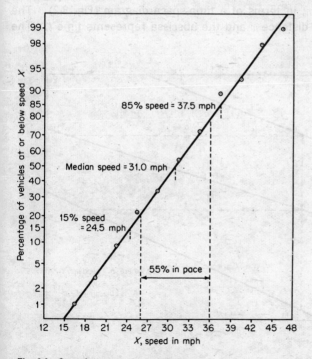

Fig. 8.1 Cumulative speed distribution.

For many years the only distribution function which scientists were accustomed to use was the normal distribution. Its principal characteristics are, as we have seen: (1) It is symmetrical, (2) it assigns a finite probability to every finite deviation, and (3) the mode and median are equal to the mean. Of course, possession of these three requisites is by no means general, especially in considering vehicular traffic characteristics. Consider the duration of time headways during successive vehicles. Here negative values are meaningless; but more than that, experience has shown that the headway of most frequent occurrence is much shorter than the mean.

8.5 DEVELOPMENT OF TRAFFIC VARIABLES

Highway traffic is concerned with the movement of discrete objects over a two-dimensional network. The control of these objects rests both with the individual drivers and the unified system. Vehicular traffic is further distinguished by the fact that the relationship between vehicles in time and the relationship between vehicles in space differ.

It is convenient to show the progress of a stream of vehicles, designated as $i = 1, 2, \ldots, n$, in terms of a time-space diagram (Fig. 8.2). The ordinate represents distance x and the abscissa represents time t. The

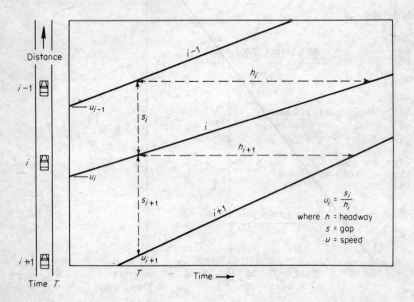

Fig. 8.2 Single-lane traffic variables.

slope of the line gives the rate of change of distance with respect to time dx/dt, or vehicular velocity u.   A curved line means a changing slope or changing velocity, acceleration.   The time and space intervals between the fronts of successive vehicles are referred to as headways h_i and gaps s_i. Their respective units are time (usually expressed in seconds) and distance (usually expressed in feet).   The reciprocal of the mean headway is the flow rate or volume, and the reciprocal of the mean gap is the overall density.

A knowledge of headways and gaps is frequently more informative than a knowledge of volume or density because it reflects much more of the essential nature of the traffic stream.   Headways and gaps are the building blocks with which the entire traffic stream is constructed.

Variation of gap length in a traffic stream ranges over slightly more than a car length to relatively great distance.   Similarly there is a wide variation in headway values.   This variability in both time and space is the most noticeable characteristic of vehicular traffic.   No matter how homogeneous traffic conditions may be, headways can be extremely dissimilar.   It is not surprising to find that attempts to understand headways and gaps have made use of the methods of mathematical statistics and probability theory, primarily by seeking to determine the theoretical distributions that account for the observed distribution.

8.6 THE NEGATIVE EXPONENTIAL DISTRIBUTION

If t represents the time between vehicular arrivals, then the probability density function for t corresponding to the negative exponential distribution is given by

$$f(t) = qe^{-qt} \tag{8.6.1}$$

where q is the mean flow rate in vehicles per unit time.   The proof of (8.6.1) is based on setting up and solving a differential difference equation relating the probability of the first arrival being before t, $G(t)$ and the probability of the first arrival being before $t + \Delta t$, $G(t + \Delta t)$.   The latter can occur in two mutually exclusive ways: (1) the first arrival not in t but in $t + \Delta t$ and (2) the first arrival in t.   This may be written

$$G(t + \Delta t) = [1 - G(t)]q\,\Delta t + G(t) \tag{8.6.2}$$

Expanding

$$\frac{G(t + \Delta t) - G(t)}{\Delta t} = [1 - G(t)]q$$

and taking the limit as Δt approaches zero gives the definition of the derivative $dG(t)/dt$

$$\frac{dG(t)}{dt} = [1 - G(t)]q \tag{8.6.3}$$

The solution of (8.6.3) is

$$G(t) = 1 - e^{-qt} \tag{8.6.4}$$

Since the probability of the first arrival being between t and $t + \Delta t$ is from (8.6.4),

$$G(t + \Delta t) - G(t) = e^{-qt} - e^{-q(t+\Delta t)}$$

and

$$\int_t^{t+\Delta t} qe^{-qt}\,dt = e^{-qt} - e^{-q(t+\Delta t)}$$

it follows that (8.6.1) is true.

What is being said, in effect, is this: The probability of a headway between t_1 and t_2 is equal to the probability that the first arrival is between t_1 and t_2. Written symbolically

$$P(t_1 < h < t_2) = e^{-qt_1} - e^{-qt_2} \tag{8.6.5}$$

Thus, the density and cumulative forms of the negative exponential distribution are given by (8.6.1) and (8.6.5), respectively.

We have seen, in the previous chapter, how the Poisson distribution is frequently used as a model to determine the distribution of vehicular traffic arrivals on a highway. If in Eq. (7.4.4) the mean number of arrivals in time t is replaced by qt, the form of the distribution becomes

$$P(x) = \frac{(qt)^x e^{-qt}}{x!} \tag{8.6.6}$$

The probability of no arrivals $x = 0$ in time t becomes

$$P(0) = e^{-qt} \tag{8.6.7}$$

But to have no arrivals in an interval t, there must be a headway greater than or equal to t. Therefore, (8.6.7) is also the probability of a headway greater than or equal to t:

$$P(h > t) = e^{-qt} \tag{8.6.8}$$

Since (8.6.8) is the "survivor" form of a cumulative distribution, the density function is the negative differential

$$\frac{-dP}{dt} = qe^{-qt}$$

and one sees how the negative exponential distribution may be established from the Poisson distribution.   Actually, the relationship is this: The distribution of time spacings between Poisson arrivals conforms to a negative exponential distribution.

The mean and variance for the negative exponential distribution are

$$\bar{t} = \frac{1}{q} \tag{8.6.9}$$

$$\sigma^2 = \frac{1}{q^2} \tag{8.6.10}$$

Naturally, $\bar{t}$ and q must be expressed in compatible units.   The mean headway in a single stream of traffic of 1,800 vph is $\frac{1}{1,800}$ hr, or 2 sec.

Although the negative exponential provides a fair model for headways in low, free-flowing traffic volumes, the headways predicted by this distribution differ greatly from observations of high-volume conditions. Vehicles possess length and obviously cannot follow at an infinitesimal headway, as the distribution predicts.   Secondly, vehicles cannot pass at will but tend to form platoons as faster vehicles overtake slower ones.

The first objection can be overcome by translating the exponential curve in (8.6.8) to the right by an amount equal to the minimum observed headway τ such that

$$P(h > t) = e^{-(t-\tau)/(\bar{t}-\tau)} \tag{8.6.11}$$

In considering the second objection regarding passing, Schuhl[2] proposed that the traffic stream be considered as composed of a combination of free-flowing and constrained vehicles

$$P(h > t) = (1 - a)e^{-t/\bar{t}_1} + ae^{-(t-\tau)/(\bar{t}_2-\tau)} \tag{8.6.12}$$

where $\bar{t}_1$ = average headway of free-flowing vehicles
$\bar{t}_2$ = average headway of constrained vehicles
τ = minimum headway of constrained vehicles
a and $1 - a$ = fractions of total volume in the two subdistributions
The form in (8.6.11) is called the translated negative exponential; the form in (8.6.12) is called the composite negative exponential.

8.7   THE PEARSON TYPE III DISTRIBUTION

Among the families of distributions which have been formulated for the purpose of enabling the statistician to deal with a wide variety of data are those developed by Karl Pearson.   Pearson noted that the binomial and

Poisson distributions satisfy the differential equation

$$\frac{dP}{dx} = \frac{(a + x)P}{b + cx + dx^2} \tag{8.7.1}$$

for some set of values of the constants a, b, c, and d. Fry[3] shows how certain solutions of this equation may be sorted out, largely because of their algebraic simplicity, to form the well-known Pearson type I and Pearson type III frequency distributions.

A random variable y is said to be distributed as the type III distribution if its density is

$$f(y) = \frac{b^a}{\Gamma(a)} y^{a-1} e^{-by} \qquad 0 < y < \infty \tag{8.7.2}$$

where $\Gamma(a)$ is called the gamma function and is defined by the formula

$$\Gamma(a) = \int_0^\infty z^{a-1} e^{-z} \, dz$$

In order to apply the type III distribution to traffic phenomena such as space headways, in which the distribution curve does not go through the origin (gaps and headways between successive vehicles in the same lane can never be zero because vehicles possess length), it is necessary to translate the distribution c units from the origin.

Recognizing that the area under $f(y)$ is unity, we obtain the desired generalized distribution by substituting $t = y + c$ for y in (8.7.2)

$$\int_0^\infty f(y) \, dy = \int_c^\infty \frac{b^a}{\Gamma(a)} (t - c)^{a-1} e^{-b(t-c)} \, dt$$

giving

$$f(t) = \frac{b^a}{\Gamma(a)} (t - c)^{a-1} e^{-b(t-c)} \qquad c < t < \infty \tag{8.7.3}$$

Taking moments about c,

$$u_k{}^c = \frac{b^a}{\Gamma(a)} \int_c^\infty (t - c)^{k+a-1} e^{-b(t-c)} \, dt$$

$$= \frac{b^a}{\Gamma(a)} \frac{\Gamma(k + a)}{b^{k+a}} \int_c^\infty \frac{b^{k+a}}{\Gamma(k + a)} (t - c)^{k+a-1} e^{-b(t-c)} \, dt$$

$$= \frac{b^a}{\Gamma(a)} \cdot \frac{\Gamma(k + a)}{b^{k+a}} \tag{8.7.4}$$

since the integral represents the area under a type III curve and must equal unity. The first moment about c is obtained by letting $k = 1$ in (8.7.4)

$$u_1{}^c = \frac{a}{b} \tag{8.7.5}$$

Adding c to (8.7.5) gives the mean of this distribution

$$\mu = \frac{a}{b} + c \tag{8.7.6}$$

The variance is given by

$$\sigma^2 = u_2{}^c - (u_1{}^c)^2$$

$$= \frac{a}{b^2} \tag{8.7.7}$$

The procedure for fitting a continuous distribution to observed data is similar to that used in the application of discrete distributions. Suppose we are measuring time headways on a freeway lane (the results are as tabulated in Table 8.2), and we want to fit a type III distribution to these data.

Table 8.2 Example headway data for fitting the type III distribution

Headway t	Number	Frequency $f(t)$	Headway t	Number	Frequency $f(t)$
$0.0 \leq t < 1.0$	16	0.027	$13.0 \leq t < 14.0$	10	0.017
$1.0 \leq t < 2.0$	74	0.123	$14.0 \leq t < 15.0$	20	0.033
$2.0 \leq t < 3.0$	65	0.108	$15.0 \leq t < 16.0$	10	0.017
$3.0 \leq t < 4.0$	51	0.085	$16.0 \leq t < 17.0$	6	0.010
$4.0 \leq t < 5.0$	42	0.070	$17.0 \leq t < 18.0$	15	0.025
$5.0 \leq t < 6.0$	46	0.077	$18.0 \leq t < 19.0$	6	0.010
$6.0 \leq t < 7.0$	46	0.077	$19.0 \leq t < 20.0$	3	0.005
$7.0 \leq t < 8.0$	40	0.067	$20.0 \leq t < 21.0$	6	0.010
$8.0 \leq t < 9.0$	27	0.045	$21.0 \leq t < 22.0$	5	0.008
$9.0 \leq t < 10.0$	26	0.043	$22.0 \leq t < 23.0$	4	0.007
$10.0 \leq t < 11.0$	33	0.055	$23.0 \leq t < 24.0$	7	0.012
$11.0 \leq t < 12.0$	21	0.035	$24.0 \leq t < 25.0$	5	0.006
$12.0 \leq t < 13.0$	17	0.028			

Mean $\bar{t} = 7.45$ sec
Variance $s^2 = 31.0$

The equation of the Pearson type III distribution is given in (8.7.3). It is apparent that the parameters a, b, and c must be estimated in order to utilize the function. Since there were 16 headways between 0 and 1 sec, the curve must go through the origin in this case. In order for $f(t)$ to

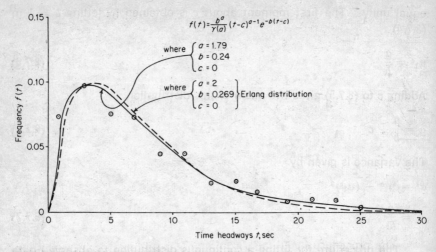

Fig. 8.3   Fitting a Pearson type III distribution to data of Table 8.1.

equal zero when t equals zero, the parameter c must equal zero. Using the method of moments, we then determine the parameters a and b from the equations

$$\bar{t} = \frac{a}{b}$$

$$s^2 = \frac{a}{b^2}$$

obtaining $a = 1.79$ and $b = 0.24$ for $\bar{t} = 7.45$ and $s^2 = 31.00$. A comparison of the theoretical curve with the observed data is shown in Fig. 8.3.

8.8   THE ERLANG DISTRIBUTION

In the previous section, we gave a brief definition of the gamma function. Interest in this formula lies in the fact that

$$\Gamma(n+1) = -x^n e^{-x} \Big|_0^\infty + n \int_0^\infty x^{n-1} e^{-x} \, dx \tag{8.8.1}$$

which may be written as

$$\Gamma(n+1) = n\Gamma(n) \tag{8.8.2}$$

From (8.8.1), if $n = 0$, we see that $\Gamma(1) = 1$. Using this result and letting $n = 1, 2, 3, \ldots$ successively in (8.8.2) we obtain by induction

$$\Gamma(n+1) = n! \qquad n = 0, 1, 2, \ldots \tag{8.8.3}$$

In most traffic applications it suffices to round off the a in (8.7.3) to the nearest integer and then calculate b from

$$b = \frac{a}{\bar{t}}$$

This special case of the type III distribution is called the Erlang distribution and its form is

$$f(t) = \frac{(qa)^a}{(a-1)!}\, t^{a-1} e^{-aqt} \qquad a = 1, 2, \ldots$$

with the rate of flow $q = \bar{t}^{-1}$ and $\Gamma(a) = (a-1)!$ from (8.8.3). The Erlang approximation with $a = 2$ has been superimposed on Fig. 8.3, showing a very satisfactory fit.

The natural distribution of vehicles in the traffic stream can be extremely useful in explaining the operations of many facilities. The Erlang frequency distribution is a logical choice because the negative exponential distribution, well established in the description of traffic headways, may be considered as a special case of the Erlang distribution for which $a = 1$. Moreover, use of the Erlang distribution for all values of a affords the opportunity of considering the distribution of vehicles for all cases between randomness ($a = 1$) to complete uniformity ($a = \infty$).

The cumulative form of the Pearson type III distribution must be evaluated by numerical methods or tables of the function. The cumulative form of the Erlang distribution may be evaluated by successive integrations by parts. The cumulative frequency curves in Figs. 8.4 and 8.5 are offered to facilitate the use of this distribution for the description of headways and gaps.

8.9 THE PEARSON TYPE I DISTRIBUTION

A random variable z is said to have a type I distribution if its density is given by

$$f(z) = \frac{z^{a-1}(1-z)^{b-1}}{\beta(a,b)} \qquad 0 < z < 1 \tag{8.9.1}$$

where $\beta(a,b)$ is called the beta function of a and b and is defined in terms of the gamma function by

$$\beta(a,b) = \frac{\Gamma(a)\Gamma(b)}{\Gamma(a+b)} \tag{8.9.2}$$

In order to apply the type I distribution to traffic phenomena, it is

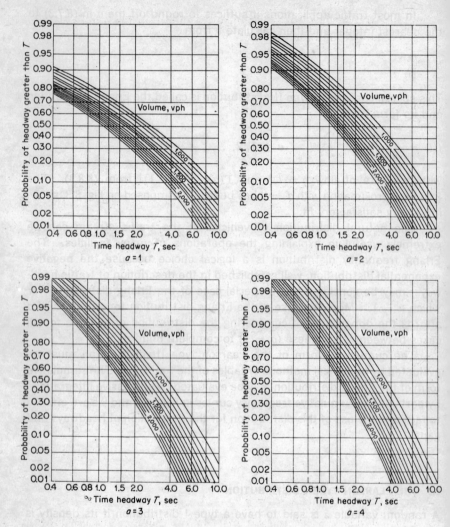

Fig. 8.4 Cumulative probability curves for the Erlang distribution.

$$\int_T^\infty \frac{(qa)^a}{(a-1)!} t^{a-1} e^{-aqt} \, dt$$

where q = flow, vps

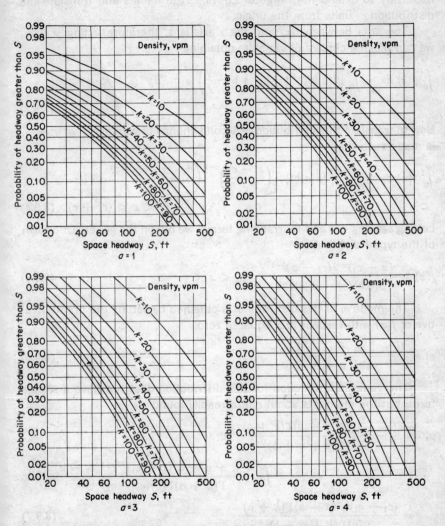

Fig. 8.5 Cumulative probability curves for the Erlang distribution.

$$\int_S^\infty \frac{(ka)^a}{(a-1)!} x^{a-1} e^{-akx}\, dx$$

where k = density, vpm

necessary to extend the range of Eq. (8.9.1) to c units and translate the distribution c_1 units from the origin.

The area between $f(z)$ and the z axis is 1, therefore letting $y = cz$, we can make a change of variable, obtaining

$$\int_0^1 f(z)\,dz = \int_0^c [\beta(a,b)]^{-1} \left(\frac{y}{c}\right)^{a-1} \left(1 - \frac{y}{c}\right)^{b-1} c^{-1}\,dy \qquad 0 < y < c$$

(8.9.3)

Now to translate the distribution, a second change of variable, $x = y + c_1$, is made such that

$$\int_0^c f(y)\,dy = \int_{c_1}^{c_1+c} [\beta(a,b)]^{-1} \left(\frac{x - c_1}{c}\right)^{a-1} \left(\frac{c - x + c_1}{c}\right)^{b-1} c^{-1}\,dx$$

$$c_1 < x < (c_1 + c) \quad (8.9.4)$$

Letting $c_2 = c_1 + c$ and rearranging terms, we obtain the generalized form of the type I distribution

$$f(x) = \frac{(x - c_1)^{a-1}(c_2 - x)^{b-1}}{\beta(a,b)(c_2 - c_1)^{a+b-1}} \qquad c_1 < x < c_2 \qquad (8.9.5)$$

The mode m of this distribution is obtained by setting the first derivative of $f(x)$ with respect to x equal to zero

$$m = \frac{(b-1)c_1 + (a-1)c_2}{a + b - 2} \qquad (8.9.6)$$

The generating function for this distribution does not have a simple form, but the moments about c_1 are readily found

$$\mu_k^{c_1} = \frac{(c_2 - c_1)^k \Gamma(a + b)\Gamma(k + a)}{\Gamma(a)\Gamma(k + a + b)}$$

$$\int_{c_1}^{c_2} \frac{\Gamma(k + a + b)(x - c_1)^{k+a-1}(c_2 - x)^{b-1}}{\Gamma(k + a)\Gamma(b)(c_2 - c_1)^{k+a+b-1}}\,dx$$

$$= \frac{(c_2 - c_1)^k \Gamma(a + b)\Gamma(k + a)}{\Gamma(a)\Gamma(k + a + b)} \qquad (8.9.7)$$

since the integral is still a type I distribution and must equal 1.   The first moment about c_1 is obtained by letting $k = 1$ in (8.9.7)

$$\mu_1^{c_1} = (c_2 - c_1) \frac{a}{a + b} \qquad (8.9.8)$$

Adding c_1 to (8.9.8) gives the mean of this distribution

$$\mu = \frac{bc_1 + ac_2}{a + b} \qquad (8.9.9)$$

The variance is given by

$$\sigma^2 = \mu_2{}^{c_1} - (\mu_1{}^{c_1})^2$$

$$= \left(\frac{c_2 - c_1}{a + b}\right)^2 \frac{ab}{a + b + 1} \tag{8.9.10}$$

The utility of the type I distribution as a traffic model can be indicated by illustrating its application to speed data (Table 8.3).   For the purposes

Table 8.3   Example speed data for fitting the type I distribution

Speed class u	Number	Frequency $f(u)$	Speed $\times$ frequency $uf(u)$
$35.0 < u \le 37.0$	2	0.01	0.36
$37.0 < u \le 39.0$	4	0.02	0.76
$39.0 < u \le 41.0$	6	0.03	1.20
$41.0 < u \le 43.0$	14	0.07	2.94
$43.0 < u \le 45.0$	14	0.07	3.08
$45.0 < u \le 47.0$	30	0.15	6.90
$47.0 < u \le 49.0$	40	0.20	9.60
$49.0 < u \le 51.0$	50	0.25	12.50
$51.0 < u \le 53.0$	30	0.15	7.80
$53.0 < u \le 55.0$	10	0.05	2.70
	200	$\Sigma f(u) = 1.00$	$\Sigma uf(u) = 47.84$

Mean $\bar{u} = 47.8$ mph   Minimum speed $c_1 = 35.0$ mph
Mode $m = 50.0$ mph   Maximum speed $c_2 = 55.0$ mph

of this discussion, the following notation will be used:

$c_1 =$ minimum observed speed sampled
$c_2 =$ maximum observed speed sampled
$m =$ most popular speed sampled
$\bar{u} =$ sample mean speed

To obtain estimates of a and b of the type I distribution, we have only to solve two of the three equations, (8.9.6), (8.9.9), and (8.9.10), simultaneously.   For (8.9.6) and (8.9.9) we get

$$b = \frac{-(c_2 - \bar{u})(c_1 + c_2 - 2m)}{(c_1 - \bar{u})(c_2 - m) - (c_1 - m)(c_2 - \bar{u})} \tag{8.9.11}$$

and

$$a = \frac{b(c_1 - \bar{u})}{\bar{u} - c_2} \tag{8.9.12}$$

For the example problem we obtained $b = 1.64$ and $a = 2.91$; the plot of

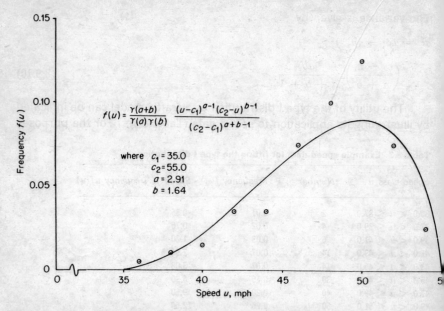

Fig. 8.6 Fitting a Pearson type I distribution to data of Table 8.3.

this theoretical curve superimposed on the observed data is illustrated in Fig. 8.6.

8.10 DISTRIBUTION RELATIONSHIPS

The relationships among the various probability distributions that have been discussed in the last two chapters are illustrated in Fig. 8.7. The discrete distributions are derived from power series, and the continuous distributions, type I and type III, are derived from the beta and gamma functions. The Poisson distribution is the limiting case of both the binomial and negative binomial distributions.

The negative exponential, translated negative exponential distribution, and Erlang distributions—often utilized as models for the distribution of gaps in the traffic stream—are special cases of the type III distribution.

The cumulative form of both the discrete and continuous distributions has wide applications in traffic engineering practice. Knowing the probability of exceeding a certain condition is usually of more significance than knowing just the probability of occurrence of the condition. Of course, in arriving at the cumulative forms, an integral sign is used in the continuous case and a summation sign in the discrete case. Considerable

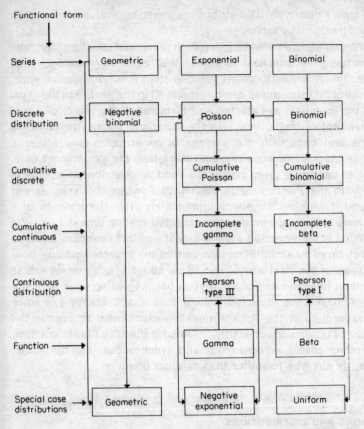

Fig. 8.7 Functional forms and organization of common probability distributions.

interest is manifest in the fact that it is possible to relate the incomplete gamma function to the cumulative Poisson distribution

$$\int_0^x \frac{b^a}{\Gamma(a)} \, t^{a-1} e^{-bt} \, dt = 1 - e^{-bx} \sum_{i=0}^{a-1} \frac{(bx)^i}{i!}$$

In a similar manner, the incomplete beta function is related to the cumulative binomial distribution

$$\int_0^x \frac{t^a (1-t)^{n-a-1}}{\beta(a+1, n-a)} \, dt = 1 - \sum_{i=0}^{a} \binom{n}{i} x^i (1-x)^{n-i}$$

The incomplete beta and incomplete gamma functions associated with use of the type I and type III distributions must be evaluated by either numerical methods or by tables.[4,5]

The correspondence between gap (continuous) and counting (discrete) distributions has great practical as well as theoretical significance.   It is, for example, much easier to count vehicles than it is to measure gaps. One of two counting procedures can be employed in the field.   In the usual procedure, traffic counts are started and terminated at given clock times quite independent of traffic flow.   In the other procedure, the counting period starts and ends with the passage of an arrival.   The former is referred to as the *asynchronous case* and the latter, the *synchronous case.*

If vehicles passing a given point on a road are distributed in time in accordance with the Poisson distribution with a mean of μ vehicles per unit time, and if we start observing immediately after the passage of a vehicle, we have already shown that the expected waiting time is $1/\mu$.   We might ask: What is the expected waiting time if we start observing at some arbitrary clock time?   Our intuition tells us that the expected waiting time for the passage of the next arrival should be about $\frac{1}{2}\mu$, since we are as likely to come in near the end of an interval between vehicles as near the beginning.   Actually, the expected waiting time is still exactly $1/\mu$ since the basic assumption in the Poisson negative exponential process is the independence of occurrence of events.   Thus for the case of random flow, the two counting cases, asynchronous and synchronous, are the same. However, this is not true for other than random flows.

8.11 FUNCTIONS AND DISTRIBUTIONS

As indicated in the previous section, continuous distributions may be formed from certain continuous functions in much the same manner as discrete distributions may be formed from certain power series.

For example, starting with the gamma function

$$\Gamma(a) = \int_0^\infty y^{a-1}e^{-y}\,dy$$

and making the change of variable $x = y/b$ gives

$$\Gamma(a) = \int_0^\infty (bx)^{a-1}e^{-bx}b\,dx$$

$$= \int_0^\infty b^a x^{a-1}e^{-bx}\,dx$$

which may be written as

$$1 = \int_0^\infty \frac{b^a}{\Gamma(a)} x^{a-1}e^{-bx}\,dx$$

proving that

$$f(x) = \frac{b^a}{\Gamma(a)} x^{a-1} e^{-bx}$$

is a valid continuous distribution called the *gamma distribution.*

A similar procedure may be used to obtain the beta distribution from the beta function.   Thus

$$\Gamma(a)\Gamma(b) = \int_0^\infty x^{a-1} e^{-x} \, dx \int_0^\infty y^{b-1} e^{-y} \, dy$$

$$= \int_0^\infty \int_0^\infty x^{a-1} y^{b-1} e^{-(x+y)} \, dx \, dy$$

Making the change of variable $u = x/(x + y)$, one obtains

$$\Gamma(a)\Gamma(b) = \int_0^\infty \int_0^\infty (uy)^{a-1} y^b e^{-y/(1-u)} (1 - u)^{-(a+1)} \, du \, dy$$

Another change of variable $v = y/(1 - u)$ gives

$$\Gamma(a)\Gamma(b) = \int_0^\infty \int_0^1 u^{a-1} (1 - u)^{b-1} v^{a+b-1} e^{-v} \, du \, dv$$

$$= \int_0^\infty v^{a+b-1} e^{-v} \, dv \int_0^1 u^{a-1} (1 - u)^{b-1} \, du$$

$$= \Gamma(a + b) \int_0^1 u^{a-1} (1 - u)^{b-1} \, du$$

Therefore

$$1 = \int_0^1 \frac{\Gamma(a + b)}{\Gamma(a)\Gamma(b)} u^{a-1} (1 - u)^{b-1} \, du$$

and the integrand is a valid continuous distribution called the *beta distribution.*

8.12 SUMMARY

As we go through examples of fitting distributions to such continuous traffic variables as headways, gaps, and speeds, we get the impression that there is quite good agreement between theory and practical experience, which may lead to false optimism concerning the methods described.   In practical work it is often difficult to obtain satisfactory agreement between the observations and the theoretical distributions.   This should not lead to a general rejection of the theory.

Fisher[6] lists five steps which should be followed in a statistical analysis.

1. *Planning* the investigation, including sample size and methods, choice of interval, and duration of study
2. *Formulation* of a mathematical-statistical model
3. *Estimation* of the parameters pertaining to the model
4. *Examination of the agreement* between the model and the observations
5. *Tests of significance* and confidence limits

In planning a traffic investigation, two important decisions required are the duration of the study and the size of increment for grouping the measurements.   Because of cyclical variations in traffic demand, we have seen that the distribution of arrivals is naturally sensitive to the duration of the study.   Intersection counts have shown that arrivals throughout the entire peak hour are not Poisson, whereas arrivals during the peak period within the peak hour are Poisson.[7]   On the other hand, Gerlough[8] has shown that traffic may be random when observed for an interval of one length but nonrandom when observed for an interval of a different length.   Sometimes the choice of interval is established by the proposed application of the study.   Thus, in signalization problems intervals equal to the phase length on an approach or the entire cycle length are logical choices.

The second step, the formulation of a mathematical-statistical model which gives a satisfactory description of the data, is not a statistical task for the statistician; but it belongs within the professional subject of traffic engineering from which the observations have been derived.   Ideally, a proper theoretical model, based on general laws, regarding the process which has generated the observations is desired.   Where the complexity of the operation studied precludes the formulation of a proper theoretical model, a purely empirical description of the observed phenomenon without any attempt at linking this description with theoretical reasoning based on traffic knowledge is necessary.   In any event, there is a demand for the greatest possible simplicity —the description should be mathematically simple and should lead to a simple test of significance.

After a distribution function is chosen for the population from which the traffic measurements have been made, the next step is the derivation of estimates of the parameters from the observations.   The method of moments, extensively illustrated in the last two chapters, is one means of estimating the parameters of the theoretical distribution chosen.   Most of the practical distributions used in traffic engineering have either one or two parameters, which can be estimated from the mean and variance of the field measurements.   In cases where the theoretical distribution has been shifted, experience has shown that it is preferable to establish the amount of this shift arbitrarily, rather than estimating it from a higher moment.   In addition to the method of moments, other methods for deriving estimates with specified properties include minimum-variance esti-mators and maximum-likelihood estimators.[9]

It is not unusual to choose the mathematical form of the distribution function on the basis of a graphical analysis of the observed data. This provides a good idea of the agreement between model and observations at the same time as the model is being constructed. After the estimates of parameters have been computed, the graphical analysis should be supple-mented by a more objective test. The expected distribution is often com-pared with the observed distribution by means of the chi-square test. Such tests are useful so that the deviations between the measurements and the estimates of the corresponding theoretical values can be computed and studied.

Lastly, it should be reemphasized that an unsatisfactory test of signifi-cance is not equivalent to rejection of the hypothesis, practically speaking, any more than a satisfactory test is equivalent to proof of the hypothesis. Statistics is the traffic engineer's helper, not his master. A great deal of work is needed to establish some ground rules for the interpretation of the chi-square test as applied to traffic phenomena. For example, the test is extremely sensitive to the sample size. In the engineer's attempt to fit a theoretical distribution to the distribution of freeway headways, the chi-square value increases with the duration of the study. Moreover, if headways are grouped into $\frac{1}{2}$-sec intervals, the chi-square value will be different than if a 1-sec interval is chosen.

PROBLEMS

8.1. If the mgf of the distribution of service times at a tollbooth is given by

$$M_T(\theta) = \frac{e^{-qT}(qT - \theta)T}{qTe^{-(qT-\theta)} - \theta}$$

find the mean of the service-time distribution.

8.2. An intersection will be closed while the sanitation department makes an improve-ment on a sewer line crossing the intersection. The time work is started until the intersection is completely repaired will be at least 12 days, no more than 30 days, and will most likely take $16\frac{1}{2}$ days (mode). A service-station owner on the corner tells you his business will be seriously affected if the intersection is in repair over 21 days. Assuming the distribution of repair time is a beta distribution $F(t)$, $12 < t < 30$, with mean of 18 days, what is the probability of the repair taking over 21 days?

8.3. The driver acceptances of time gaps on the main street were observed for the traffic on the minor street which is controlled by a stop sign. For the following observed distribution, test the hypothesis that these data represent a random sample from a normal population.

Midpoint of interval, sec	Observed frequency
1.0	0
2.0	6
3.0	34
4.0	132
5.0	179
6.0	218
7.0	183
8.0	146
9.0	69
10.0	30
11.0	3
12.0	0

8.4. In a 176-ft speed trap the following frequency of times were recorded:

Time, sec	No. of vehicles observed
1.0–1.4	2
1.4–1.8	1
1.8–2.2	5
2.2–2.6	36
2.6–3.0	54
3.0–3.4	52
3.4–3.8	25
3.8–4.2	12
4.2–4.6	3
4.6–5.0	2
>5.0	0

a. Convert the time to speeds.
b. Calculate the mean and modal speed of the sample.
c. Plot the cumulative frequency (percentage) on probability paper and measure the 15th, 50th, and 85th percentile speeds.
d. What would you recommend as a speed limit on this facility? Explain your reasoning in making your choice.

8.5. The following are tabulated gaps in a lane of traffic. Test the observed gaps for fit to the following distributions using the chi-square test:
a. Translated negative exponential
b. Negative exponential
c. Erlang with $a = 2$
d. Erlang with $a = 3$

Gap, sec	No. of observed gaps
0.0–1.0	0
1.0–2.0	21
2.0–3.0	34
3.0–4.0	17
4.0–5.0	15
5.0–6.0	6
6.0–7.0	2
7.0–8.0	3
8.0–9.0	1
9.0–10.0	1
>10.0	0

8.6. If the distribution of space headways conforms to a composite exponential distribution for a density of 66 vph evenly divided between free-flowing and constrained vehicles, find the probability of a headway between 10 and 60 ft. The minimum headways for free-flowing and constrained vehicles are 60 and 20 ft, respectively.

8.7. What characteristics of the beta (Pearson type I) distribution make it preferable to the normal distribution as a description of the distribution of speeds?

8.8. Develop the moment-generating function for $f(x) = qe^{-qx}$ (negative exponential distribution). Check your development by evaluating the mean and variance based on the mgf vs. those given in this chapter.

8.9. The distribution of gaps in a pedestrian stream entering a department store is found to be distributed as a translated negative exponential distribution. During the rush hour the probability of a gap greater than 4.5 sec is 0.368. What is the mean interval at which shoppers enter the store?

8.10. Arrivals into an oversaturated intersection from one approach are found to conform to an Erlang $a = 4$ distribution

$$f(x) = \frac{b^a}{(a-1)!} x^{a-1} e^{-bx}$$

during the rush period. The approach volume is 1,620 vph. An upstream intersection will be blocked if, during a 60-sec red phase, more than 12 cars arrive. What is the probability of blocking the intersection?

8.11. If the expected distance between cars is 100 ft, what is the probability that a 325-ft city block is empty?

REFERENCES

1. Kennedy, N., J. H. Kell, and W. S. Homburger: "Fundamentals of Traffic Engineering," Institute of Transportation and Traffic Engineering, University of California, Berkeley, 1963.
2. Schuhl, Andre: The Probability Theory Applied to the Distribution of Vehicles on Two-lane

Highways, in "Poisson and Traffic," The Eno Foundation for Highway Traffic Control, Saugatuck, Conn., 1955.

3. Fry, Thorton C.: "Probability and Its Engineering Uses," D. Van Nostrand Company, Inc., Princeton, N.J., 1928.
4. Pearson, Karl: Tables of the Incomplete Gamma-function, *Biometrika*, London, 1922.
5. Pearson, Karl: Tables of the Incomplete Beta-function, *Biometrika*, 1948.
6. Hald, A.: "Statistical Theory With Engineering Applications," John Wiley & Sons, Inc., New York, 1952.
7. Drew, D. R., and C. Pinnell: A Study of Peaking Characteristics of Signalized Urban Intersections as Related to Capacity and Design, *Highway Res. Board Bull.* 352, pp. 1–54, 1962.
8. Gerlough, D. L.: Use of Poisson Distribution in Highway Traffic, in "Poisson and Traffic," The Eno Foundation for Highway Traffic Control, Saugatuck, Conn., 1955.
9. Kendall, M. G., and A. Stuart: "The Advanced Theory of Statistics," vol. 2, Hafner Publishing Company, Inc., New York, 1961.

CHAPTER NINE
GAP ACCEPTANCE

So geographers, in Afric maps,
With savage pictures fill their gaps.

Jonathan Swift

9.1 INTRODUCTION

One of the most important aspects of traffic operation is the interaction of vehicles within a single stream of traffic or the interaction of two separate traffic streams. This interaction takes place when a driver changes lanes, merges into a traffic stream, or crosses a traffic stream. Inherent in the traffic interaction associated with these basic maneuvers is the concept of gap acceptance.

Gap acceptance is best illustrated in the analysis of ramp-freeway connections. This is true because the unsignalized intersection, another location in which gap acceptance provides the basis of operation, is essentially just a special case of the ramp-freeway connection in which there is no acceleration lane.

The subject of ramp vehicles merging into the freeway stream has deservedly been treated by a number of researchers. Most of the research has been devoted to empirical studies, leading to design and operational procedures. Mathematical treatment of the merging maneuver has been attempted, too, with somewhat limited success because of the complexity of the vehicle interactions. Computer technologists have contributed several digital-computer simulation programs, but lack of detailed criteria on gap acceptance and merging logic has hampered progress.

The three fundamental maneuvers performed by vehicles in the vicinity of this connection area may be identified as follows:

Lane Change   The transfer of a vehicle from one traffic lane to the next adjacent traffic lane.

Merging   The process by which vehicles in two separate traffic streams moving in the same general direction combine or unite to form a single stream.

Weaving   The oblique crossing of one stream of traffic by another, accomplished by the merging of the two streams into one and then the diverging of this common stream into separate streams again.

The freeway elements associated with the above maneuvers are:

Acceleration lane   An added width of pavement adjacent to the main roadway traffic lanes, enabling vehicles entering the main roadway to adjust their speed to the speed of through traffic before merging.

Ramp   A connecting roadway between two intersecting or parallel roadways, one end of which joins in such a way as to produce a merging maneuver.

Frontage road   A roadway paralleling a freeway so as to intercept traffic entering or leaving the facility and to furnish access to abutting property.

Basic to the description of traffic interaction in the merging process are the following variables:

Headway   The interval of time between successive vehicles moving in the same lane and measured from head to head as they pass a point on the road.

Space headway   The distance between successive vehicles moving in the same lane and measured from head to head at a given instant in time.

Gap   A major-stream headway that is evaluated by a minor-stream vehicle desiring to either merge into or cross the major stream. The units may be those of either time or distance (space gap).

Lag   The interval of time between the arrival of a minor-stream vehicle and the arrival of a major-stream vehicle at a reference point or points in the vicinity of the area where the streams either cross or merge.

Space lag   The distance between a minor-stream vehicle and a reference point where the minor stream crosses or joins a major stream sub-

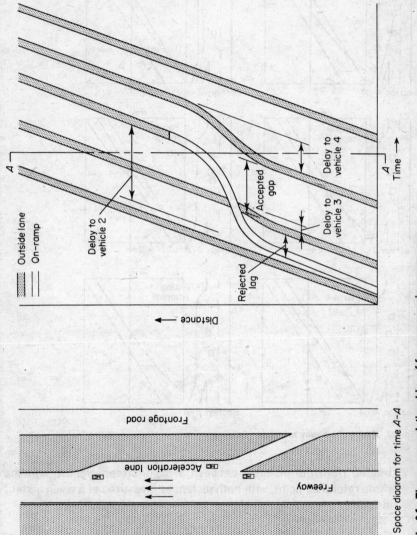

Fig. 9.1 Time-space relationships of freeway merging maneuver.

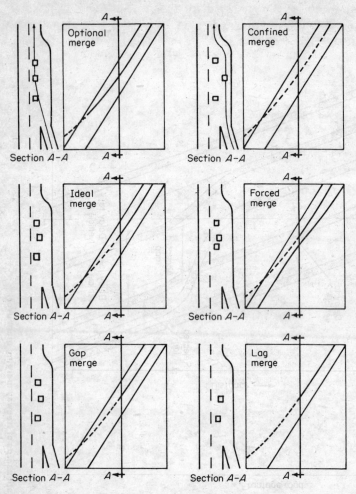

Fig. 9.2 Types of freeway merges.

tracted from the distance between a major-stream vehicle and the same reference point, with both distances measured at a given instant in time.

Figure 9.1 is a time-space diagram prepared to illustrate the relationship between the geometric elements comprising the merging area and the movements of mainstream and ramp vehicles within the area. Distance is plotted as the ordinate, time is the abscissa, and the slope of the traces denotes speed. The traces for freeway and ramp vehicles are identified in the figure. A vehicle is regarded as a ramp vehicle as long as it remains at least partially on the acceleration lane.

It is convenient to subdivide merges into their basic types and classifications, then to determine the kind and amount of information needed for intelligent analysis of the movements.   Following is a listing of merge maneuvers, defined for use in this report (see Fig. 9.2):

Optional merge   The merging vehicle voluntarily moves from the acceleration lane into the outside traffic lane.

Confined merge   The merging vehicle is forced into the outside traffic lane by the presence of the end of the acceleration lane.

Ideal merge   The merging vehicle is able to enter the freeway stream without causing a freeway vehicle to reduce its speed or change lanes.

Forced merge   The ramp vehicle effects the merging maneuver into the freeway stream so that an oncoming freeway vehicle or vehicles must either slow down or change lanes.

Single-entry merge   One ramp vehicle moves into a single freeway time gap.

Multiple-entry merge   Two or more ramp vehicles merge into a single freeway time gap.

From Figs. 9.1 and 9.2 and the above classification, it is apparent that a merge must be qualified as being either single or multiple entry, optional or confined, ideal or forced, and gap or lag.   For example, a given merge might be described as being a "single-entry, optional, ideal-gap merge."

9.2 MERGING PARAMETERS

In the previous section, a few traffic variables —headways, gaps, and lags— were identified.   In addition to these variables, the speed of the major-stream vehicles, speed of the merging vehicles, relative speed, major-stream flow, and minor-stream flow are additional variables that must be considered in any rational description of the merging process.

Whereas a variable is a quantity which can assume any value or number, a parameter is a term used in identifying a particular variable or constant other than the coordinate variables.   For example, *volume* is a traffic variable and *capacity*—the maximum volume that a facility can accommodate—is the corresponding traffic parameter.   Some important parameters for describing the gap acceptance phenomenon in freeway merging are the critical gap, percent of ramp vehicles delayed, mean duration of static delay accepting a gap, mean length of queue, and total waiting time on the ramp.

Several "critical" values have been discussed in the literature.   Greenshields[1] defined the "acceptable average-minimum time gap" as a gap accepted by half the drivers.   Raff[2] used a slightly different parameter, the "critical lag."   The critical lag is the size of lag whose number of accepted lags shorter than it is equal to the number of rejected lags longer than it.   The Greenshields and Raff parameters are median values and will be referred to as the *median critical gap* and *median critical lag* in this chapter.

The principal use of gap acceptance parameters is to simplify the computation of the delay duration by permitting the assumption that all intervals shorter than the critical value (lag or gap) are rejected, and all intervals longer are accepted.   It has been suggested[3,4] that the mean of the critical-gap distribution, not the median, should be used in delay computations.   This shall be referred to as the *mean critical gap* in this chapter.   Nevertheless, the median critical gap remains a practical parameter because, as shall be shown, it is more readily obtained.

In evaluating any critical gap it is apparent that a given gap must be either accepted or rejected by a given driver.   Each driver can accept only one gap, but he can reject several of them.   This means that if all rejected gaps are given the same weight as accepted gaps, then the percentage of intervals accepted for a particular size will not be a true measure of the percentage of drivers who find such an interval acceptable.   If the percentage of intervals accepted is to be used to determine the percentage of drivers who are willing to accept them, then the same number of intervals must be counted for each driver.   Raff[2] accomplished this by counting only lags and ignoring the gaps.

Table 9.1   Accepted and rejected gaps at Dumble ramp

	Stopped vehicles		Moving vehicles		All vehicles	
Length of gap t, sec	Accepted gaps $< t$	Rejected gaps $> t$	Accepted gaps $< t$	Rejected gaps $> t$	Accepted gaps $< t$	Rejected gaps $> t$
0.0	0	100	0	89	0	189
$\Delta t = 0.5$	0	100	0	89	0	189
1.0	0	95	0	80	0	175
1.5	0	71	1	52	1	123
2.0	2	49	7	27	9	76
2.5	11	34	$a = 13$	$c = 16$	$a = 24$	$c = 50$
3.0	$a = 15$	$c = 20$	$b = 26$	$d = 7$	$b = 41$	$d = 27$
3.5	$b = 23$	$d = 10$	38	4	61	14
4.0	32	5	46	3	78	8
4.5	41	4	55	3	96	7
5.0	48	2	63	2	111	4
5.5	57	0	70	1	127	1
10.0	100	0	106	0	206	0

Critical gap $T = t + \dfrac{(c - a)\Delta t}{(b + c) - (a + d)}$

$T \text{ (stopped)} = 3.1$ $T \text{ (moving)} = 2.5$ $T \text{ (all)} = 2.8$

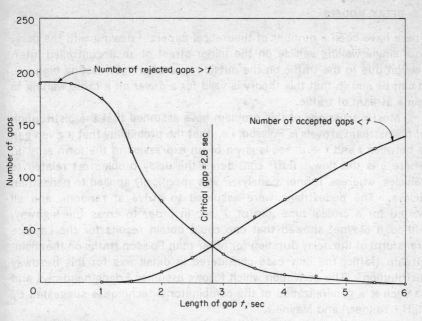

Fig. 9.3   Accepted and rejected gaps for all vehicles at Dumble ramp.

At the Dumble entrance ramp of the Gulf Freeway the number of gaps accepted and rejected have been tabulated in cumulative form in Table 9.1 for stopped and moving vehicles and then for all vehicles.   The critical gap may be determined algebraically, as indicated in the table, or graphically, as illustrated in Fig. 9.3.   Two cumulative distribution curves are shown which depict the number of accepted gaps shorter than t and the number of rejected gaps longer than t.   The value of t for which these two curves intersect is the critical gap.

This arbitrary classification of stopped and moving merges affords the means of eliminating any bias due to the inclusion of all rejected gaps. For example, in the case of stopped vehicles only two gaps were considered for each vehicle—the *largest* rejected gap and the gap finally accepted. To evaluate the gap acceptance characteristics for moving vehicles, only the first gap available to those ramp vehicles not delayed by previous ramp vehicles was considered, much in the same manner that Raff treated lags.

The effect of relative speed on the critical gap for merging ramp vehicles is evident in Table 9.1.   At the Dumble entrance ramp, the critical gap for moving vehicles (based on the first gap evaluated) is seen to be 2.5 sec as against a critical gap of 3.1 sec for stopped vehicles (vehicles ejecting at least one gap before finally accepting a gap).

9.3 DELAY MODELS

There have been a number of theoretical papers[5,6] dealing with the delay to a single waiting vehicle on the minor street of an uncontrolled intersection due to the traffic on the outside lane of an intersecting highway. It can be shown that this theory is valid for a driver on a ramp waiting to join a stream of traffic.

Most discussions of this problem have assumed that the distribution of mainstream arrivals is Poisson, i.e., that the probability that a given gap is between t and $t + dt$ sec is given by an expression of the form $qe^{-qt}\,dt$, where q is the flow. Raff[2] considered the delay problem as related to vehicles, whereas Tanner's analysis[5] was specifically applied to pedestrian delays. The pedestrians were assumed to arrive at random, and all waited for a critical time gap of T sec in order to cross the highway. Although Mayne[6] showed that one could obtain results for the Laplace transform of the delay duration for other than Poisson traffic on the mainstream traffic, the only case considered in detail was for this headway distribution. The derivation which follows assumes Erlang headways and as such is a generalization of the combinatorial techniques suggested by Raff,[2] Tanner,[5] and Mayne.[6]

It may be assumed that a ramp driver waiting to merge measures each time gap t in the traffic on the outside lane of the freeway until he finds an acceptable gap T, which he believes to be of sufficient length to permit his safe entry. If he accepts the first gap $(t > T)$, his waiting time is zero. If he rejects the first gap $(t < T)$, but accepts the second gap, his expected waiting time would be one gap interval. If it is assumed that the driver's gap acceptance does not change with time, then by induction the individual waiting periods form a geometric distribution; and the probability $P(n)$ of any driver having a wait for n intervals, each less than T sec, before merging is

$$P(n) = p^n(1 - p) \qquad n = 1, 2, \ldots \tag{9.3.1}$$

where

$$p = P(t < T) = \int_0^T f(t)\,dt \tag{9.3.2}$$

and $f(t)$ is the distribution of gaps in the major stream. It is convenient to define a normalized random variable w such that

$$w = \frac{t}{T} \tag{9.3.3}$$

or

$$t = Tw \tag{9.3.4}$$

and

$$dt = T\, dw \tag{9.3.5}$$

Substituting (9.3.4) and (9.3.5) in (9.3.2) yields

$$p = P(w < 1) = \int_0^1 f(Tw)T\, dw \tag{9.3.6}$$

The density of w may be defined as

$$g(w) = f(Tw)T \tag{9.3.7}$$

which, when inserted in (9.3.6), gives

$$p = \int_0^1 g(w)\, dw \tag{9.3.8}$$

If the distribution of gaps on an outside freeway lane of flow q can be described by the Erlang distribution

$$f(t) = \frac{(aq)^a}{\Gamma(a)}\, t^{a-1} e^{-aqt} \tag{9.3.9}$$

then it follows from (9.3.3) to (9.3.8) that

$$p = \int_0^1 \frac{aqT}{\Gamma(a)}\, w^{a-1} e^{-aqTw}\, dw \tag{9.3.10}$$

Equation (9.3.10) is of the form of the incomplete gamma distribution, which is of course equivalent to the cumulative Poisson:

$$p = 1 - e^{-aqT} \sum_{i=0}^{a-1} \frac{(aqT)^i}{i!} \tag{9.3.11}$$

and

$$1 - p = e^{-aqT} \sum_{i=0}^{a-1} \frac{(aqT)^i}{i!} \tag{9.3.12}$$

Placing (9.3.11) and (9.3.12) in (9.3.1) gives the probability that the first n time gaps between vehicles in the major stream are all less than the normalized critical gap $w = 1$ but that the $n + 1$th is greater than the normalized critical gap. This probability $P(n)$, averaged over the distribution of gaps less than the normalized critical gap $g(w)$ where $0 < w < 1$, gives the distribution of waiting time, or delay, $f(\mu)$ to a ramp vehicle in position to merge. In terms of the moment-generating functions, this may be expressed[7] as

$$M_\mu(\theta) = \sum_{n=0}^{\infty} P(n)M_w(\theta)^n \tag{9.3.13}$$

where $M_\mu(\theta)$ is the mgf of the distribution of delay to a ramp vehicle waiting for a gap greater than the normalized critical gap. From (9.3.1), (9.3.11), and (9.3.12), it is apparent that

$$M_\mu(\theta) = (1 - p) \sum_{n=0}^{\infty} [pM_w(\theta)]^n \tag{9.3.14}$$

$$= \frac{1 - p}{1 - pM_w(\theta)} \tag{9.3.15}$$

In order to use (9.3.15), $M_w(\theta)$ must be found. It is apparent that $g(w)$, where $0 < w < 1$, is a conditional probability, which by definition becomes

$$g(w)(0 < w < 1) = P(w < W | w < 1)$$

$$= \frac{\int_0^W g(w)\,dw}{\int_0^1 g(w)\,dw} \tag{9.3.16}$$

Since the denominator of (9.3.16) is given in (9.3.8), one obtains the density of gaps less than the normalized critical gap as

$$g(w)(0 < w < 1) = \frac{g(w)}{p} \tag{9.3.17}$$

and the mgf as

$$M_w(\theta) = \frac{1}{p} \int_0^1 e^{\theta w} g(w)\,dw \tag{9.3.18}$$

Changing the limits of summation and substituting for $g(w)$ yields

$$M_w(\theta) = \frac{(aqT)^a}{p\Gamma(a)} \left(\int_0^\infty w^{a-1} e^{-w(aqT-\theta)}\,dw - \int_1^\infty w^{a-1} e^{-w(aqT-\theta)}\,dw \right)$$

If a change of variable is made such that $u = w(aqT - \theta)$ in the first integral and $u + 1 = w$ in the second integral, then

$$M_w(\theta)^\cdot = \frac{(aqT)^a}{p\Gamma(a)} \left[(aqT - \theta)^{-a} \int_0^\infty u^{a-1} e^{-u}\,du \right.$$

$$\left. - e^{-(aqT-\theta)} \int_0^\infty (u+1)^{a-1} e^{-u(aqT-\theta)}\,du \right]$$

Using Newton's binomial identity[7]

$$(u+1)^{a-1} = \sum_{i=0}^{a-1} \binom{a-1}{i} u^{a-i-1}$$

in the second integral and noting that the first integral is a gamma function gives

$$M_w(\theta) = \frac{1}{p}\left(1 - \frac{\theta}{aqT}\right)^{-a}$$

$$- \frac{(aqT)^a}{p\Gamma(a)} e^{-(aqT-\theta)} \sum_{i=0}^{a-1} \binom{a-1}{i} \int_0^\infty u^{a-i-1} e^{-u(aqT-\theta)} \, du \quad (9.3.19)$$

Since the second integral is a gamma function with parameters $a - i$ and $aqT - \theta$, the second term may be written as

$$\frac{(aqT)^a}{p\Gamma(a)} e^{-(aqT-\theta)} \sum_{i=0}^{a-1} \frac{\Gamma(a)}{i!\Gamma(a-i)} \frac{\Gamma(a-i)}{(aqT-\theta)^{a-i}}$$

or

$$\left(\frac{aqT}{aqT-\theta}\right)^a \frac{e^{-(aqT-\theta)}}{p} \sum_{i=0}^{a-1} \frac{(aqT-\theta)^i}{i!} \quad (9.3.20)$$

Substituting (9.3.20) for the second term in (9.3.19) and collecting terms, one obtains

$$M_w(\theta) = \frac{1}{p}\left(1 - \frac{\theta}{aqT}\right)^{-a} \left[1 - e^{-(aqT-\theta)} \sum_{i=0}^{a-1} \frac{(aqT-\theta)^i}{i!}\right] \quad (9.3.21)$$

Now it is possible to find the moments of the delay to a ramp vehicle in position to merge by evaluating the derivatives of (9.3.15)

$$\frac{dM_\mu(\theta)}{d\theta} = \frac{p(1-p)}{[1 - pM_w(\theta)]^2} \frac{dM_w(\theta)}{d\theta} \quad (9.3.22)$$

and

$$\frac{d^2M_\mu(\theta)}{d\theta^2} = \frac{p(1-p)[dM_w^2(\theta)/d\theta^2]}{[1 - pM_w(\theta)]^2} + \frac{2(1-p)[pdM_w(\theta)/d\theta]^2}{[1 - pM_w(\theta)]^3} \quad (9.3.23)$$

at $\theta = 0$, where the derivatives of $M_w(\theta)$ in the expressions are given by

$$\frac{dM_w{}^n(\theta)}{d\theta^n} = \frac{\Gamma(a+n)}{p\Gamma(a)} \frac{(aqT)^a e^{-(aqT-\theta)}}{(aqT-\theta)^{a+n}} \left[e^{aqT-\theta} - \sum_{i=0}^{a+n-1} \frac{(aqT-\theta)^i}{i!}\right]$$

$$(9.3.24)$$

Expansion of (9.3.22) and (9.3.23) leads to the following expressions for the mean and variance of the delay distribution:

$$\mu(w)_a = \frac{e^{aqT} - \sum_{i=0}^{a} \dfrac{(aqT)^i}{i!}}{qT \sum_{i=0}^{a-1} \dfrac{(aqT)^i}{i!}} \tag{9.3.25}$$

$$\sigma^2(w)_a = \frac{(a+1)\left[e^{aqT} - \sum_{i=0}^{a+1} \dfrac{(aqT)^i}{i!} \right]}{a(qT)^2 \sum_{i=0}^{a-1} \dfrac{(aqT)^i}{i!}} + \mu^2(w)_2 \tag{9.3.26}$$

Converting from the normalized parameter w to the original variable t, it is seen that the mean and variance for values $a = 1, 2, 3,$ and 4 become:

$$\mu(t)_1 = q^{-1}(e^{qT} - 1 - qT) \tag{9.3.27}$$

$$\mu(t)_2 = \frac{e^{2qT} - 1 - 2qT - 2(qT)^2}{q(1 + 2qT)} \tag{9.3.28}$$

$$\mu(t)_3 = \frac{e^{3qT} - 1 - 3qT - 4.5(qT)^2 - 4.5(qT)^3}{q[1 + 3qT + 4.5(qT)^2]} \tag{9.3.29}$$

$$\mu(t)_4 = \frac{e^{4qT} - 1 - 4qT - 8(qT)^2 - 10.7(qT)^3 - 10.7(qT)^4}{q[1 + 4qT + 8(qT)^2 + 10.67(qT)^3]} \tag{9.3.30}$$

$$\sigma^2(t)_1 = \frac{e^{2qT} - 2qTe^{qT} - 1}{q^2} \tag{9.3.31}$$

$$\sigma^2(t)_2 = \frac{2e^{4qT} - e^{2qT} - 2qTe^{2qT} - 8(qT)^2e^{2qT} - 1 - 4qT - 2(qT)^2}{2q^2(1 + 2qT)^2} \tag{9.3.32}$$

$$\sigma^2(t)_3 = \frac{4e^{3qT} - 4 - 12qT - 18(qT)^2 - 18(qT)^3 - 13.5(qT)^4}{3q^2[1 + 3qT + 4.5(qT)^2]} + \mu^2(t)_3 \tag{9.3.33}$$

$$\sigma^2(t)_4 = \frac{5e^{4qT} - 5 - 20qT - 40(qT)^2 - 53.3(qT)^3 - 53.3(qT)^4 - 42.67(qT)^5}{4q^2[1 + 4qT + 8(qT)^2 + 10.67(qT)^3]} + \mu^2(t)_4 \tag{9.3.34}$$

Equations (9.3.27) to (9.3.34) are plotted in Figs. 9.4 and 9.5.

A simpler derivation[4] of Eqs. (9.3.27) to (9.3.30) holds if $f(t)$ is restricted to the Erlang distribution. If we assume again that the driver's gap acceptance policy does not change with time, then by induction the indi-

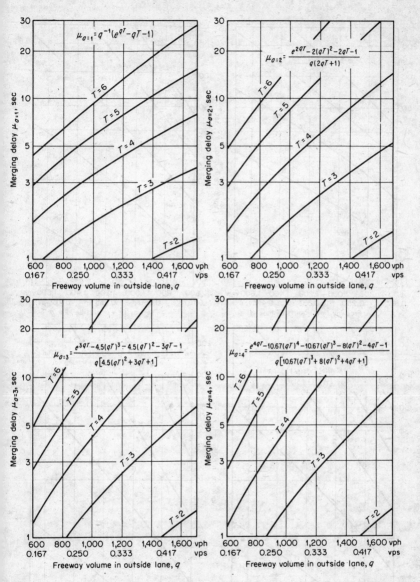

Fig. 9.4 Merging delay in terms of the freeway flow q, critical gap T, and Erlang constant a.

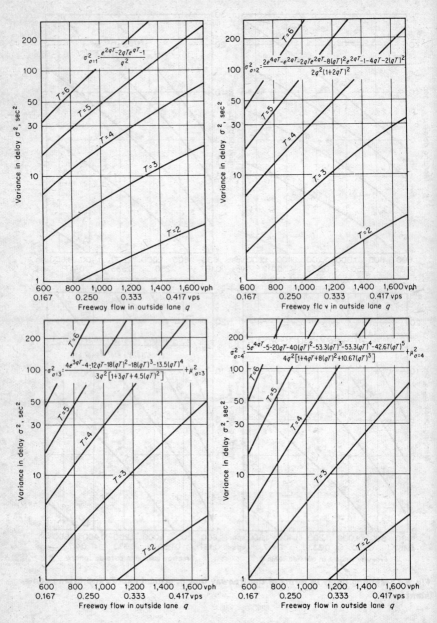

Fig. 9.5   Variance in delay in terms of the freeway flow q, critical gap T, and Erlang constant a.

vidual waiting periods form a geometric distribution and the probability P_n of any driver having to wait for n intervals, each less than T sec. before merging is

$$P_n = p^n(1 - p) \qquad n = 1, 2, \ldots \qquad (9.3.35)$$

where

$$p = P(t < T) = \int_0^T f(t)\, dt \qquad (9.3.36)$$

and $f(t)$ is the distribution of gaps in the major stream. The expected number of intervals for which a driver has to wait is given by

$$E(n) = \frac{p}{1 - p} = \frac{\int_0^T f(t)\, dt}{\int_T^\infty f(t)\, dt} \qquad (9.3.37)$$

The average time for a ramp vehicle which is in position to merge to find an acceptable gap in the freeway traffic stream will be the product of the expected number of intervals less than T and the average length of interval less than T. The average length of interval less than T is, in turn, equal to the total time less than T sec divided by the number of intervals less than T sec:

$$\text{Average length of intervals} < T = \frac{q \int_0^T t f(t)\, dt}{q \int_0^T f(t)\, dt} \qquad (9.3.38)$$

where q is the rate of flow. Multiplying Eqs. (9.3.37) and (9.3.38) yields the average waiting time for a ramp vehicle in position to merge:

$$\mu = \frac{\int_0^T t f(t)\, dt}{\int_T^\infty f(t)\, dt} \qquad (9.3.39)$$

Recalling that the proportion of ramp vehicles actually delayed is given by (9.3.36), it is apparent that the average waiting time of those who suffer delay is

$$\mu' = \frac{\mu}{\int_0^T f(t)\, dt} \qquad (9.3.40)$$

9.4 CRITICAL–GAP DISTRIBUTIONS

The theoretical delay values of the previous section are based on a fixed critical gap for all drivers. A more realistic description of delays can be obtained by replacing the fixed critical gap with a distribution of critical gaps $f(T)$. The forms for the mean and variance of the distribution of delay become

$$M(T) = \int_0^\infty \mu(T)f(T)\,dT \tag{9.4.1}$$

and

$$S^2(T) = \int_0^\infty \sigma^2(T)f(T)\,dT \tag{9.4.2}$$

If one assumes that the headway distribution on the freeway is a negative exponential, for example, substitution of (9.3.27) in (9.4.1) yields

$$M(T) = q^{-1}\int_0^\infty e^{qT}f(T)\,dT - \int_0^\infty Tf(T)\,dT - q^{-1}\int_0^\infty f(T)\,dT \tag{9.4.3}$$

Realizing that the second term defines the mean of the critical-gap distribution and that the integral in the last term must equal unity, one sees that

$$M(T) = q^{-1}\int_0^\infty e^{qT}f(T)\,dT - \bar{T} - q^{-1} \tag{9.4.4}$$

Some representative forms for critical-gap distributions are shown in Fig. 9.6. If one assumes that the critical gaps of drivers are distributed uniformly between c and c_1, then

$$f(T) = (c_1 - c)^{-1} \tag{9.4.5}$$

The translated negative exponential distribution

$$f(T) = (\bar{T} - c)^{-1}e^{-(T-c)/(\bar{T}-c)} \tag{9.4.6}$$

has also been suggested[8] as a gap distribution function as has the Erlang distribution[4]

$$f(T) = \frac{(a/\bar{T})^a}{(a-1)!}\,T^{a-1}e^{-aT/\bar{T}} \tag{9.4.7}$$

Substitution of (9.4.5) to (9.4.7) in (9.4.4) gives, respectively,

$$M(T) = (c_1 - c)^{-1}q^{-2}(e^{qc_1} - e^{qc}) - \bar{T} - q^{-1} \tag{9.4.8}$$

$$M(T) = [q(1 - q\bar{T} + qc)]^{-1}e^{qc} - \bar{T} - q^{-1} \tag{9.4.9}$$

and

$$M(T) = \frac{1}{q}\left(\frac{a}{a - q\bar{T}}\right)^a - \bar{T} - q^{-1} \tag{9.4.10}$$

which correspond to the distribution of delay for the first three critical-gap distributions in Fig. 9.6.

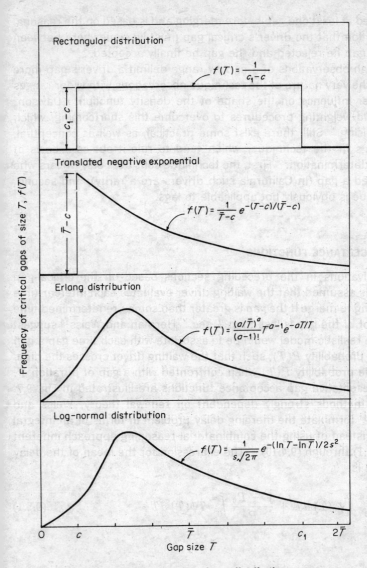

Fig. 9.6 Representative forms for critical-gap distributions.

The derivations of this section [Eqs. (9.4.1) to (9.4.10)] are based on utilization of a distribution of critical gaps for all drivers $f(T)$. This is a concession to the obvious fact that not all drivers have the same critical gap. The difficulty lies, however, in measuring the critical gap for individual drivers in order to obtain a distribution of critical gaps. One technique[4] to obtain such a frequency distribution samples only drivers who have rejected at least one gap before merging and is based on the reasonable assumption that the driver's critical gap must lie somewhere between the largest gap he rejected and the gap he finally accepted.

Although observations of a narrow range delimit a driver's gap more closely, by the very nature of the technique observations with wider ranges have greater influence on the shape of the density function. Dawson[9] suggests two weighting procedures to overcome this shortcoming, which seem promising. Still, there exist some practical, as well as conceptual, shortcomings in the technique, which tend to rule it out as a basis of parameter determination. First, the technique considers only drivers who have rejected a gap (in California such drivers are a rarity); and second, the technique is obviously not applicable to lags.

LIBRARY

11.1 USING GAP SEQUENCES TO

9.5 GAP ACCEPTANCE FUNCTIONS   *ESTIMATE ACCEPTANCE FUNCTION*

MIT   HANI MAHMASSANI & YOSEF SHE

In the derivations in the preceding sections based on the critical-gap concept, it is assumed that the waiting driver evaluates each intercar time gap, choosing to merge if the gap is greater than some predetermined time gap T or not, if the gap is less than T sec. Herman and Weiss[10] suggest that a more realistic model would be to associate with each time gap a gap acceptance probability $P(T)$, such that the waiting driver crosses the highway with the probability $P(T)$ when confronted with a gap of duration T. Some representative gap acceptance functions are illustrated in Fig. 9.7.

Using methods strongly dependent on renewal theory, Weiss and Maradudin[11] formulate the merging delay problem in terms of an integral equation, instead of using the combinatorial reasoning approach inherent in Eqs. (9.3.1) through (9.4.10). The expression for the mean of the delay distribution is

$$M(T) = \int_0^\infty T\psi_0(T)\, dT + \frac{1 - P_0}{P} \int_0^\infty T\psi(T)\, dT \qquad (9.5.1)$$

where

$$P_0 = \int_0^\infty P(T)f_0(T)\, dT \qquad (9.5.2)$$

$$P = \int_0^\infty P(T)f(T)\, dT \qquad (9.5.3)$$

$$\psi_0(T) = f_0(T)[1 - P(T)] \qquad (9.5.4)$$

$$\psi(T) = f(T)[1 - P(T)] \qquad (9.5.5)$$

and $f(T)$ is the distribution of gaps on the outside freeway lane and $P(T)$ is the gap acceptance function. The subscript 0 refers to the gap availability distribution and gap acceptance function for the first gap.

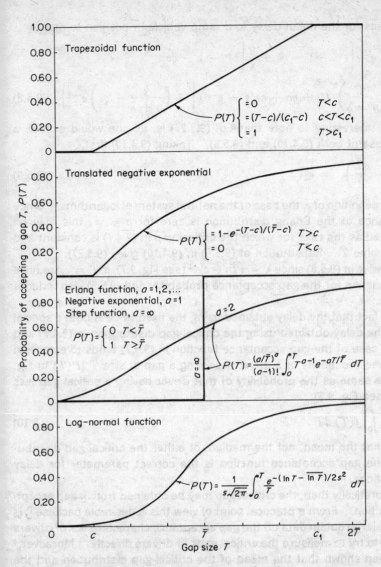

Fig. 9.7 Representative forms for gap acceptance functions.

Weiss[10,14] has applied this theory to some specific distributions. For example, if the distribution of gaps on the outside lane of the freeway is taken as

$$f(T) = f_0(T) = qe^{-qT} \qquad (9.5.6)$$

and the gap acceptance function is

$$P(T) = P_0(T) = 1 - e^{-\lambda(T-c)} \qquad (9.5.7)$$

one obtains for the mean delay to a ramp vehicle in position to merge

$$M(T) = q^{-1} \left\{ e^{qc} - 1 - qc + \frac{q}{\lambda} \left[e^{qc} - 1 - qc \right. \right.$$

$$\left. \left. + \left(\frac{q}{q+\lambda} \right)^2 (1 + qc + c)(1 - e^{-qc}) \right] + \left(\frac{q}{q+\lambda} + qc \right) e^{-qc} \right\} \quad (9.5.8)$$

It is interesting to note that Eq. (9.3.27) is, as one would expect, a special case of both (9.4.10) and (9.5.8). Taking (9.4.10) first,

$$\lim_{a \to \infty} \left(1 - \frac{q\bar{T}}{a} \right)^{-a} = e^{q\bar{T}} \quad (9.5.9)$$

from the definition of e, the base of the natural system of logarithms. Since the variance of the Erlang distribution is zero for $a = \infty$, this may be interpreted as the case for which the critical gap in (9.4.7) is constant and has the value $\bar{T}$. Substitution of (9.5.9) in (9.4.10) gives (9.3.27).

Likewise in (9.5.8) since $\lambda = (\bar{T} - c)^{-1}$ (see Fig. 9.7), if $\bar{T} = c$, we have a step function for the gap acceptance probability and (9.5.8) also reduces to (9.3.2).

The fact that the delay obtained using the fixed value is both a special case of the delay obtained using the critical-gap distribution in (9.4.7) and a special case of the gap acceptance function in (9.5.7) tends to establish: (1) that the probability of a driver accepting a gap of size T $[P(T)$ in Fig. 9.7] is the same as the probability of that driver having a critical gap less than T (see Fig. 9.6)

$$P(T) = \int_0^T f(T) \, dT \quad (9.5.10)$$

and (2) that the mean, not the median, of either the critical-gap distribution or the gap acceptance function is the correct parameter for delay computations.

Theoretically then, the critical gap may be obtained from a gap acceptance function. From a practical point of view this is desirable because it is much easier to obtain data on the gap acceptance characteristics of drivers than it is to try to measure the critical gaps of drivers directly. Moreover,[4] it has been shown that the mean of the critical-gap distribution and the mean of the gap acceptance function for a given ramp do in fact exhibit close agreement.

9.6 THE BINOMIAL RESPONSE

If an entrance ramp driver, selected at random from a population, is given an opportunity to merge, the probability he will accept is p; the probability of rejecting the opportunity is $1 - p$, or q. The opportunity to merge here may be measured by the time gap, space gap, relative speed, or some combination of these or other factors. If two drivers are given the same opportunity and if their reactions are completely independent, the probability that both accept is p^2, and the probability that both reject is q^2; the probability that only the first accepts is pq, and the probability that only the second accepts is qp. Thus, the total probabilities of two, one, and zero accepting are p^2, $2pq$, and q^2, respectively, the successive terms in the expansion of $(p + q)^2$. In a similar manner it may be seen that if a group of n drivers is exposed to the same merging conditions, and all react independently, the probabilities of $n, n - 1, n - 2, \ldots, 2, 1, 0$ responding are the $n + 1$ terms in the binomial expansion $(p + q)^n$. The probability of exactly x acceptances is therefore

$$P(x) = \binom{n}{x} p^x q^{n-x} \tag{9.6.1}$$

Equation (9.6.1) of course represents the binomial distribution of probabilities. The average number of gaps accepted in n opportunities is np and the average number rejected is nq. The variance of the distribution is npq.

It is apparent that the gap acceptance phenomenon involves a *stimulus* (available time gap) applied to a *subject* (the driver of a ramp vehicle). Variation of the stimulus is followed by a change in some measurable characteristic—the acceptance of a certain gap, in the case at hand— referred to as the *response* of the subject.

Methods employed for the estimation of the nature of a process by means of the reaction that follows its application to living matter are called *biological assays.* In its widest sense the term should be understood to mean the measurement of any stimulus (physical, chemical, biological, physiological, or psychological) by means of the reactions which it produces. One type of assay which has been found useful in many different disciplines is that dependent on the all-or-nothing response. The decision to accept or reject a gap is the type of response which permits no graduation and which can only be expressed as occurring or not occurring. The statistical treatment of this particular assay has been greatly facilitated by the development of probit analysis.

9.7 THE PROBIT METHOD

Probit analysis is a well-established technique used widely in toxicology and bioassay work. Reference is made to two books by D. J. Finney.[12,13] In the first, the technique is described and a rather lengthy computational method is set forth; in the second, many examples covering a wide variety of cases are presented, and the underlying statistical theory is presented in some detail.

Normally, in an experiment to which a probit analysis is applied, the magnitude of the stimulus is controlled by the experimenter. A number of subjects (usually insects) are administered a stimulus (usually a poison) to which they either do or do not exhibit a certain response (usually dying). Each subject is assumed to have a tolerance for the stimulus; if the stimulus is greater, the subject responds (i.e., insect dies).   Typically what is desired is a relationship expressing the percent killed as a function of the dose. Since this is not a linear relationship, the estimation of the equation of the curve and tests of significance are severely complicated.   The transformation from percentages to *probits* forces the curve into a linear relationship.

In a study of lag and gap acceptances at stop-controlled intersections, Solberg and Oppenlander[14] showed that the probit of the percentage accepting a time gap is related to the logarithm of the time gap x by the equation

$$Y = a + bx$$

By means of the probit transformation the study data were used to obtain an estimate of this equation.   The parameters of the tolerance distribution, mean and variance, were also determined.   In particular the median gap and lag acceptance times were easily estimated as that value of x when $Y = 5$ (acceptance is 50 percent).

Whereas Solberg and Oppenlander[14] tabulated the data into groups at 1-sec intervals in order to obtain an estimate of drivers accepting this interval in the time series, such groupings are usually reserved for experiments in which the magnitude of the stimulus is controlled.   In traffic studies such control is not possible, though the magnitude of the stimulus (available gap size) may be measured and recorded along with the response (accepted or rejected).   Finney[13] terms such data *individual records*.   The probit method is just as applicable to individual records as to data from a controlled (grouped) experiment.

It is important to remember that what is desired is the form of the gap (or lag) acceptance function for the freeway merging maneuver.   Many theoretical forms of this function were described in the section on theory and are illustrated in Fig. 9.7.   Several researchers have shown that the log-normal function provides the best description of the gap acceptance phenomena.   The purpose of the probit analysis is to transform this log-normal function into the linear form

$$Y = a + bx \tag{9.7.1}$$

where $x = \log t$, t being the time interval (either lag or gap as the case may be) and Y is the probit of P, P being the probability of accepting a gap or lag of t (see Fig. 9.7).

If x and Y are plotted on a grid, Eq. (9.7.1) has the form of a straight line.   For convenience then, log-probability graph paper has been used in the following sections so that the relationship between t and P from Fig. 9.7 may be viewed as a straight line.   The abscissa of this graph paper (see Fig. 9.8) is in the form of a logarithmic scale denoting the gap (or lag) interval t in seconds; the right ordinate is a probability scale denoting

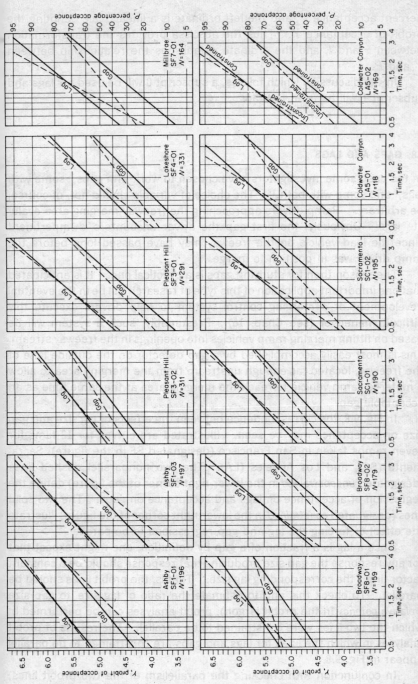

Fig. 9.8 Lag and gap acceptance regressions for U.S. ramps.

percent acceptance P, and the left ordinate is a linear scale denoting the probit of acceptance Y. Thus the abscissa provides the transformation between x and t, and the ordinate establishes the relationship between Y and P.

In the discussion which follows, a subscript 0 will be used for lags and subscript 1 will be used for gaps.

9.8 GAPS AND LAGS

In the past, some investigators have chosen to work with gaps,[1] some with lags,[2] and some with both.[14] The choice of variable should not, of course, be arbitrary. Ramp drivers approaching the freeway merging area evaluate lags. Obviously, the first gap the driver encounters is impertinent since the lead vehicle on the freeway may have passed long before the ramp driver was in position to merge.

The reason then for studying gaps is a practical one. It is anticipated that one important application of this type of research will be in the eventual development of a merging control system to help drivers execute this difficult maneuver (see Chap. 16). Presumably, such a system will be based on fitting merging ramp vehicles into openings in the freeway stream. This is most easily accomplished by a gap detector on the outside lane of the freeway, located far enough upstream from the merging area to allow a metered ramp vehicle to reach the merging area at the same time as an acceptable gap.

In Fig. 9.8, the correspondence between the effects of lag size and gap size on acceptance may be seen. Illustrated are curves for 12 films for seven ramps taken in San Francisco (designated SF in the figure), Sacramento (SC), and Los Angeles (LA). The two nonparallel dashed lines (designated *unconstrained* in the lower-right corner of the figure) define the regression of acceptance on lags and acceptance on gaps. The two solid lines (designated *constrained* in the lower-right corner of the figure) illustrate the effect of forcing the lag and gap regressions to be parallel.

Since a lag is a fraction of a gap by definition, one might expect that for any response the gap size producing that response should be a constant factor times the corresponding lag size. That is, the probit lines should be parallel. To check this and to estimate the ratio of lag size to gap size (*relative potency* in the probit jargon), probit analyses can be performed in which the two lines were constrained to be parallel. The results of probit analyses in which the lines for lags and gaps are constrained to be parallel appear in Fig. 9.8.

In conjunction with checking the parallelism of the two probit lines, the primary purpose of the analyses is to estimate the *relative potencies* of lags to gaps, that is, to estimate the effectiveness of lags relative to gaps in inducing drivers to merge. This is estimated as the antilogarithm of the horizontal distance between the two parallel probit lines with the ratio R being greater than unity for the case when the probit line for lags lies to the left of the probit line for gaps.   Thus a value of $R = 2$ implies that in order to induce the same percentage of response as a particular lag, the gap must be twice the size of the lag.

9.9   MULTIPLE ENTRIES

A multiple entry occurs when two or more drivers accept the same gap. For example, three probit lines may be constructed considering r_i the responses and x_i the stimuli.   The three lines are first fitted separately as

$$Y_i = a_i + b_i x_i \qquad i = 1, 2, 3$$

With these lines, it is possible to estimate for any given gap size the probability that one, two, or three cars will accept that available gap.

For each ramp there is a tendency for the b's (slopes) to become successively larger: that is, b_1, b_2, b_3 (see Fig. 9.9).   As the sensitivity of such platoons of cars to differences in gap size is affected by the decision of the last car in the platoon, this trend indicates the reasonable conclusion that the $i + 1$th car in line for a given gap is more sensitive to differences in gap size than the ith car.   Furthermore, these analyses do not take account of ramp speed: It is expected that the $i + 1$th vehicle in line must slow down more than the ith in line.   Barring any interaction of ramp speed and gap size, the probit line for three or more cars per gap would be shifted to the right of the probit line for two or more and the line for two or more to the right of one or more.   This would aggravate the shifting of the lines because of the simple fact that it takes a bigger gap for two cars than for one and for three than for two.

9.10   THE IDEAL MERGE

In the first part of this chapter, it was suggested that a merge must be qualified according to the terminology illustrated in Fig. 9.2.   The most desirable type of gap merge would be both *optional* and *ideal* in that it would not be made because the ramp driver had run out of acceleration lane, nor would it cause turbulence in the freeway stream.   The requirements of such a merge serve to document the interaction of the basic traffic elements—the driver and the vehicle—and the basic traffic characteristics—headways and vehicular speeds.

If it is assumed that the average driver's normal acceleration of the merging vehicle is inversely proportional to his speed,[15] the following differential equation may be written

$$\frac{du}{dt} = a - bu \qquad (9.10.1)$$

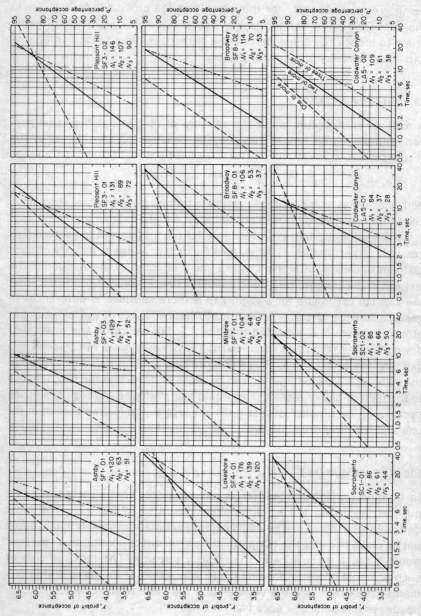

Fig. 9.9 Gap acceptance for multiple vehicle merges.

where u is the speed of the merging vehicle, t is time, and a and b are constants.   If the merging vehicle is moving at a speed u_r at the beginning of the merge, then the speed-time relationship is

$$u = \frac{a}{b}(1 - e^{-bt}) + u_r e^{-bt} \tag{9.10.2}$$

Since $u = dx/dt$, integration of (9.10.2) provides the equation of the time-space curve

$$x = \frac{a}{b}t - \frac{a}{b^2}(1 - e^{-bt}) + \frac{u_r}{b}(1 - e^{-bt}) \tag{9.10.3}$$

Substitution of (9.10.2) in (9.10.1) gives the acceleration-time relationship for the merging maneuver:

$$\frac{du}{dt} = (a - bu_r)e^{-bt} \tag{9.10.4}$$

The units of the constants a and b^{-1} are those of acceleration and time, respectively, where a is the maximum acceleration and a/b is the free speed in the merging area.   The forms of Eqs. (9.10.1) to (9.10.4) are illustrated in Fig. 2.2.

The time-space relationship of Fig. 2.2 has been reproduced in Fig. 9.10 along with the procedure for determining the theoretical minimum ideal gap for merging.   Such a gap is made up of three time intervals: (1) a safe time headway between the merging vehicle and the freeway vehicle ahead T_r, (2) the time lost accelerating during the merging maneuver T_L, and (3) a safe time headway between the second freeway vehicle and the merging vehicle T_f.   The *safe headway* referred to is that headway between two vehicles in a lane which will allow the following vehicle to stop safely even if the vehicle in front makes an emergency stop.   If reaction times τ, braking capabilities, and speeds u are assumed to be equal, then the safe headway is $(L/u) + \tau$, where L is the length of the vehicle in front.

The time necessary for the merging ramp vehicle to accelerate from a speed u_r at the beginning of the merge to attain the speed of the freeway traffic u may be obtained by solving Eq. (9.10.2) for time.

$$T_2 = -\frac{1}{b}\ln\frac{a - bu}{a - bu_r} \tag{9.10.5}$$

Subtracting the travel time T_1 to travel the same distance covered during merging, only at a constant speed u, from (9.10.5) gives the time lost during merging T_L.   The theoretical minimum ideal gap for merging $T_r + T_f + T_L$ is

$$T = \frac{L_f + L_r}{u} + 2\tau + \frac{u + u_r}{bu} + \frac{(a/b) - u}{bu}\ln\frac{a - bu}{a - bu_r} \tag{9.10.6}$$

in which L_f and L_r are the lengths of the freeway and ramp vehicles,

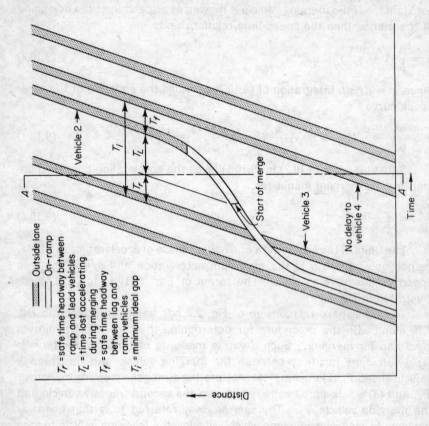

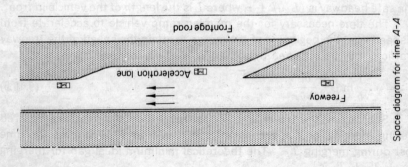

Fig. 9.10 Theoretical minimum ideal gap for merging.

respectively. It is apparent from (9.10.6) that a merging truck would require a larger gap than a merging passenger car by virtue of its longer length L_r and its reduced accelerating capabilities a. Similarly, the first of the two freeway vehicles influences the size gap as indicated in L_f in (9.10.6).

9.11 SPEED OF THE MERGING VEHICLE

It is apparent from the previous section that the ramp driver's problem in executing the merging maneuver is more than one of simply evaluating successive time headways in the freeway traffic stream until he finds a large enough gap. For example, Eq. (9.10.6) shows that the ideal gap is based on the ramp driver's reaction time τ; the vehicle characteristics L_f, L_r, a, and b; and the speeds of the mainstream vehicles and ramp vehicle u and u_r. The influence of speeds on merging comes as no surprise; we know, for example, that under conditions of forced flow on the freeway, comparatively large time headways are rejected. Theoretically, infinitely large time headways would necessarily be rejected as the concentration increases and movement ceases.

In delineating the effect of speeds on merging, consider a simple model consisting of two vehicles A and B separated by a space headway s and traveling at a constant velocity u on the outside lane of the freeway, and a vehicle C traveling at speed u_r on a corner intersecting the freeway at an angle δ. It follows that the gap between A and B is

$$t = \frac{s}{u} \qquad\qquad (9.11.1)$$

However, since the driver in vehicle C is moving, he views the time gap between A and B as

$$t' = \frac{s}{u - u_r \cos \delta} \qquad\qquad (9.11.2)$$

where $u_r \cos \delta$ is vehicle C's speed component along the freeway. Solving for s in (9.11.2) and substituting in (9.11.1) yields

$$t = \frac{u - u_r \cos \delta}{u} t' \qquad\qquad (9.11.3)$$

If $t' > T'$ where T' is a constant, the gap will be accepted. Therefore, the minimum acceptable gap T

$$T = \frac{u - u_r \cos \delta}{u} T' \qquad\qquad (9.11.4)$$

is clearly a function of the speed of the freeway traffic u and the speed of the ramp vehicle u_r.

The effect of ramp geometry on merging operation is indicated in Eq. (9.11.4). The equation suggests that the difference between traffic operation at an ordinary intersection and that at a freeway ramp terminal is primarily in the angle at which entering traffic and through traffic converge. For an intersection $\delta = 90°$, one obtains $T = T'$ from (9.11.4); for a lane change $\delta = 0°$

$$T = \frac{u - u_r}{u} T' \tag{9.11.5}$$

Whereas for an intersection gap acceptance is independent of the speed of the merging vehicle, the speed of a vehicle merging from an entrance ramp has a profound effect on gap acceptance.

9.12 ANGULAR VELOCITY MODEL

Several researchers have hypothesized the use of angular velocity as the basis for the gap acceptance decision. The advantage of this parameter is that it takes into consideration the distance of the approaching freeway vehicle and its speed of approach.

The angular velocity model for gap acceptance is an application of the calculus of related rates (see Sec. 3.9). Referring to the moving coordinate system in Fig. 9.11, we can write the equations for the relative speed of the freeway vehicle and merging vehicle and for their space lag:

$$\frac{dx}{dt} = u - u_r \tag{9.12.1}$$

$$-x = w \cot \theta \tag{9.12.2}$$

Differentiating (9.12.2) with respect to time, equating it to (9.12.1), and solving for the angular velocity $\dot{\theta}$ yields

$$\dot{\theta} = \frac{w}{s^2} (u - u_r) \tag{9.12.3}$$

The basic aspect of the theory is that a ramp driver rejects a lag if he detects an angular velocity but accepts the lag if he cannot perceive any motion. Michaels and Weingarten[16] utilize this criterion to develop the equation for the minimum acceptable gap T.

$$T = \frac{s}{u} = \left[\frac{w}{\dot{\theta} u^2} (u - u_r) \right]^{\frac{1}{2}} \tag{9.12.4}$$

Although the direct application of Eq. (9.12.4) to determine gap acceptance times depends on the evaluation of drivers' thresholds of angular velocity, the role of speed in gap acceptance is again evidenced. The

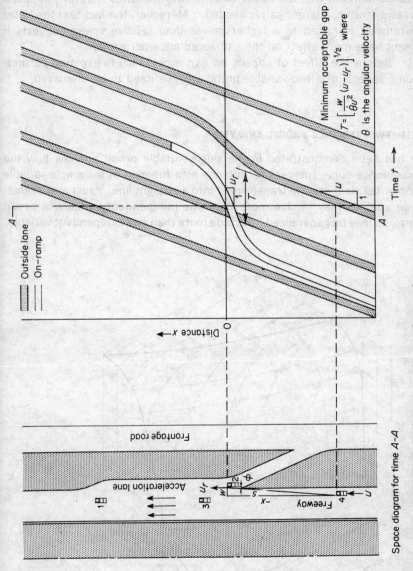

Fig. 9.11 Theoretical minimum acceptable gap for merging.

Minimum acceptable gap

$$T = \left[\frac{w}{\theta u^2}(u - u_r)\right]^{1/2} \text{ where}$$

θ is the angular velocity

relative speed $u - u_r$ appearing in the numerator of (9.12.4) suggests that the lower the relative speed, the smaller the gap needed. The speed of the mainstream traffic u in the denominator indicates that for low-speed freeway traffic, a larger gap is needed. Moreover, the fact that the latter parameter is raised to a higher power than relative speed suggests it might have more effect on the gap acceptance performance.

Before the effect of speeds on gap acceptance is explored further, some aspects of a two-variable probit analysis need to be discussed.

9.13 TWO–VARIABLE PROBIT ANALYSIS[17]

It has been demonstrated in the single-variable probit analysis how the acceptance curve (percent acceptance as a function of the single-variable gap or lag size) may be transformed into a straight line. Just as a regression analysis may involve more than one independent variable, a probit analysis may be generalized to include more than one independent variable.

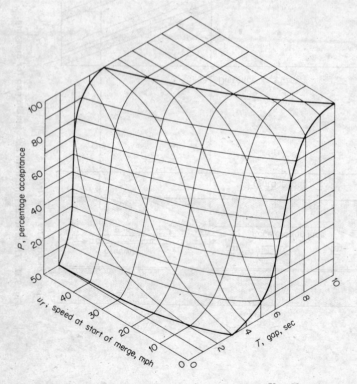

Fig. 9.12 Gap acceptance surface for freeway speed, $U = 50$ mph.

Based on the theory of the preceding sections, the two obvious independent variables are gap (or lag) and some speed parameter (u, u_r or $u - u_r$). Figure 9.12 illustrates a hypothetical gap acceptance surface for a freeway speed $u = 50$ mph.   A vertical plane parallel to the T axis yields a gap acceptance curve of the form shown in Fig. 9.7 and discussed in the single-variable analysis.   It should be apparent that the equation of the surface in Fig. 9.12 would be difficult to write—the value of the parameters of the equation from raw data, even more difficult to estimate.

A probit analysis for gap acceptance can be presented in which a combination of x_0 and u_r is considered the stimulus and r, the response. In such an analysis there is obtained an estimate of a probit plane defined

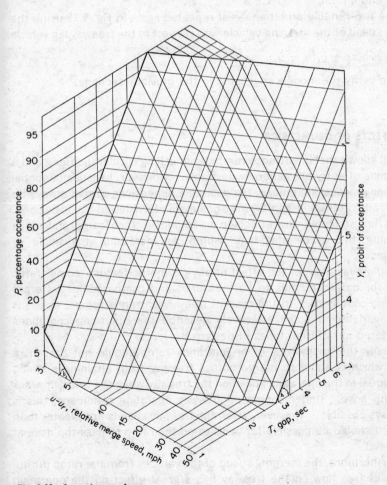

Fig. 9.13 Acceptance plane.

by $Y = a + b_1 x_0 + b_2 u_r$. Thus for any combination of x_0 and u_r, it is possible to estimate Y and the corresponding percentage of all ramp drivers who would accept that combination. Or, going in reverse, it is possible to estimate for any given percentage acceptance the corresponding combinations of x_0 and u_r producing that percentage. The effect of the two-variable probit analysis is to change the acceptance surface of Fig. 9.12 to an acceptance plane such as is shown in Fig. 9.13. Again, Figs. 9.12 and 9.13 are meant to be illustrative rather than conclusive, the actual independent variables used being the lag x_0 and the speed of the ramp vehicle u_r (see Fig. 9.14).

The three lines on each graph of Fig. 9.14 depict, from left to right, a speed at the start of the merge of 50 mph, the mean speed for the study period, and 0 mph.

The two-variable probit analysis, repeated again in Fig. 9.15 using the relative speed of the merging vehicle with respect to the freeway lag vehicle

$$Y = a + b_1 x_0 + b_2 (u - u_r)$$

tends to verify the models suggested in the previous sections.

9.14 EFFECTS OF GEOMETRICS

It is well known that in entrance-ramp design a long acceleration lane and a small angle of entry are generally desirable. However, because of spatial and other practical limitations, the desirable dimensions cannot always be attained and often some compromise or substandard design has to be used. Little is known about the effect on operation of such compromises as decreasing the acceleration-lane length or increasing the angle of convergence.

In the evaluation of the effect of any geometric feature, some traffic variable or parameter has to be measured and compared for different geometric conditions. In the case of freeway entrance ramps, speed is such a variable and can effectively describe the operating conditions encountered at different types of designs.

Ideally the geometrics of an entrance ramp should not prevent a vehicle which is about to merge onto a freeway from attaining a speed nearly equal to the operating speed on the freeway when the vehicle passes the ramp nose. Under congested freeway operating conditions, this is not always possible. However, when the freeway speeds are greater than 40 mph it should be possible for vehicles on the ramp to pass the nose at high speed.

Furthermore, the merging speed of the vehicles from the ramp during periods of free flow on the freeway has a great effect on the operation

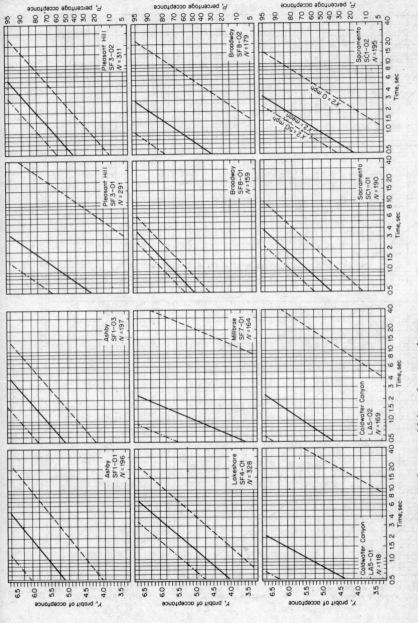

Fig. 9.14 Effect of speed of ramp vehicle on lag acceptance.

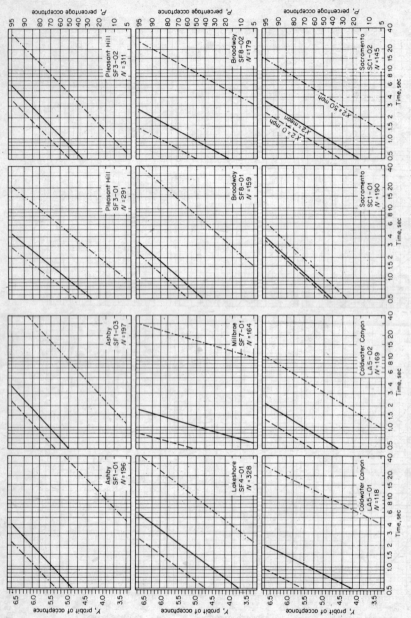

Fig. 9.15 Effect of relative speed on lag acceptance.

in the merging area.   Normally, higher merging speeds of ramp vehicles indicate a better operation.

Similarly, the speed change between the nose and the merge point should be a good indicator of the effect of different ramp geometrics. It seems reasonable to expect that at poorly designed ramps, many vehicles will be forced to stop or travel very slowly at the nose while selecting a gap and would therefore have to accelerate rapidly in an attempt to approach the freeway speed.   Under these conditions, better designs would generally exhibit lower-speed changes between the nose and merge point.

For the purposes of this discussion, the speed of each vehicle traversing the ramp while freeway speeds were in excess of 40 mph is displayed at the nose and at the merge point in the form of cumulative distributions in order to illustrate the effects of acceleration-lane length and of angle convergence.

The gap acceptance characteristics of drivers can also be used to illustrate the effect of different geometric features, bearing in mind that a better design should result in acceptance of smaller gaps by drivers.

9.15 ANGLE OF CONVERGENCE

The effect of the angle of convergence on ramp speed is illustrated by Fig. 9.16, which shows cumulative plots of ramp vehicle speeds at the nose and at the merge point and of the speed change between the nose and the merge point.[18]   Different plots are shown for ramps having angles of con-

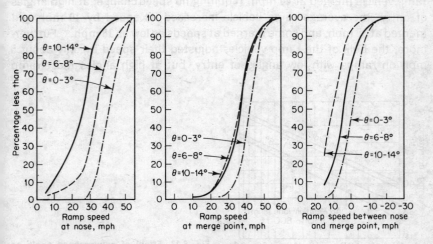

Fig. 9.16 Effect of convergence angle on ramp speed.

vergence falling into the groups indicated. All these ramp speeds were observed under freeway speed conditions in excess of 40 mph. The effect of the angle of convergence is immediately evident from the figures.

At the nose, small angles of convergence resulted in speeds generally 15 to 20 mph higher than speeds at the nose of ramps with angles of convergence in the 10 to 14° group. The percent of slow ramp vehicles increased quite rapidly with an increase in the angle of convergence. For example, in the 0 to 3° group, only about 2 percent of ramp vehicles traveled slower than 30 mph at the nose; in the 10 to 14° group, this was true of about 85 percent of all ramp vehicles. The operation was also considerably more uniform at the ramps with low angles of entry. In the 0 to 3° group, the bulk of the vehicles passed the nose at speeds between 35 and 40 mph; at higher angles, most ramp vehicles passed the nose at speeds ranging from 12 to 30 mph. The advantage of lower angles of entry is thus clearly demonstrated in terms of higher ramp speeds at the nose. For example, if the angle of entry of an on-ramp could be reduced from about 12 to about 7°, seemingly not a major change, it could be expected that about 50 percent more of the ramp vehicles would pass the nose at ramp speeds exceeding 30 mph. The median relative speeds at the nose, although not shown in a figure, were found to vary from less than 10 mph on ramps with small convergence angles to median relative speeds generally exceeding 25 mph on ramps with convergence angles exceeding 10°.[18]

The speeds at the merge point and the speed changes between the nose and the merge point further substantiate the findings revealed by the speeds at the nose. At small angles of convergence, the average ramp vehicle merged at 40 mph, requiring no speed change; at high angles of entry, the average ramp vehicle increased its speed by 10 mph and merged at 35 mph, and some merged at speeds as low as 10 mph. Furthermore, the bulk of the ramp vehicles adjusted their speed by only about 5 mph on ramps with low angles of entry, but at high angles most ramp

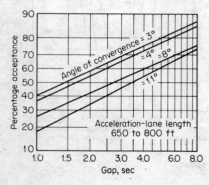

Fig. 9.17 Effect of convergence angle on gap acceptance characteristics.

vehicles had to increase their speed from 5 to 20 mph.   The relative speeds at the merge point were found to vary from about a median of 2 mph for ramps with small convergence angles to about 16 mph for ramps with high angles.[18]

Since vehicles generally merge at lower speeds when the angle of convergence is high, it was expected that they would require bigger gaps to merge into.   This was confirmed by a study of the gap acceptance behavior of drivers at ramps with different angles of entry, as illustrated by the gap acceptance curves in Fig. 9.17.[19]   These curves apply to entrance ramps having acceleration lanes between 650 and 800 ft to reduce the effect of acceleration-lane length on the gap acceptance behavior of drivers.   It can be seen that 50 percent of the drivers accepted gaps less than 1.5 sec at the entrance ramp with a 3° angle of convergence and less than 3.5 sec at the entrance ramp with an 11° angle.   At the lower angle, 58 percent of the drivers accepted gaps less than 2 sec; only 33 percent of the drivers on the ramp with an 11° angle of convergence were willing to accept this gap.

As expected then, a lower angle of convergence was found to be more desirable in the design of entrance ramps.   Although this is nothing new, the significance of the last two figures is that they put quantitative values on these effects, so that a designer can see what is gained or lost by using a certain angle of convergence as opposed to another.

9.16 ACCELERATION–LANE LENGTH

The effect of the acceleration-lane length on the speeds of ramp vehicles is illustrated by means of the cumulative speed distributions shown in Fig. 9.18.   Different curves are shown for ramps with acceleration-lane lengths that fall into the groups indicated.   The different figures are for ramp-speeds at the nose, ramp speeds at the merge point, and speed changes between the nose and the merge point.   All the ramp speeds included were observed under freeway speed conditions in excess of 40 mph.[18]

The median ramp speed at the nose was about 18 mph for ramps with short acceleration lanes (less than 350 ft) to about 35 mph for acceleration lanes exceeding 950 ft.   On ramps with acceleration lanes exceeding 950 ft, about 85 percent of all drivers passed the nose at speeds greater than 30 mph.   This was true of only about 15 percent of drivers on ramps with acceleration lanes less than 350 ft.   From the plots, it can also be seen that speeds were much more uniform on the ramps with longer acceleration lanes.   The median relative speed at the nose varied from about 10 to 15 mph on ramps with long acceleration lanes to median relative speeds exceeding 35 mph on ramps with short acceleration lanes.

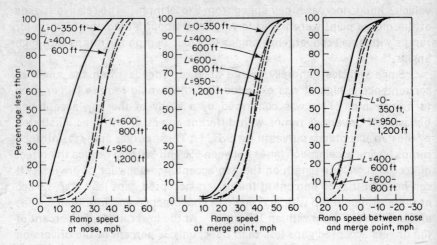

Fig. 9.18   Effect of acceleration-lane length on ramp speed.

Ramp speeds at the merge point were generally higher than at the nose but required considerably less adjustment in speed on ramps with long acceleration lanes.   On short acceleration lanes, the average driver increased his speed by 10 mph and merged at 30 mph; on larger acceleration lanes, the average driver merged at 40 mph without adjusting his speed.   This resulted in median relative speeds at merge of more than 10 mph on short acceleration lanes and less than 5 mph on long acceleration lanes.

The effect on relative speeds of using a certain design, as opposed to another, may not be too meaningful to a designer.   However, when the effect of this relative speed is related to the gap acceptance behavior of drivers, which can be directly tied to the capacity of the entrance ramp, the designer will generally have a better feel for the gains or losses associated with a certain change in the design.

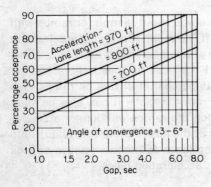

Fig. 9.19   Effect of acceleration-lane length on gap acceptance characteristics.

The effect of the length of acceleration lane on the gap acceptance behavior of drivers is shown in Fig. 9.19 for ramps having convergence angles between 3 and 6°. As expected, in view of the effects revealed by the study of speed distributions, drivers on long acceleration lanes were generally more willing to accept smaller gaps than they were on short acceleration lanes.[19]

A stepdown regression analysis with the slope of the gap acceptance line as the dependent variable yielded the following equation:

$$B = 1.394 + 0.289\theta - 0.027L\theta$$

where θ = angle of convergence in degrees
L = length of acceleration lane in stations
B = slope of the gap acceptance lines as expressed by $Y = A + B_1X$, Y being the probit and X the logarithm of the gap size

An increase in the angle increases the slope of the gap acceptance line and thus decreases the variance of the critical-gap distribution. The acceleration-lane length has the opposite effect for a fixed angle of entry. The gap acceptance characteristics based on the above two regression equations are shown in Fig. 9.20 for different angles of convergence and different lengths of parallel acceleration lane.[20]

9.17 ACCEPTED-GAP NUMBER

The mean and standard deviations of the number of the gap accepted by a ramp vehicle offer valuable figures of merit for describing entrance ramp operation.   In establishing the notation, gap 1 for each ramp vehicle is the gap which is at the physical nose; gap 2 is behind (upstream of) gap 1; and gap 0, −1, etc., are downstream of gap 1.

Figure 9.21 presents the mean and standard deviation of the accepted-gap number at each ramp.[18]   Each point represents a ramp with a particular set of geometric design features.   The first number in parentheses is the mean accepted gap for that ramp, and the second number is the standard deviation of the accepted-gap number.   Only periods in which the freeway operating speed exceeded 40 mph were considered, so the data reflect freeway operation at each ramp.   If each ramp vehicle accepted the gap that was adjacent to the nose when the ramp vehicle reached the nose, the mean accepted-gap number would be 1.0 and the standard deviation would be zero.

It can be seen in Fig. 9.21 that ramps with low convergence angles and long acceleration lanes have associated with them a low mean and standard deviation of accepted-gap number.   This reflects on the ability of ramp drivers to select their gap at or before the nose on this type of

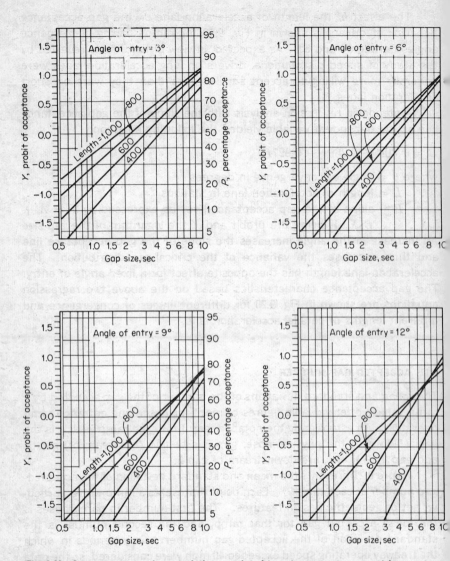

Fig. 9.20 Gap acceptance characteristics as related to entrance-ramp geometrics.

ramp and indicates the consistency of operation on these ramps with good geometric designs.   The consistency of behavior of ramp drivers is important in the achievement of a high degree of safety, so that ramps with low standard deviations of accepted-gap number probably have a lower accident record than the ramps on which the operation is more erratic. Many ramp accidents are caused by one driver assuming that the vehicle

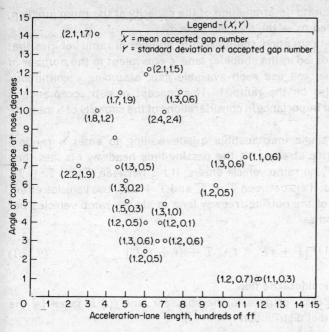

Fig. 9.21 Mean and standard deviation of accepted-gap numbers in seconds, freeway operating speeds over 40 mph.

in front will accept a particular gap. If this assumption proves true, all is well. If not, the following vehicle may find it necessary to decelerate rapidly, especially at ramps with large convergence angles and short acceleration lanes.

It is the object of the freeway merging control system discussed in Sec. 16.11 to automatically fit all ramp vehicles into gap 1. Preliminary experiments with a prototype of this system (Chap. 16) indicate that the mean gap number does in fact approach 1.0 and the standard deviation does approach zero. Accident analyses indicate that since the installation of this system, rear end-type ramp accidents due to false starts have been virtually eliminated.

9.18 APPLICATION TO DESIGN

The efficiency of traffic movement on the through lanes of an urban free-way is directly affected by the adequacy of the associated ramps. The proper design and placement of ramps on high-volume freeways is there-fore imperative if those facilities are to afford fast, efficient, and safe operation. The development of such suitable designs depends to a large extent on the accurate determination of the capacity at the ramp junction, heretofore referred to as the *merging capacity*.

In the merging situation, the maximum number of ramp vehicles that can be accommodated in the shoulder lane is equivalent to the number of ramp vehicles that will use each available gap, assuming a continuous backlog of vehicles on the ramp.[21]   The concept of gap acceptance is therefore of major importance in considerations of the capacity of a merging area.

Consider a single inexhaustible queue waiting to enter a random shoulder-lane traffic stream.   If the passing-time headway t is less than the critical gap T, no ramp vehicle enters; if t is between T and $T + T'$, one vehicle enters; if t is between $T + T'$ and $T + 2T'$, two vehicles enter, etc.   The ability of the outside freeway lane to absorb ramp vehicles per unit of time becomes

$$q_r = q_i \sum_{i=0}^{\infty} (i + 1)P[T + iT' < t < T + (i + 1)T'] \qquad (9.18.1)$$

where q is the shoulder-lane flow.

If the distribution of gaps in the shoulder lane $f(t)$ is given by the negative exponential distribution

$$f(t) = qe^{-qt}$$

then it follows from (9.18.1) that

$$q_r = \frac{qe^{-qT}}{1 - e^{-qT'}} \qquad (9.18.2)$$

In Sec. 9.9 dealing with multiple entries it was concluded that double entries are more sensitive than single entries and triple entries more sensitive than double entries to differences in gap sizes.   Here *sensitivity* means the data shows that the percent acceptance curves are steeper for triple entries than for double entries and steeper for double than for single entries.   The percent acceptance curves show that at the 50 percentile acceptance level (the critical gap), T and T' are approximately equal.   Therefore, the expression for ramp capacity in (9.18.2) may be simplified, becoming

$$q_r = \frac{qe^{-qT}}{1 - e^{-qT}} \qquad (9.18.3)$$

Equation (9.18.3) is illustrated in Fig. 9.22.   In order to use the graph, it is necessary that the shoulder-lane volume q and the critical gap T be known.   If an existing design is being evaluated, q, q_r, and T can be measured.   In the case of a proposed design, methods of estimating the percent of the total freeway volume in the shoulder lane are well documented. Variables which have been found to significantly affect q are the total freeway volume, the entrance ramp volume q_r, upstream ramp volume, downstream exit ramp volume, and distance to downstream exit ramp. Figure 5.4 is illustrative of the relationship between q and two of these variables—total freeway volume and entrance ramp volume q_r.

As defined above, the merging capacity is the maximum number of vehicles that can be accommodated with a continual backlog or queue of ramp vehicles.   As such, it is a possible capacity.   Whenever the opportunity occurs for r vehicles to enter the shoulder-lane stream there must, of course, be at least r vehicles queued on the ramp to utilize this capacity potential.   Although the delay or queue lengths associated with such a traffic condition could be excessive, they were not considered in the capacity analysis.   However, in order to provide a certain level of service, the determination of merging service volumes must take delays or queue lengths into account.   This is discussed in Sec. 10.8 in the next chapter.

In the design of a new facility, however, the engineer is confronted with a set of assigned volumes and has no knowledge of the value of T and a; he is therefore at a loss regarding which curve or even which set of curves to use for determining the service volume of a proposed design.   Collected data have been analyzed with the objective of formulating relationships useful to the designer in that they should allow the prediction of the Erlang

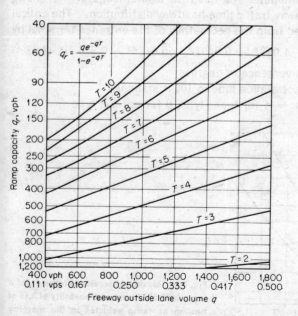

Fig. 9.22 Possible capacity of merging areas.

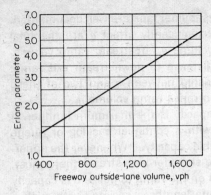

Fig. 9.23 Approximate value of Erlang a as related to the freeway outside lane volume.

parameter a and the initial gap T from information that would normally be available.[20]

The Erlang parameter is largely affected by the volume level. To be sure, there are several other variables such as alignment, grade, and other environmental elements which affect the value of a. However, in the absence of any knowledge of these variables, the curve shown in Fig. 9.23 has been found to approximate the value of a as related to the freeway volume. This curve can be used in conjunction with the relationships shown in Fig. 10.8 to develop the set of curves shown in Fig. 9.24, which relates the ramp service volume to the outside freeway lane volume and the critical gap T. This eliminates the need for a knowledge of the Erlang parameter a of the freeway traffic time-headway distribution. The critical gap T can be estimated from the geometrics of the entrance terminal by

$$T = 5.547 + 0.828\theta - 1.043L + 0.045L^2 - 0.042\theta^2 - 0.874S \quad (9.18.4)$$

where θ = angle of convergence in degrees
L = length of acceleration lane in 100-ft stations
S = shape factor = 1 for taper type, 0 for parallel type

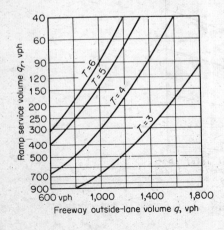

Fig. 9.24 Maximum service volume q_r for a ramp arrival to have a probability of 0.67 of finding no ramp vehicles in the merging area.

From the equation, it can be seen that increasing the angle increases the critical gap and that increasing the length decreases the critical gap. As expected then, a lower angle of convergence and a longer acceleration lane are desirable in the design of the ramp junction. Although this idea is not new, the significance of the above equation is that it quantifies these effects so that the designer can evaluate the trade offs between θ and L. Curves for estimating the critical gap from the length of acceleration lane and angle of entry are shown in Fig. 9.25.

Another interesting aspect of this analysis is that it shows a difference between a tapered and a parallel-lane acceleration lane—a topic that has aroused considerable controversy. According to the analysis, which is based on observations of the operation of 13 tapered and 16 parallel-lane acceleration lanes, a tapered entrance terminal will, on the average, have a critical gap that is about 0.9 sec smaller than that of a parallel-lane acceleration lane with the same length and the same angle of convergence.[20]

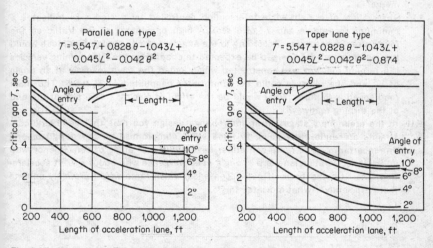

Fig. 9.25   Critical gap T based on entrance ramp geometrics.

PROBLEMS

9.1. Define each of the following terms, identifying the distinctiveness of each:
   a. Lane change
   b. Weaving maneuver
   c. Merging maneuver

9.2. Ramp drivers with a critical gap of 3 sec enter the merging area when the outside freeway lane flow is 800 vph. If the freeway gaps are described by an Erlang $a = 1$ distribution, what is the expected mean and variance of such drivers' delay.

9.3. For the same ramp and group of drivers in Prob. 9.2, what is the expected mean and variance of merging delay if the outside freeway lane is described by 1,200 vph and $a = 2$?

9.4. A driver's gap acceptance function is taken as

$$P(T) = 1 - e^{-\lambda(T-c)}$$

If the freeway lane flow is 0.4 vps and $c = 1.0$ sec with $\lambda = 3.0$ sec^{-1}, what is the mean delay while in position to merge?

9.5. Is it easier to determine *gap acceptance* or *critical-gap* characteristics? Discuss why this is so.

9.6. For a vehicle with a maximum acceleration rate of 10 ft/sec^2 entering a merging area at an initial speed of 25 mph where the free speed in the merging area is 70 mph, plot the speed-time relationship.

9.7. Plot the space-time curve for the driver-vehicle element of Prob. 9.6.

9.8. An acceleration lane 800 ft long is available for use by the ramp driver of Probs. 9.6 and 9.7. Using the curves plotted in the previous problems determine the maximum speed the ramp driver could attain by merging at the end of the acceleration lane.

9.9. Is it better to face twice the volume of traffic than to require a gap twice as long to merge?

9.10. Observation of a right-angle intersection controlled by a stop sign indicates a minimum acceptable gap T' of 5 sec. A high percentage of the traffic on the minor street turns right. According to the angular velocity model, how much would the minimum acceptable gap be expected to change for the right-turning vehicles if a turning roadway was constructed which joined the main roadway at an angle of 15° and the roadway could be comfortably negotiated at 20 mph? The main road speed is 45 mph. What would be the effect of placing a stop sign at the end of the turning roadway?

9.11. In the preliminary stages of an interchange design you note that geometric constraints are quite likely going to force some compromise on the design of one entrance ramp. The ramp can either have about a 6° angle of convergence with an 800-ft acceleration lane or a 3° angle of convergence with about a 600-ft acceleration lane. On the basis of gap acceptance, which ramp would probably operate more efficiently? What indicates this?

Gap data

Gap	No. observed	No. accepted	No. rejected
0–1	150	1	91
1–2	249	27	122
2–3	200	59	61
3–4	153	68	21
4–5	110	59	7
5–6	74	40	2
6–7	31	21	1
7–8	12	7	0
8–9	15	11	1
9–10	5	3	0
>10	10	6	0

9.12. Perform the following with respect to the gap data shown on page 220:

a. Compare the observed distribution of gaps to the Erlang $a = 2$ distribution using the chi-square test.

b. Determine the median acceptable gap, T, by plotting the percent accepted vs. gap size on log-probability paper. Percent accepted $= 100 \times$ number accepted $\div$ (number accepted $+$ number rejected).

c. Assuming that the observed distribution may be deserved by the Erlang $a = 2$ distribution, calculate the average merging delay to a ramp vehicle.

9.13. Given the following information about traffic flow on a major street through an intersection with a minor street: The minor street is stop-sign controlled. Out of every 100 gaps, 65 are acceptable. When a vehicle arrives on the minor street:

a. What is the probability that the first gap is not an acceptable one?

b. What is the probability that the second gap is acceptable with the first one being unacceptable?

c. What is the probability that the first two gaps will both be acceptable?

d. What is the probability that the driver will not have to wait for more than two gaps to find an acceptable one?

REFERENCES

1. Greenshields, B. D., D. Shapiro, and E. L. Erickson: "Traffic Performance at Urban Street Intersections," Technical Report 1, Bureau of Highway Traffic, Yale University, New Haven, Conn., 1947.
2. Raff, M. S., and J. W. Hart: "A Volume Warrant for Urban Stop Signs," The Eno Foundation for Highway Traffic Control, Saugatuck, Conn., 1950.
3. Blunden, W. R., C. M. Clissold, and R. B. Fisher: Distribution of Acceptance Gaps for Crossing and Turning Maneuvers, *Australian Road Res. Board Proc.*, vol. 1, pt. 1, 1962.
4. Drew, D. R.: "Gap Acceptance Characteristics for Ramp-freeway Surveillance and Control," presented at the annual meeting of the Highway Research Board, Washington, D.C., January, 1966.
5. Tanner, J. C.: The Delay to Pedestrians Crossing a Road, *Biometrika*, vol. 38, pp. 3–4, December, 1951.
6. Mayne, A. J.: Some Further Results in the Theory of Pedestrians and Road Traffic, *Biometrika*, vol. 41, no. 375, 1954.
7. Feller, W.: "An Introduction to Probability Theory and Its Applications," vol. 1, John Wiley & Sons, Inc., New York, 1950.
8. Garwood, F.: An Application of the Theory of Probability to the Operation of Vehicular-controlled Traffic Signals, *J. Roy. Statist. Soc.*, vol. 7, no. 65, 1940.
9. Dawson, R.: "Comments on 'Gap Acceptance Characteristics for Ramp-freeway Surveillance and Control,'" presented at the annual meeting of the Highway Research Board, Washington, D.C., January, 1966.
10. Herman, R., and G. Weiss: Comments on the Highway-crossing Problem, *Operations Res.*, vol. 9, no. 6, pp. 828–840, November–December, 1961.
11. Weiss, G., and A. Maradudin: Some Problems in Traffic Delay, *Operations Res.*, vol. 10, no. 1, pp. 74–104, January–February, 1962.
12. Finney, D. J.: "Probit Analysis," Cambridge University Press, New York, 1947.
13. Finney, V. J.: "Statistical Method in Biological Assay," Hafner Publishing Company, Inc., New York, 1964.
14. Solberg, Per, and J. C. Oppenlander: Lag and Gap Acceptances at Stop-controlled Intersections, *Highway Res. Board Rec.* 118, 1965.

15. Matson, T. M., W. S. Smith, and F. W. Hurd: "Traffic Engineering," McGraw-Hill Book Company, New York, 1955.
16. Michaels, R. M., and H. Weingarten: "Driver Judgment in Gap Acceptance," unpublished.
17. Drew, D. R., L. R. LaMotte, J. H. Buhr, and J. A. Wattleworth: Gap Acceptance in the Freeway Merging Process, *Texas Transportation Inst. Res. Rept. RF* 430-2, Texas A&M University, College Station, 1966.
18. Wattleworth, J. A., J. H. Buhr, D. R. Drew, and F. Gerig: Operational Effects of Some Entrance Ramp Geometrics on Freeway Merging, *Texas Transportation Inst. Res. Rept. RF* 430-3, Texas A&M University, College Station, 1966.
19. Drew, D. R., J. A. Wattleworth, and J. H. Buhr: Gap Acceptance and Traffic Interaction in the Freeway Merging Process, *Proc. Traffic Systems Div.*, U.S. Bureau of Public Roads, 1966.
20. Drew, D. R., J. H. Buhr, and R. H. Whitson: The Determination of Merging Capacity and Its Application to Freeway Design and Control, *Texas Transportation Inst. Rept.* 430-4, Texas A&M University, College Station, 1967.
21. Major, N. G., and N. J. Buckley: Entry to a Traffic Stream, *Australian Road Res. Board Proc.*, vol. 1, pt. 1, pp. 206–228, 1962.

CHAPTER TEN
QUEUEING PROCESSES

They also serve who only stand and wait.

Milton

10.1 INTRODUCTION

A primary objective in operational problems involving flow is to ensure that
the average capacity can handle the average flow, so that persistent traffic
jams do not occur.   However, because flow fluctuates, merely guaranteeing
that highway capacity can handle traffic demand on the average does not
preclude the formation of transient or even permanent bottlenecks.

Queueing theory was developed in order to describe the behavior
of a system providing services for randomly arising demands.   It originated
in a paper by A. K. Erlang written in 1909 on the problem of congestion
in telephone traffic.   In later works he observed that the telephone system
was characterized by Poisson inputs, exponential holding times, and
multiple service channels.   In recent years much research into the theory
has been carried out, particularly in the field of operations research.

The fundamental idea of the theory is that congestion is manifest
through delay caused by an interruption in a flow pattern.   The flow is
usually discrete such as the arrivals of vehicles at a parking lot; but it

may be continuous such as the flow of traffic into and out of a freeway bottleneck. Since this general explanation of queueing offers a plausible description of vehicular traffic congestion, it is not surprising that queueing models have received wide consideration in traffic theory.

To specify a queueing system completely, the following must be given: (1) the input function, that is, the distribution of arrivals; (2) the input source, whether finite or infinite; (3) the queue discipline, whether first come, first served, priority, or random selection for service; (4) the channel configuration, that is, the number of channels and, if more than one, whether the channels are in series or parallel; and (5) the distribution of service times for each channel.

The collection of arrivals waiting to be served is known as the queue; the number of arrivals waiting for service at a time t is called the queue length. A queue is judged differently by different people. A driver on arrival will observe the length of queue at a parking lot and will be interested in his waiting time. The owner of the parking lot will be interested in slack periods when there are no parkers, to know how many attendants he needs. In general, the questions of importance in the study of queueing are (1) the distribution of the length of queue, (2) the distribution of the waiting time in the queue, and (3) the percent of time during which the service system is idle. The answers to these questions depend directly on the nature of the input and service-time distributions.

Input and service-time distributions may be of any form. However the science of queueing theory has been developed around three special types of distributions: regular D, random M, and Erlang E_a. In addition to these some models utilize an undefined or general distribution denoted by G. $M/M/1$ denotes a single-channel queue with a random (Poisson) input and a random (negative exponential) service. The number 1 stands for the number of channels in parallel.

A queueing system is said to be in a state n if it contains exactly n items $(n > 0)$, including those items in the line and in service. If the flow of arrivals q is less than the service rate Q, stability exists in the system and there is a finite time-independent probability of the queue being in any state. If, however, the ratio of flow rate to service rate is greater than unity, the state of the waiting line increases in length and is no longer independent of time.

10.2 THE SINGLE CHANNEL

An appreciation of the characteristics of queueing can be gained through the derivation of a simple case, that of a single channel with Poisson-distributed arrivals and an exponentially distributed service rate. Letting

$P_n(t + \Delta t)$, $n > 0$ be the probability that the queueing system has n items at the time $t + \Delta t$, consider the three ways in which the system could have reached this state, assuming Δt is so small that only one unit could have entered or departed.

1. The system did not change from time t to $t + \Delta t$.
2. The system was in state $n - 1$ at t and one arrived in Δt.
3. The system was in state $n + 1$ at t and one departed in Δt.

Since the probability of one arrival in Δt equals $q \, \Delta t$ and the probability of one departure in Δt is $Q \, \Delta t$, the corresponding probabilities of zero arrivals and zero departures are $1 - q \, \Delta t$ and $1 - Q \, \Delta t$.   If higher orders of Δt are neglected, the mathematical description of the three ways for the system to reach n in $t + \Delta t$ is:

$$P_n(t + \Delta t) = P_n(t)[1 - (q + Q) \, \Delta t] + P_{n-1}(t)q \, \Delta t + P_{n+1}(t)Q \, \Delta t$$
$$n = 1, 2, \ldots \quad (10.2.1)$$

$$P_0(t + \Delta t) = P_0(t)(1 - q \, \Delta t) + P_1(t)Q \, \Delta t \qquad n = 0 \qquad (10.2.2)$$

Passing to the limit with respect to Δt, these equations become

$$P_n'(t) = -(q + Q)P_n(t) + qP_{n-1}(t) + QP_{n+1}(t) \qquad n = 1, 2, \ldots$$
$$(10.2.3)$$

$$P_0'(t) = -qP_0(t) + QP_1(t) \qquad n = 0 \qquad (10.2.4)$$

If the time derivatives are set equal to zero and time is eliminated from the above equations, there results

$$1 + \frac{q}{Q} \, P_n = P_{n+1} + \frac{q}{Q} \, P_{n-1} \qquad n = 1, 2, \ldots \qquad (10.2.5)$$

$$P_1 = \frac{q}{Q} \, P_0 \qquad n = 0 \qquad (10.2.6)$$

where P_n is the steady-state (time-independent) probability that there are n units in the system.

Letting $n = 1$ in (10.2.5) and substituting for P_1 from (10.2.6) in (10.2.5), one sees that Eq. (10.2.5) becomes

$$P_2 = \left(\frac{q}{Q}\right)^2 P_0 \qquad (10.2.7)$$

Then solving (10.2.5) and (10.2.7) by induction one sees that the queue length probabilities take the form of the geometric distribution.

$$P_n = \left(\frac{q}{Q}\right)^n \left(1 - \frac{q}{Q}\right) \qquad (10.2.8)$$

The mean of this distribution is by application of (7.3.1)

$$E(n) = \frac{q}{Q}\left(1 - \frac{q}{Q}\right)^{-1}$$

or

$$E(n) = \frac{q}{Q - q} \tag{10.2.9}$$

where $E(n)$ is the expected number in the system.

Use of (10.2.9) will be illustrated by an example. Consider the operation of a single lane of freeway traffic at a tollbooth. If the lane flow is $q = 800$ vph and the tollbooth operator can process an automobile in 4 sec on the average, then

$$Q = \frac{3{,}600 \text{ sec/hr}}{4 \text{ sec/vehicle}} = 900 \text{ vph} \tag{10.2.10}$$

Thus Q, the service rate in a queueing system in general, takes on an added physical significance in the special case of a highway facility—that of capacity. In most traffic engineering problems, an important consideration is the solution of a capacity-demand relationship. Although it is recognized that capacity on the average must exceed demand on the average, the importance of the distributions of these parameters is not generally appreciated. Using the tollbooth example this will be illustrated (see Fig. 10.1a).

If it is assumed that the flow of 800 vph, representing the demand on the tollbooth, conforms to a Poisson distribution and that the rate at which

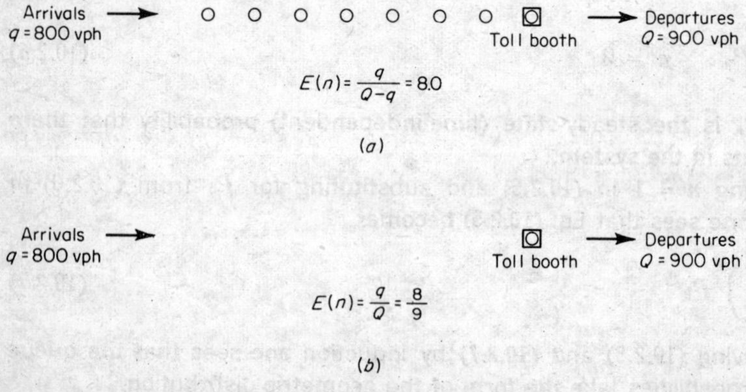

Fig. 10.1 Comparison of the effect of the distribution of arrivals and the distribution of service on congestion. (a) Poisson (random) arrivals and exponential service; (b) regular (metered) arrivals and regular service.

the operator works is exponentially distributed, the expected number of vehicles at the tollbooth is obtained from (10.2.9).

$$E(n) = \frac{800}{900 - 800} = 8 \tag{10.2.11}$$

On the other hand, if arrivals at the tollbooth are regular (accomplished perhaps through metering) and some uniform automatic servicing device is utilized, the expected number of vehicles at the tollbooth becomes (Fig. 10.1b)

$$E(n) = \frac{q}{Q} = \frac{8}{9} \tag{10.2.12}$$

Two important conclusions may be drawn:

1. The distribution of demand and the distribution of service (capacity) are important.
2. The expected number in the system $E(n)$ is a measure of the inefficiency of a facility, or a quantitative measure of traffic congestion.

Because there is a definite probability of the system being empty (P_0), it should be apparent that the mean queue length $E(m)$ will never be exactly one less than the mean or expected number in the system $E(n)$. The expected number in the queue is

$$E(m) = \sum_{n=1}^{\infty} (n - 1)P_n = E(n) - \rho \tag{10.2.13}$$

where $\rho = q/Q$.

10.3 APPLICATION OF THE GENERATING FUNCTION

A useful tool in the analysis of infinite-queue systems, which become more important as the systems become more formidable, is the probability-generating function discussed in Sec. 7.6. The form of Eq. (7.6.1) used in this case is

$$Z_n(\theta) = \sum_{n=0}^{\infty} \theta^n P_n \tag{10.3.1}$$

It is convenient to establish the following identities both to illustrate this application of pgf as well as to simplify the taking of the inverse after the pgf of the single exponential channel is obtained:

$$P_n = 1 \qquad Z_n(\theta) = (1 - \theta)^{-1} \tag{10.3.2}$$

$$P_n = C_1 \qquad\qquad Z_n(\theta) = C_1(1 - \theta)^{-1} \qquad\qquad\qquad (10.3.3)$$

$$P_n = C_2{}^n \qquad\qquad Z_n(\theta) = (1 - C_2\theta)^{-1} \qquad\qquad\qquad (10.3.4)$$

$$P_n = C_1 C_2{}^n \qquad Z_n(\theta) = C_1(1 - C_2\theta)^{-1} \qquad\qquad\qquad (10.3.5)$$

$$P_n = P_{n+k} \qquad Z_{n+k}(\theta) = \theta^{-k} \sum_{n=0}^{\infty} \theta^{n+k}P_{n+k} = \theta^{-k}Z_n(\theta)$$

$$- \sum_{n=0}^{k-1} \theta^{n-k}P_n \qquad (10.3.6)$$

Letting $\rho = q/Q$, one can write Eq. (10.2.5) as

$$(1 + \rho)P_{n+1} - \rho P_n - P_{n+2} = 0 \qquad\qquad\qquad (10.3.7)$$

Taking the pgf of each term gives

$$(1 + \rho)Z_{n+1}(\theta) - \rho Z_n(\theta) - Z_{n+2}(\theta) = 0 \qquad\qquad (10.3.8)$$

Applying (10.3.6) to (10.3.8) and solving for $Z_n(\theta)$, one obtains

$$Z_n(\theta) = P_0(1 - \rho\theta)^{-1} \qquad\qquad\qquad (10.3.9)$$

From (10.3.1) it is evident that since $Z_n(\theta) = 1$ for $\theta = 1$,

$$1 = P_0(1 - \rho)^{-1}$$

or

$$P_0 = 1 - \rho \qquad\qquad\qquad (10.3.10)$$

Substitution of (10.3.10) in (10.3.9) yields the pgf of the $M/M/1$ queue:

$$Z_n(\theta) = \frac{1 - \rho}{1 - \rho\theta} \qquad\qquad\qquad (10.3.11)$$

To obtain the inverse of the pgf in (10.3.11), we note that (10.3.11) is the same form as (10.3.5) with $C_1 = 1 - \rho$ and $C_2 = \rho$; therefore

$$P_n = (1 - \rho)\rho^n \qquad\qquad\qquad (10.3.12)$$

The moments of (10.3.12) are easily obtained by taking the derivatives of the pgf according to (7.6.2) and (7.6.4):

$$E(n) = Z'_n(1) = \frac{\rho}{1 - \rho} \qquad\qquad\qquad (10.3.13)$$

and

$$\sigma^2 = Z''_n(1) + \mu - \mu^2 = \frac{\rho}{(1 - \rho)^2} \qquad\qquad\qquad (10.3.14)$$

10.4 FINITE QUEUES

It is well to start with Eq. (10.2.5) in order to remove the restrictions concerning an infinite input source and a ρ value less than unity. If the maximum number in the system is limited to N such that vehicles arriving when $n = N$ will not join the queue, then (10.2.5) and (10.2.6) become

$$P_1 = \rho P_0$$

$$(1 + \rho)P_n = P_{n+1} + \rho P_{n-1} \qquad 0 < n < N$$

$$P_n = \rho P_{n-1} \qquad n = N$$

Now, to obtain the probability of some number in the system for a finite queue length, one uses

$$\sum_{n=0}^{N} P_n = 1 = P_0 + \rho P_0 + \rho^2 P_0 + \cdots + \rho^N P_0$$

which, from (7.9.3), gives

$$1 = P_0 \frac{1 - \rho^{N+1}}{1 - \rho} \tag{10.4.1}$$

Solving for P_0 in (10.4.1) and substituting in $P_n = \rho^n P_0$, one obtains the queue length probability for the truncated case:

$$P_n = \frac{1 - \rho}{1 - \rho^{N+1}} \rho^n \tag{10.4.2}$$

The expected number in a system in which the maximum number is N is

$$E(n) = \sum_{n=0}^{N} n P_n$$

which can be shown to be equal to

$$E(n) = \frac{\rho}{1 - \rho} \frac{1 - (N + 1)\rho^N + N\rho^{N+1}}{1 - \rho^{N+1}} \tag{10.4.3}$$

Finite or truncated queues have an important application in traffic as models for ramp metering (see Chap. 16). If the merging rate is reduced by placing a traffic signal on the ramp, some of the ramp traffic will be diverted. It is important to know what this diversion is since it may create problems at another entrance ramp downstream or on the surface streets.

This diversion is the difference between the ramp volume observed and the ramp demand q. The ramp demand q, or arrival rate, can be estimated by measuring the queue lengths on the ramp. One merely checks the ramp drivers' actions by measuring the P_n's in the ramp queue and sees whether they are a constant times some fraction to the nth power. Morse[6] suggests that the P_n be plotted against n on semilog paper. The curve should be a straight line if ramp traffic stays or diverts, depending on queue length; the slope of the curve enables one to obtain P and, knowing Q, to calculate q.

10.5 WAITING TIMES

This distribution of such random variables as the waiting time of an arrival before being taken into service, w, and the total time of an arrival in the system, v, are important parameters in such traffic applications as merging and crossing.

The waiting-time distribution from arrival until start of service is considered in two parts. First there is a finite probability that the waiting time will be zero, which is the same as the probability of the system being empty [see Eq. (10.3.10)].

$$P(w = 0) = P_0 = 1 - \rho \qquad n = 0 \tag{10.5.1}$$

Second, there is the probability that the waiting time is between time w and $w + dw$:

$$P(w < \text{wait} < w + dw) = f(w)\,dw \qquad n > 0 \tag{10.5.2}$$

The waiting-time density function described by Eqs. (10.5.1) and (10.5.2) is illustrated in Fig. 10.2.

What is desired are all the possible combinations that will yield a waiting time between w and $w + dw$. Theoretically, such a delay to an arrival A is possible as long as there is a unit in service. This may be

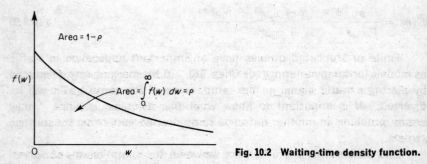

Fig. 10.2 Waiting-time density function.

expressed symbolically as

$$P(n \geq 1) = \sum_{n=1}^{\infty} P_n \tag{10.5.3}$$

But in order for the waiting time of A to be exactly between w and $w + dw$, all the units in the system ahead of A except the one immediately ahead must depart in time w and the one immediately ahead must be served in dw. This is the product of two probabilities:

$$P_{n-1}(w) = \frac{(Qw)^{n-1}e^{-Qw}}{(n-1)!} \tag{10.5.4}$$

and

$$P_1(dw) = Q \, dw \tag{10.5.5}$$

These probabilities, summed over (10.5.3) and substituted in (10.5.2), yield

$$\begin{aligned}
f(w)dw &= \sum_{n=1}^{\infty} P_n P_{n-1}(w) P_1(dw) \\
&= \sum_{n=1}^{\infty} \frac{\rho^n(1-\rho)(Qw)^{n-1}e^{-Qw}Q \, dw}{(n-1)!} \\
&= Q\rho(1-\rho)e^{-Qw} \, dw \sum_{n=1}^{\infty} \frac{(\rho Qw)^{n-1}}{(n-1)!}
\end{aligned} \tag{10.5.6}$$

The desired probability density function for waiting times is, therefore,

$$f(w) = \rho(Q-q)e^{-w(Q-q)} \qquad w > 0 \tag{10.5.7}$$

The cumulative forms of (10.5.7) are

$$\begin{aligned}
P(\text{wait} > w) &= \int_w^{\infty} f(w) \, dw \\
&= \rho e^{-(Q-q)w}
\end{aligned} \tag{10.5.8}$$

and

$$\begin{aligned}
P(0 < \text{wait} < w) &= \int_0^w f(w) \, dw \\
&= \rho - \rho e^{-(Q-q)w}
\end{aligned} \tag{10.5.9}$$

Again, attention is called to the fact that the integral from 0 to ∞ is ρ, not unity (see Fig. 10.2).

The mgf of the density function for waiting times (10.5.7) is

$$M_w(\theta) = \frac{\rho(Q-q)}{(Q-q) - \theta} \tag{10.5.10}$$

The average waiting time of an arrival is therefore

$$M'_w(0) = E(w) = \frac{q}{Q(Q-q)} \tag{10.5.11}$$

The conditional density function for waiting time, that is, the density of an arrival that has to wait, is by definition

$$f(w|w > 0) = \frac{f(w)}{P(w > 0)}$$

which becomes

$$f(w|w > 0) = (Q - q)e^{-w(Q-q)} \tag{10.5.12}$$

The average waiting time of an arrival that waits is

$$E(w|w > 0) = \frac{1}{Q-q} \tag{10.5.13}$$

The density function for the total time of an arrival v in the system may be found in the same manner as was w [Eqs. (10.5.1) to (10.5.7)]. It is

$$f(v) = (Q - q)e^{-v(Q-q)} \tag{10.5.14}$$

The average time an arrival spends in the system is given by

$$E(v) = \frac{1}{Q-q} \tag{10.5.15}$$

which is, as one would expect, exactly $1/Q$ longer than $E(w)$

$$E(v) = E(w) + \frac{1}{Q} \tag{10.5.16}$$

The relationship between the expected delays $E(w)$ and $E(v)$ and the expected number in the system $E(n)$, as evidenced in Fig. 10.3, are for the $M/M/1$ case given by

$$E(w) = E(n)\frac{1}{Q} \tag{10.5.17}$$

and

$$E(v) = E(n)\frac{1}{q} \tag{10.5.18}$$

A more general relationship between the parameters studied is given by

$$q = \frac{E(n)}{E(v)} = \frac{E(m)}{E(w)} \tag{10.5.19}$$

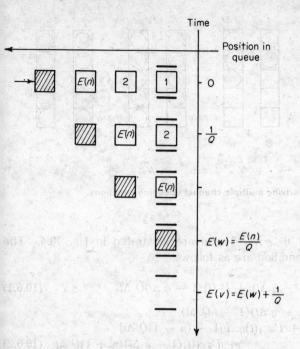

Fig. 10.3 **Relations between expected values for single-channel queues.**

10.6 MULTIPLE CHANNELS

Problems involving parallel channels arise, for example, in the study of telephone traffic. Other examples are restaurants with counter seats, bank tellers, and supermarket check-out counters. Traffic applications of this sort of system include operations at tollbooths, truck warehouse terminals, and automobile parking facilities.

The assumptions of Poisson arrivals at an average rate q and exponential service at an average rate Q for each of the N channels will be utilized as was done for the single-channel case. An arrival may enter any free channel; if none are available, he joins the queue. The chance of a departure in a single channel in Δt is $Q\,\Delta t$; the chance of a departure from the system in time Δt is $nQ\,\Delta t$ for $n < N$ and $NQ\,\Delta t$ for $n \geq N$.

The derivation of the equations for this, the $M/M/N$ case, proceeds in much the same manner as for the single channel, or $M/M/1$, case. However, for multiple channels there are four conditions at time t which must be considered in the formulation of the transient equations: $n = 0$,

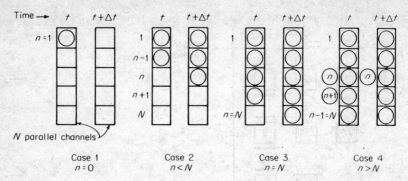

Fig. 10.4   Conditions for deriving multiple channel queueing equations.

$n < N$, $n = N$, and $n > N$. These are illustrated in Fig. 10.4. The equations for each condition are as follows:

$$P_0(t + \Delta t) = P_0(t)(1 - q\,\Delta t) + P_1(t)(1 - q\,\Delta t)Q\,\Delta t \qquad (10.6.1)$$

$$P_n(t + \Delta t) = P_n(t)(1 - q\,\Delta t)(1 - nQ\,\Delta t)$$
$$+ P_{n-1}(t)q\,\Delta t[1 - (n - 1)Q\,\Delta t]$$
$$+ P_{n+1}(t)(1 - q\,\Delta t)(n + 1)Q\,\Delta t \qquad (10.6.2)$$

$$P_n(t + \Delta t) = P_n(t)(1 - q\,\Delta t)(1 - NQ\,\Delta t)$$
$$+ P_{n-1}(t)q\,\Delta t[1 - (N - 1)Q\,\Delta t]$$
$$+ P_{n+1}(t)(1 - q\,\Delta t)(NQ\,\Delta t) \qquad (10.6.3)$$

$$P_n(t + \Delta t) = P_n(t)(1 - q\,\Delta t)(1 - NQ\,\Delta t) + P_{n-1}(t)q\,\Delta t(1 - NQ\,\Delta t)$$
$$+ P_{n+1}(t)(1 - q\,\Delta t)(NQ\,\Delta t) \qquad (10.6.4)$$

Equations (10.6.3) and (10.6.4) become the same when higher orders of Δt are neglected, and so the transient equations become

$$P_0'(t) = -qP_0(t) + QP_1(t) \qquad (10.6.5)$$

$$P_n'(t) = -(q + nQ)P_n(t) + qP_{n-1}(t) + (n + 1)QP_{n+1}(t) \qquad (10.6.6)$$

$$P_n'(t) = -(q + NQ)P_n(t) + qP_{n-1}(t) + NQP_{n+1}(t) \qquad (10.6.7)$$

The steady-state equations are obtained as before, only letting $\rho = q/Q < N$ such that $q/NQ < 1$:

$$P_1 = \rho P_0 \qquad n = 0 \qquad (10.6.8)$$

$$(\rho + n)P_n = \rho P_{n-1} + (n + 1)P_{n+1} \qquad n < N \qquad (10.6.9)$$

$$(\rho + N)P_n = \rho P_{n-1} + NP_{n+1} \qquad n \geq N \qquad (10.6.10)$$

The solution of these steady-state equations is obtained by induction to be

$$P_n = \frac{\rho^n}{n!} P_0 \qquad\qquad n < N \tag{10.6.11}$$

$$P_n = \frac{\rho^n}{N! N^{n-N}} P_0 \qquad n \geq N \tag{10.6.12}$$

where P_0 is found from

$$\sum_{n=0}^{\infty} P_n = \sum_{n=0}^{N-1} P_n + \sum_{n=N}^{\infty} P_n = 1$$

$$= P_0 \sum_{n=0}^{N-1} \frac{\rho^n}{n!} + P_0 \frac{N^N}{N!} \sum_{n=N}^{\infty} \frac{\rho^n}{N^n} \tag{10.6.13}$$

Solving (10.6.13) for P_0 gives

$$P_0 = \frac{1}{\displaystyle\sum_{n=0}^{N-1} \frac{\rho^n}{n!} + \frac{\rho^N}{N![1 - (\rho/N)]}} \tag{10.6.14}$$

The expected number in the queue is calculated from

$$E(m) = \sum_{n=N}^{\infty} (n - N) P_n = \frac{P_0}{N!} \sum_{n=N}^{\infty} (n - N) \frac{\rho^n}{N^{n-N}}$$

$$= \frac{P_0 \rho^{N+1}}{N!N} \left[1 + 2\frac{\rho}{N} + 3\left(\frac{\rho}{N}\right)^2 + \cdots \right]$$

$$= \frac{P_0 \rho^{N+1}}{N!N} \frac{1}{[1 - (\rho/N)]^2} \tag{10.6.15}$$

The other expected values of interest calculated as before are

$$E(n) = E(m) + \frac{q}{Q} \tag{10.6.16}$$

$$E(v) = \frac{E(n)}{q} \tag{10.6.17}$$

$$E(w) = E(v) - \frac{1}{Q} \tag{10.6.18}$$

If the number of channels is very large $(N \to \infty)$, we see from (10.6.14) that

$$\lim_{N \to \infty} P_0 = e^{-\rho} \tag{10.6.19}$$

Equation (10.6.19) provides a good model for an off-street parking operation. The expected number of vehicles parked would be

$$E(n) = \sum_{n=0}^{\infty} nP_n = P_1 + 2P_2 + 3P_3 + \cdots$$

$$= \rho P_0 \left(1 + \rho + \frac{\rho^2}{2!} + \cdots \right)$$

$$= \rho \qquad\qquad (10.6.20)$$

For curb parking, a more realistic model assumes that cars are not allowed to queue to park, that is, $P_n = 0$ if $n > N$. The probability of no cars parked on the street is

$$P_0 = \frac{1}{\sum\limits_{n=0}^{N} \rho^n/n!} \qquad\qquad (10.6.21)$$

Since

$$P_n = \frac{\rho^n/n!}{\sum\limits_{i=0}^{N} \rho^i/i!} \qquad\qquad (10.6.22)$$

it follows that the probability that a car cannot park is

$$P_N = \frac{\rho^N/N!}{\sum\limits_{i=0}^{N} \rho^i/i!}$$

10.7 MOVING QUEUES

As traffic volumes increase, vehicles tend to form platoons, or moving queues. The criteria for determining when two moving vehicles are queued is arbitrary. And, for that matter, the criteria for determining when two stationary units are queued in classical queueing systems is equally arbitrary since the concept of distance did not appear in the queueing derivations in the previous sections. Borrowing from car-following theory, a line of moving vehicles could be considered to be in a single queue if each must react instantly to the speed reductions of its predecessor. For the purposes of this discussion, it will be assumed that a vehicle is queued to the vehicle ahead when its headway is less than S, if space is the parameter, and less than T, if time is the parameter.

There are several traffic characteristics that can indicate congestion

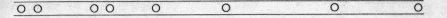

$$E(n) = \frac{8}{6} = 1.33$$

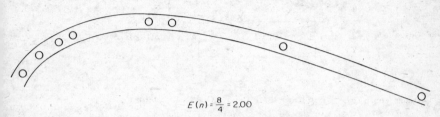

$$E(n) = \frac{8}{4} = 2.00$$

Fig. 10.5 Comparison of two highway facilities with equal densities and different congestion indices.

on a highway facility: low speeds, high flow/capacity ratios, high space densities, and high time densities (lane occupancy). These various parameters tend to ignore the distribution of traffic and therefore give incomplete descriptions of congestion and the state of a system, as exemplified in the tollbooth illustration if only q and Q are considered. It has been suggested that $E(n)$ (which, in the case of moving queues, shall be called the *queue length* or *number queued*) is a logical measure of congestion on a highway facility, just as it is in conventional queueing systems.   Figure 10.5 illustrates how the moving queue length might be a more sensitive indicator of congestion than density.

The formulation of the moving queues model is based on performing a Bernoulli test with probabilities p and $1 - p$ on each headway (either time headway or space headway). If headways between successive vehicles are assumed to be independent, then the probability of having a queue of exactly one vehicle is

$$P_1 = 1 - p \tag{10.7.1}$$

where $1 - p$ is the probability that the headway of vehicle 2 (Fig. 10.6a) is greater than the arbitrary queueing headway S.   Similarly the probability of a queue of exactly two vehicles is obtained by a combination of one "success" followed by a "failure," or

$$P_2 = p(1 - p) \tag{10.7.2}$$

where p is the probability that the headway of vehicle 2 (Fig. 10.6b) is

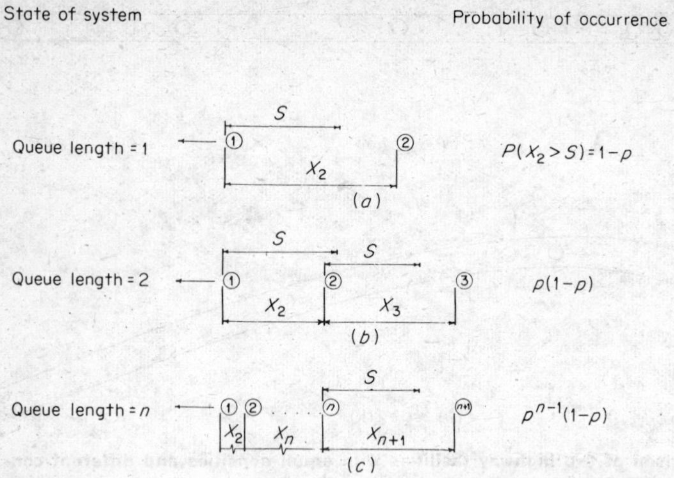

State of system   Probability of occurrence

Fig. 10.6   Bernoulli test for determining the probability of individual queue lengths.

greater than the arbitrary queueing headway S.   By induction, the individual queue lengths form a geometric distribution (Fig. 10.6c):

$$P_n = p^{n-1}(1 - p) \qquad n = 1, 2, \ldots \tag{10.7.3}$$

The expected queue length $E(n)$ is given by

$$E(n) = \sum_{n=1}^{\infty} nP_n$$

which yields

$$E(n) = 1P_1 + 2P_2 + 3P_3 + \cdots$$
$$= (1 - p)(1 + 2p + 3p^2 + \cdots) \tag{10.7.4}$$

The second term of (10.7.4) is a series whose sum is $1/(1 - p)^2$; therefore

$$E(n) = (1 - p)^{-1} \tag{10.7.5}$$

The probability $1 - p$ of any vehicle headway x being greater than the arbitrary queueing headway S is of course dependent on the distribution of vehicles in space on the highway lane:

$$1 - p = P(x > S) = \int_S^{\infty} f(x;k,a) \, dx \tag{10.7.6}$$

The two-parameter probability distribution implied in (10.7.6) is the Erlang distribution.   Substituting in (10.7.6),

$$P(x > S) = \int_S^{\infty} \frac{(ka)^a}{(a - 1)!} \, x^{a-1}e^{-akx} \, dx \tag{10.7.7}$$

where k = average concentration of vehicles

$\quad x$ = distance

$\quad a = 1, 2, 3, \ldots$

Substituting (10.7.7) and (10.7.6) in (10.7.5) yields

$$E(n) = \left[\int_S^{\infty} \frac{(ka)^a}{(a-1)!} \, x^{a-1} e^{-akx} \, dx \right]^{-1} \qquad (10.7.8)$$

which is the fundamental relation between the moving queue length $E(n)$, concentration k, the arbitrary queueing headway S, and the distribution of concentration a.

Integration of (10.7.8) for $a = 1, 2, 3$, and 4 yields

$$E(n)_{a=1} = e^{kS} \qquad (10.7.9)$$

$$E(n)_{a=2} = \frac{e^{2kS}}{2kS + 1} \qquad (10.7.10)$$

$$E(n)_{a=3} = \frac{e^{3kS}}{4.5(kS)^2 + 3kS + 1} \qquad (10.7.11)$$

$$E(n)_{a=4} = \frac{e^{4kS}}{10.67(kS)^3 + 8(kS)^2 + 4kS + 1} \qquad (10.7.12)$$

In Fig. 10.7, these equations have been plotted using a space headway of 100 ft as a basis for the queueing criteria. The observed points numbered

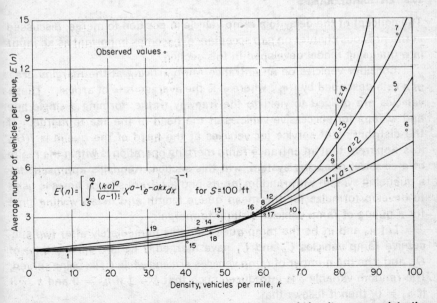

Fig. 10.7 **Observed relationship between queue length and density compared to the theoretical relationship.**

1 to 19 refer to consecutive $\frac{1}{3}$-mile lengths of the shoulder lane of the Gulf Freeway in Houston.   It is apparent that the observed relationship between queue length and density compares very favorably to the theoretical relationship.

Where the smooth flow of traffic is interrupted by some form of bottleneck, the downstream flow of traffic will usually take the form of platoons or moving queues.   Because the motorist recognizes congestion as compromising his right to choose his speed and his traffic lane, the natural tendency for groups to form and increase in length affords a logical index of congestion.

It should also be noted that the moving-queue length parameter $E(n)$ was formulated in such a way that it is the reciprocal of the probability of receiving a gap larger than the queueing criteria.   Thus, since

$$P(x > S) = \frac{1}{E(n)}$$ (10.7.13)

$E(n)$ is actually a measure of gap availability.   This property makes it extremely important as a tool in rational geometric design, freeway capacity analyses, and entrance ramp control.   These applications will be discussed in more detail in a later chapter.

10.8   ENTRANCE RAMPS

The concept of the delay to a ramp vehicle in position to merge, discussed in the previous chapter on gap acceptance, becomes important as an input in a queueing model developed in this section.

Consider vehicles on an entrance ramp arriving at the merging area at a rate described by $f(q_r)$, where q_r is the average rate of arrival.   These vehicles are obliged to yield to the freeway traffic, forming a single line and waiting for successive vehicles at the head of the line to merge.   If the distribution of service for vehicles at the head of the queue is $f(Q)$, it is apparent that an entrance-ramp merging operation is within the realm of a classical queueing system.   In this section, Kendall's approach[2] for a queueing system with random inputs and arbitrary service times is used to develop formulas for the mean queue length and mean waiting time for a queue of ramp vehicles waiting to merge.

Let n_0 and n_1 be the ramp queue lengths immediately after two successive ramp vehicles C_0 and C_1 have merged, t be the service time of C_1, and r be the number of ramp vehicles arriving while C_1 is being served.   If a random variable δ is introduced such that $\delta = 1$ if $n_0 = 0$ and $\delta = 0$ if $n_0 \neq 0$, then it follows that

$$n_1 = n_0 + r - 1 + \delta$$ (10.8.1)

It is to be noted from the definition of δ that

$$\delta^2 = \delta$$

and

$$n_0 \delta = 0$$

and hence, from (10.8.1), on taking expected values, we obtain

$$E(n_1) = E(n_0) + E(r) - 1 + E(\delta) \tag{10.8.2}$$

If the system is assumed to be in a state of statistical equilibrium, $E(n_i) = E(n_0)$ and

$$E(r) = q_r E(t) = \frac{q_r}{Q} = \rho$$

Thus substituting in (10.8.2)

$$E(\delta) = 1 - \rho$$

Squaring both sides of (10.8.1) and taking expected values as before,

$$E(r - 1)^2 + E(\delta^2) + 2E[n_0(r - 1)] + 2E[\delta(r - 1)] = 0$$

which reduces to

$$E(n_0) = \rho + \frac{E(r^2) - \rho}{2(1 - \rho)} \tag{10.8.3}$$

It is now necessary to calculate $E(r^2)$, the second moment of the number of arrivals in the service time t, making use of its relationship to the mean and variance in arrivals. Assuming that ramp arrivals are Poisson and remembering that "averaging" here must be carried out with respect to both r and the service time t, we have

$$E(r^2) = q_r E(t) + q_r^2 E(t^2)$$

Since $E(t) = Q^{-1}$ and

$$E(t^2) = \sigma^2 + E(t)^2$$

then

$$E(r^2) = \rho + \rho^2 + q_r^2 \sigma^2 \tag{10.8.4}$$

Substituting (10.8.4) in (10.8.3) gives the expected queue length on the ramp as

$$E(n)_r = \frac{q_r}{Q} + \frac{Q q_r^2 (Q^{-2} + \sigma^2)}{2(Q - q_r)} \tag{10.8.5}$$

If w is the waiting time (before merging) of C_1, then n_1 ramp vehicles arrive in time $t + w$. Thus since the mean arrival rate is q_r,

$$E(n)_r = q_r E(t + w)$$

It follows that the average waiting time for a ramp vehicle before merging is

$$E(w)_r = \frac{E(n)_r}{q_r} - Q^{-1} \qquad (10.8.6)$$

and the mean wait in the system for a ramp vehicle is

$$E(v)_r = \frac{E(n)_r}{q_r} \qquad (10.8.7)$$

Major and Buckley[3] have interpreted the service time for the queue Q^{-1} as identical to the summation of the rejected gaps for a ramp vehicle in position to merge (what we have referred to as the *merging delay* in the previous chapter). The Pearson type III distribution was fitted to the observed distribution of delays at various ramps, and the results are summarized in Table 10.1. Based on the assumptions that the average service time is equal to the average merging delay and the distribution of service times is of the form of the Pearson type III distribution, we obtain from (10.8.5)

$$E(n)_r = \mu q_r + \frac{(\mu q_r)^2(1 + a^{-1})}{2(1 - \mu q_r)} \qquad (10.8.8)$$

where μ is given by Eqs. (9.3.27) to (9.3.30), and a is the parameter describing the distribution of service times.

If the a value in (10.8.8), calculated in Table 10.1, is rounded off to unity, Eqs. (10.8.5) to (10.8.8) take the form of the conventional Poisson negative exponential queueing formulas of Sec. 10.2. Figure 10.8 illustrates the relationship between the freeway outside-lane volume q, the headway

Table 10.1 Summary of service-time moments and Pearson type III parameter a

Ramp	Service-time moments		Type IIi a value
	Mean	Standard deviation	
Cullen	4.0	5.2	0.6
Dumble	3.5	5.5	0.4
Wayside	6.2	8.8	0.5
Griggs	1.5	2.8	0.3
Mossrose	4.4	6.2	0.5
Woodridge	4.6	6.0	0.6

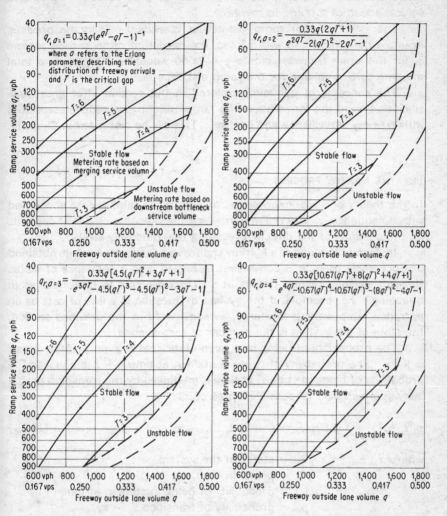

Fig. 10.8 Maximum service volume q_r, for a ramp arrival to have a probability of 0.67 of finding no ramp vehicles in the merging area.

distribution of the freeway volume a, the critical gap T, and the ramp volume q_r. Consider an entrance ramp operating with a critical gap of 4.0 sec and with the distribution of freeway traffic conforming to an Erlang distribution with $a = 2$. It is apparent that the sum of the coordinates of any point on the line $T = 4$ in the graph $a = 2$ describes the merging service volume for that ramp. For example, the point described by $q = 1,500$ and $q_r = 120$ tells us that the merging service volume is 1,620 and that under these operating characteristics, a ramp arrival has a 67 percent chance of finding the ramp empty or a 33 percent of finding a vehicle ahead of it trying to merge.

The effect of poor ramp geometrics is evident. If in the previous example the critical gap was 5.0 sec, the maximum ramp service volume q_r for the same freeway volume $q = 1{,}500$ would be 50 vph, or a total merging service volume for the ramp of 1,550 vph.

The concepts of queueing explored in this section offer great possibilities toward the formulation of a ramp metering procedure associated with freeway surveillance and control. This will be discussed in Chap. 16.

10.9 THE MARKOV PROCESS

In the theory of a Markov process, or a Markov chain, we are concerned with a sequence of experiments in which the outcome of a trial depends on the outcome of the directly preceding trial (and only on it).[4]

Instead of thinking of a Markov chain as a sequence of events obtained on successive trials, it is convenient to regard it as a sequence of states through which a system passes at successive points in time, just as in queueing. For example, if the system in question is a line of cars at an entrance ramp waiting to merge, as considered in the previous section, the state of the system is the number of cars queued.

The probability of being in a given state x_i at a given point in time t is $p_i(t)$. The probability of being in state x_j at time $t + 1$ is equal to the probability of being in state x_i at time t, multiplied by the transition probability summed over all possible states of x_i:

$$p_j(t + 1) = \sum_i^m p_i(t)p_{ij} \qquad j = 1, 2, \ldots, n \tag{10.9.1}$$

where $x_1, x_2, \ldots, x_n$ = states of the system
$t_1, t_2, \ldots, t_n$ = points in time
$\qquad p_i(t)$ = probability of being in state x_i at time t
$\qquad p_{ij}$ = probability of transition $x_i \rightarrow x_j$

To make the algebra clear, the state probability at time $t + 1$ [Eq. (10.9.1)] can be written as follows:

$$
\begin{aligned}
p_1(t + 1) &= p_1(t)p_{11} + p_2(t)p_{21} + \cdots + p_n(t)p_{n1} \\
p_2(t + 1) &= p_1(t)p_{12} + p_2(t)p_{22} + \cdots + p_n(t)p_{n2} \\
&\cdots\cdots\cdots\cdots\cdots\cdots\cdots\cdots\cdots\cdots\cdots \\
p_n(t + 1) &= p_1(t)p_{1n} + p_2(t)p_{2n} + \cdots + p_n(t)p_{nm}
\end{aligned}
\tag{10.9.2}
$$

In the case of the ramp example, Eqs. (10.9.1) and (10.9.2) would mean that the probability of one car waiting at time $t + 1$ equals the probability of there being one car at time t and no arrivals and no departures in Δt, plus two cars at time t and no arrivals and one departure in Δt, etc.

Equation (10.9.2) is a system of equations which can be represented in matrix form as

$$P(t+1) = TP(t) \tag{10.9.3}$$

where P represents the vector of state probabilities at time $t+1$ and t, and T is the matrix of transition probabilities.   It can be shown by induction that

$$P(t+1) = T^{(t+1)}P(0) \tag{10.9.4}$$

and that once the system becomes stationary,

$$\lim_{t \to \infty} P(t) = \lim_{t \to \infty} TP(t) = \lim_{t \to \infty} T^t P(0) \tag{10.9.5}$$

Consider the table for the hypothetical case of the origin and destination of lane changes for 200-ft sections of a three-lane freeway (Fig. 10.9).

	From lane		
To lane	1	2	3
1	0.1	0.3	0.2
2	0.4	0.3	0.5
3	0.5	0.4	0.3

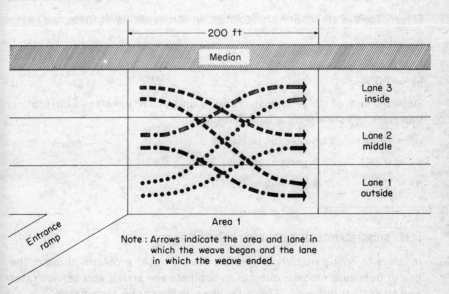

Note : Arrows indicate the area and lane in which the weave began and the lane in which the weave ended.

Fig. 10.9 Origin-termination distribution of weaving maneuvers.

The matrix of transition probabilities can be written directly from the table

$$T = \begin{pmatrix} 0.1 & 0.3 & 0.2 \\ 0.4 & 0.3 & 0.5 \\ 0.5 & 0.4 & 0.3 \end{pmatrix}$$

The probability vector at the merging area for the ramp vehicles is

$$P(0) = \begin{pmatrix} 1 \\ 0 \\ 0 \end{pmatrix}$$

since all vehicles must enter on the outside lane.   By applying Eq. (10.9.4), we can find $P(t+1)$ for consecutive 200-ft sections.   For example letting $t = 1$ gives

$$P(2) = \begin{pmatrix} 0.1 & 0.3 & 0.2 \\ 0.4 & 0.3 & 0.5 \\ 0.5 & 0.4 & 0.3 \end{pmatrix}^2 \begin{pmatrix} 1 \\ 0 \\ 0 \end{pmatrix} = \begin{pmatrix} 0.23 \\ 0.41 \\ 0.36 \end{pmatrix}$$

The interpretation is that 400 ft downstream only 23 percent of the vehicles that entered are still in the outside lane.   If the lane distribution becomes constant in the limit as t (number of 200-ft sections) becomes very large, then we say the process becomes stationary:

$$\begin{pmatrix} p_1 \\ p_2 \\ p_3 \end{pmatrix} = \begin{pmatrix} 0.1 & 0.3 & 0.2 \\ 0.4 & 0.3 & 0.5 \\ 0.5 & 0.4 & 0.3 \end{pmatrix} \begin{pmatrix} p_1 \\ p_2 \\ p_3 \end{pmatrix}$$

Solving for the stationary probabilities, in terms of one of them, we get

$$p_1 = 0.569 p_3$$

$$p_2 = 1.039 p_3$$

$$p_3 = 1.000 p_3$$

Since the sum of the probabilities equals unity, $2.608 p_3 = 1$.   Substituting this gives the steady-state lane distribution:

$p_1 = 0.218$ 21.8 percent in lane 1

$p_2 = 0.399$ 39.9 percent in lane 2

$p_3 = 0.384$ 38.4 percent in lane 3

10.10 SIGNIFICANCE OF QUEUEING SERVICE RATE

In order to obtain numerical answers to real traffic problems through the use of queueing models, one has to estimate the arrival and service rates and their distributions.   These requirements may be formidable.

First, the problem of fitting any data to mathematically tractable frequency distributions is an obstacle in itself and may preclude the use of queueing as a model.   In a comprehensive study of traffic delays at tunnel tollbooths, Edie[5] was forced to resort to empirical measures of delay which, when compared with the appropriate queueing theories, showed that the average booth servicing times at a given volume were more nearly constant than exponential in distribution.   Moreover, for flows above 600 vph, the normal distribution provided a better approximation than the Poisson distribution.

Second, in applying queueing to traffic, obtaining the actual rates may be difficult.   Although counting arrivals poses little problem, measuring or even identifying the service times can be perplexing.   This is true because the limitations of the vehicle, the characteristics of the driver, and the requirements of the traffic stream serve to distinguish vehicular traffic queueing.   Vehicular speed changes are not achieved instantaneously, and sluggishness in starting can represent a high percentage of the total service time, leading to appreciable errors.   On the other hand, drivers tend to apportion their delay over space, rather than accepting it at a definite point or service channel.   Thus, a driver approaching a red traffic signal will try to adjust his speed to avoid having to stop behind the waiting queue.

Last, vehicles must follow at some finite headway which though small (say in the neighborhood of 2 sec) is still relatively high when compared to many congestion-causing traffic maneuvers.   Some typical discrepancies between actual and apparent service times are illustrated for three basic traffic maneuvers in Fig. 10.10.

In promotion of the idea that traffic capacity may be defined in terms of service times such as those illustrated in Fig. 10.10, the notation Q for service rate was employed throughout the development of the classical queueing concepts in this chapter.   The definition of service rate Q as the number of units a (service) facility can handle in a unit of time, assuming no shortage of demand, is not alien to the commonly accepted definition of highway capacity as the maximum number of vehicles that can cross a point on a road in some unit of time.   We know that capacity is influenced by the geometrics of the highway just as surely as it is by a traffic signal.   Thus, the maximum rate at which a given traffic flow element can service traffic is a unique, stable property.   The problem of determining the traffic service rate is essentially the problem of determining capacity since the reciprocal of mean service time is the maximum flow rate.

In Sec. 10.8, the concept of variable capacity arising from a stream of ramp traffic that is merging with a priority freeway stream is discussed. The absorption rate represents the service rate and establishes ramp

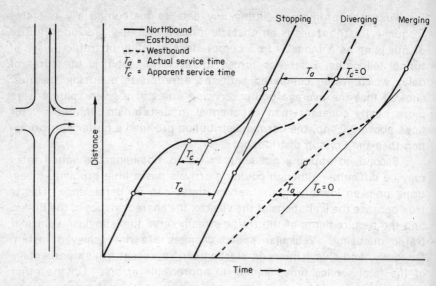

Fig. 10.10 Actual and apparent queueing service times for some basic maneuvers.

capacity.   The importance of the variance of the service-time distribution is illustrated in Eqs. (10.8.5) to (10.8.7).   Thus, it is apparent from (10.8.5) that a fundamental property of a queueing system is that the delay is reduced as the variance is reduced, even though the maximum flow rate remains constant.   Ramp metering, to be discussed in more detail in a later chapter, seems to offer one such variance reduction technique.

For barrier-type traffic elements, capacity determinations are quite straightforward.   In Chap. 7, we saw how intersection approach capacities can be determined from mean starting headways.   Blunden[6] has found that the distributions of starting headways—that is, the distribution of service times in queueing-theory terminology—may be approximated by the Erlang distribution.

10.11 NONEXPONENTIAL DISTRIBUTIONS

In Chap. 7, the concepts of randomness, less random than random, and more random than random were established and were identified as being described, in the case of discrete events, by the Poisson binomial and negative binominal distributions.   Morse[7] shows how the same variations of randomness can be simulated by a suitably chosen set of exponential facilities leading to the Poisson distribution, compound Poisson distribution and hyper-Poisson distribution.   Their counterparts for continuous phenomena such as service-time distributions are: the negative exponential distribution for "random," the Erlang distribution for less random than random, and the hyperexponential distribution for more random than random.

The schematic diagrams for simulating negative exponential, Erlang, and hyperexponential service times appear in Fig. 10.11. Assuming a continuous queue exists at the entrance to the service area, the times between departures will be at rate Q for each of the three cases. However, the distribution of the times between departures will be different: negative exponential for the single channel, hyperexponential for the parallel channels, and Erlang for the channels in series.

Consider the series arrangement in more detail: One unit at a time is allowed to enter the service area and must go through each phase which has an exponential time distribution with a mean rate of completion aQ. A classic example is a car wash with the first stage washing and the second, wiping the car dry.   It is important to remember that only one unit is allowed in the service area at a time; therefore although the area is compounded in the schematics (Fig. 10.11) it can be considered as a single area.   Because of this, even when the physical realization of the Erlang

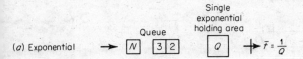

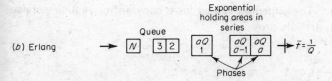

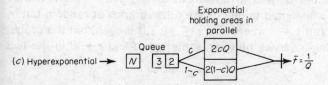

Fig. 10.11 Schematic diagrams for simulating exponentially, Erlang, and hyperexponentially distributed service times.

distribution is not present, some member of the family may give adequate fit to the observed distribution of service time.

The equations of expected values for the case of Poisson arrivals with mean arrival rate q and service times following the ath Erlang distribution with mean service rate Q are

$$E(m) = \frac{a+1}{2a} \frac{q^2}{Q(Q-q)} \tag{10.11.1}$$

$$E(n) = E(m) + \frac{q}{Q} \tag{10.11.2}$$

$$E(v) = \frac{E(n)}{q} \tag{10.11.3}$$

$$E(w) = E(v) - \frac{1}{q} \tag{10.11.4}$$

It is evident that the previously derived form in Eq. (10.8.8)

$$E(n) = \frac{q}{Q} + \frac{(q/Q)^2(1 + a^{-1})}{2(1 - q/Q)}$$

is the same as Eq. (10.11.2)

$$E(n) = \frac{a+1}{2a} \frac{q^2}{Q(Q-q)} + \frac{q}{Q}$$

If a is set equal to 1, the $M/M/1$ case is obtained. If a is set equal to infinity, the variance of the Erlang distribution is zero and the service time becomes a constant Q^{-1}. Thus, both are special cases of Erlang service-time distribution. The expected values for a constant or regular service time and Poisson arrivals are

$$E(m) = \frac{q^2}{2Q(Q-q)}$$

etc., for Eqs. (10.11.2) to (10.11.4).

Now look at the parallel arrangement in Fig. 10.11: When a unit enters service it is assigned to one or the other branch at random but in the proportion c and $1 - c$ where $0 < c \leq \frac{1}{2}$. It is seen that the probability of completing a service between some time t and $t + dt$ in the two phases is

$$P(t < t_1 < t + dt) = 2cQe^{-2cQt}\, dt$$
$$P(t < t_2 < t + dt) = 2(1 - c)Qe^{-2(1-c)Qt}\, dt$$

It follows that the probability that a service is completed between t and $t + dt$ is the probability that a given phase is used times the probability that a service is completed in that phase

$$f(t)\ dt = 2c^2 Q e^{-2cQt}\ dt + 2(1 - c)^2 Q e^{-2(1-c)Qt}\ dt \qquad (10.11.5)$$

Equation (10.11.5) establishes the probability density fraction for the distribution of service times. The mean of this distribution is $1/Q$ and the variance is $1/aQ^2$ where

$$a = \left[1 + \frac{(1 - 2c)^2}{2c(1 - c)} \right]^{-1}$$

10.12 STATUS OF QUEUEING

Most traffic situations are soluble using queueing theory only if some simplifying assumptions can be made. Typical assumptions are that the inputs are separated by constant times or are Poisson distributed and that the service times are constant or exponentially distributed. Note that the case in which both inputs and service are constant is considered a deterministic, rather than a queueing, approach.

In traffic control, even greater complexity enters the problem. Each street intersection is comparable to a servicing area such as we have considered in the previous section. Complicating features include the tendency to change lanes if the waiting line in one traffic approach lane becomes excessively large. Added to this are the driver-vehicle limitation we have already discussed.

However, the benefits that can be derived from queueing theory approaches to complex traffic problems are not limited to numerical answers. Of perhaps more importance is the recognition of the parameters and their sensitivity, leading to a more comprehensive understanding of the problem and its components.

Simulation methods, which will be discussed in the next chapter, are extremely useful in queueing problems that are difficult or impossible to solve. Advantages of this approach are (1) the ability to increase the realism by eliminating simplifying assumptions and (2) being able to take a much more microscopic look at the system. In other words, from an operational point of view, the maximum queue length on an intersection approach during the peak period is probably of more significance than the average queue length.

PROBLEMS

10.1. Arrivals at a rural entrance tollbooth to the Lower High Valley Parkway are considered Poisson with mean arrival rate of 20 vph. The time to process an arrival (pay toll or issue ticket) is approximately exponentially distributed with a mean time of 1 min.

a. What percentage of the time is the tollbooth operator free to work on operational reports?

b. How many cars are expected to be waiting to be processed on the average?

c. What is the average time a driver waits in line before paying his toll?

d. Whenever the average number of vehicles waiting to pay tolls or get tickets reaches five, a second tollbooth will be opened during the day shift. How much will the average hourly rate of arrivals have to increase to require the addition of a second operator?

10.2. A bank has two reserved on-street parking spaces. Measurements indicate an average of six random (Poisson) arrivals per hour and an average of six exponentially distributed departures per hour per space. What is the average number parked in front of the bank? When the city reserved the spaces, it was agreed that if the bank traffic kept the two spaces filled more than 50 percent of the time, on the average, a third space would be reserved. Does the bank have grounds to ask for another space?

10.3. A freeway on-ramp will be metered experimentally in the near future. Present demand on the ramp is 320 Poisson arrivals during the peak hour. The ramp metering station will be located such that six cars is the maximum number that can queue up behind the car being metered. Initially a constant metering rate of 6 vpm will be used (regular service time). Assuming that the demand does not change and that the probability of the queueing system being full is the fraction of demand turned away, how many vehicles will probably be diverted during the peak hour?

10.4. Derive the form for the expected waiting time of an arrival that waits, $E(w|w > 0)$, for multiple channels in parallel, $M/M/N$, given the following information:

$$P(0 < w < T) = \int_0^T f(w)\ dw$$

$$f(w) = \frac{(\rho N)^N}{N!} P_0 e^{-(N\mu - \lambda)w} N\mu$$

$$f(w|w > 0) = \frac{f(w)\ dw}{P(w > 0)}$$

where w = time in queue before service
N = number of channels in parallel
μ = service rate of one channel
λ = arrival rate
$\rho = \lambda/N\mu$

10.5. A parking lot has 100 parking stalls; 200 vehicles enter the lot per day according to a Poisson distribution, and the length of time a stall is occupied is negative exponentially distributed with a mean of 0.4 day. Determine the following:

a. Probability of the lot being empty

b. Average number of cars parked

10.6. A university data processing center designated two on-street curb-parking stalls for student use only for convenience in picking up and depositing computer programs in inclement weather. Students arrive randomly (Poisson) with average time between arrivals of 15 min. Time in the DPC is exponentially distributed with a mean of 30 min. If on a rainy day all students drive, what is the probability a student can park? What is the expected number of cars parked?

10.7. Flow on a roadway is found to be described by Erlang $a = 2$ distribution when the density is in the vicinity of 25 vpm and $a = 3$ when the density reaches 35 vpm. The respective queueing headways are taken to be

$$S = 140 \text{ ft} \big|_{a=2} \qquad S = 120 \text{ ft} \big|_{a=3}$$

Compare the expected number of moving queues at the two levels of flow. Which is the more congested level?

10.8. A random sample of merging delay to ramp drivers in position to merge at ramp x is as follows in seconds of delay: 0, 3, 2, 0, 5, 1, 4, 3, 3, 0, 4, 1, 4, 3, 0, 7, 0, 3, 1, 2. If the ramp has Poisson arrivals which arrive at average rate of 360 vph and $a = 2$, estimate the following:

a. Average queue length on the ramp
b. Average waiting time before merging
c. Mean wait in ramp system

10.9. On-ramp 1 is 2,000 ft upstream of on-ramp 2. If the matrix below is the transition matrix of the probability of lane change, what is the probable lane distribution of traffic entering the freeway at on-ramp 1 when the traffic reaches on-ramp 2?

Probability of weaving in a 500-ft section

To lane	From lane		
	1	2	3
1	0.2	0.3	0.2
2	0.5	0.4	0.4
3	0.3	0.3	0.4

10.10. Discuss the possible applicability (and limitations) of queueing theory to the operational analysis of a four-way stop-controlled intersection.

10.11. The main entrance to a state park is regulated by two gatekeepers. The park visitors arrive in a Poisson manner at the rate of one car per 2 min, and the service time is exponentially distributed with a mean of 2 min.

a. What is the probability of an arrival finding a gate empty?
b. How long is the average waiting line?
c. What percentage of the time is each gatekeeper expected to be free to observe the birds and the bees?
d. What is the average length of waiting line if the arrival rate goes up to 60 vph on the Independence Day weekend?

REFERENCES

1. Drew, D. R., and C. Pinnell: Some Theoretical Considerations of Peak-hour Control for Arterial Street Systems, in "Traffic Control Theory and Instrumentation," Plenum Press, New York, 1965.
2. Kendall, D. G.: Some Problems in the Theory of Queues, *J. Roy. Statist. Soc.*, ser. B, vol. 13, no. 151, 1951.
3. Major, N. G., and D. S. Buckley: Entry to a Traffic Stream, *Australian Road Res. Board Proc.*, vol. 1, pt. 1, 1962.
4. Clough, D. J.: "Concepts in Management Science," Prentice-Hall, Inc., Englewood Cliffs, N.J., 1964.
5. Edie, L. C.: Traffic Delays at Toll Booths, *J. Operations Research Soc. Am.*, vol. 2, pp. 107–138, 1954.
6. Blunden, W. R.: "Capacity of Highways: Some Fundamental Considerations," presented at the 35th annual meeting of the Institute of Traffic Engineers, Boston, Oct. 21, 1965.
7. Morse, P. M.: "Queues, Inventories and Maintenance," John Wiley & Sons, Inc., New York., 1963.

CHAPTER ELEVEN
SIMULATION

*You have only to take in what you please and leave out what you
please; to select your own conditions of time and place; to
multiply and divide at discretion; and you can pay the National
Debt in half an hour.*

Lord Brougham

11.1 INTRODUCTION

A penny is tossed until it comes down heads.   If this happens on the first
toss, the player receives $1 from the bank.   If heads appears for the first
time on the second toss, the player receives $2, on the third toss, $4, etc.,
doubling each time.   What should the player pay the bank for the privilege
of playing this game if the game is fair?

If one has a coin handy, the simplest way to get some insight into
how much a player should pay is to play say 1,000 games and determine
the average winnings per game.   This seemingly unscientific approach to
the "gambler's ruin" or "St. Petersburg paradox," as it is also called,
represents a simple illustration of simulation.

Simulation is essentially a working analogy.   It involves the construction of a working model presenting similarity of properties or relationships to the real problem under study.   Simulation is a technique which
permits the study of a complex traffic system in the laboratory rather
than in the field.

In a more general sense, simulation may be defined as a dynamic representation of some part of the real world, achieved by building a computer model and moving it through time.[1] The term *computer model* is used to denote a special kind of formal mathematical model, namely, a model which is not intended to be solved analytically but rather to be simulated on an electronic computer. Thus, simulation consists in using a digital or analog computer to trace the time paths, with the distinction that the digital device counts and the analog device measures. This distinction is actually a fundamental one, being essentially the mathematical distinction between the discrete variable (digital) and the continuous variable (analog). The differences in capabilities between the digital and analog computers are manifest in the mathematical distinctions between summation and integration, or between difference equations and differential equations.

11.2 MONTE CARLO METHODS

Stochastic:
Involving probability or chance

Often an equation arises in the formulation of a model which cannot be solved by standard numerical techniques. It may then be more efficient to construct an analogous stochastic model of the problem. Thus, essentially, an experiment is set up to duplicate the features of the problem under study. The calculation process is entirely numerical and is carried out by supplying *random numbers* into the system and obtaining numerical answers.

One of the simplest and most powerful applications of this idea is the evaluation of a multidimensional integral. Consider the simplest case of evaluating the area of a bounded area. Surround the area with a square, normalizing to make its side of unit length. Taking a point in the area at random, the probability that it lies in area A is simply A. If a large number of points are taken at random, it follows that the proportion of these lying in the area is an estimate of A (see Fig. 11.1). The general equation for evaluating a definite integral, using the *point-distribution method*, as it is called, is

$$\int_a^b f(x)\ dx = (b - a)y_{\max}P[Y < f(X)] \tag{11.2.1}$$

where

$$X = (b - a)RN + a \tag{11.2.2}$$

and

$$Y = y_{\max}RN \tag{11.2.3}$$

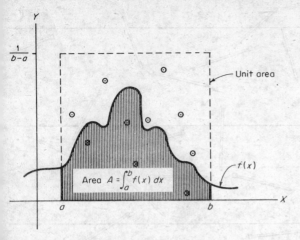

Note : $A = \dfrac{4}{4+6} = 0.4$ based on 10 pairs of random numbers

Fig. 11.1   Monte Carlo evaluation of an integral.

Thus, the point-distribution method consists of choosing two random numbers, the first between a and b and the second between 0 and some number which is greater than or equal to the maximum of $f(x)$ in (a,b), or y_{max}.   If the first is regarded as the abscissa and the second as the ordinate, the probability that the point will fall below the curve is equal to the ratio of the area below the curve to the area of the indicated rectangle (see Fig. 11.1).   This provides the basis for obtaining the value of the definite integral using Eq. (11.2.1), which is of course the area under the curve.   This method yields an accuracy proportional to $n^{1/d}$, where n is the number of points and d is the number of dimensions.   The whole idea can be generalized to higher dimensions and the result remains true.

A scheme based on the mean-value theorem appears to be more efficient.   The theorem states that if $f(x)$ is continuous in (a,b), there exists a point T in (a,b) such that

$$\int_a^b f(x)\,dx = (b-a)f(T) \tag{11.2.4}$$

This is accomplished by generating a sequence of random numbers X between a and b and calculating the average of $f(X)$:

$$f(T) = n^{-1} \sum_{i=1}^{n} f(X_i) \tag{11.2.5}$$

where X is given by Eq. (11.2.2) as before.

Throwing nails on a floor made of narrow boards (Fig. 11.2) is another

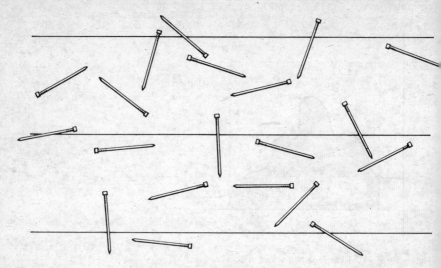

Fig. 11.2 Variation of the Buffon needle problem.

illustration of the Monte Carlo method.[1a]   The fraction P of nails intersecting the cracks in the floor provide an estimate of pi, π:

$$\hat{\pi} = \frac{2L}{PW} \tag{11.2.6}$$

where L is the length of nail and W is the width of board.   Of course by
repeated throws of the nails it is possible to predict just how often a nail
will touch on a crack.   Atomic scientists adapted the principle to figure
out the chances that a neutron produced by the fission of an atomic
nucleus would be stopped or deflected by another nucleus in the shielding
around it.   This was instrumental in developing the proper shielding for
atomic reactors.

11.3 RANDOM WALKS

A second origin of simulation lies in the demands of applied mathematicians
for methods of solving problems involving partial differential equations.[2]
A typical problem was the solution of the diffusion equation arising in the
diffusion of gases as well as in the conduction of heat in a medium.   The
characteristic of many of these physical systems was that the actual
mechanism for the movement of the gas or heat involved a large number
of particles behaving in a partly regular and partly irregular manner.
Averaging over the particles enabled the random element to be eliminated
and a deterministic description to be given.

Fig. 11.3 Taking a random walk.

As it turned out, mathematical techniques for dealing with a system of particles moving in this partly regular, partly random manner had been studied under the name of random walks. For example, a blindfolded girl who walks away from a lamppost, changing direction now and then according to whim, moves in a wholly irregular fashion, illustrating the *random-walk concept* (see Fig. 11.3). If the girl moves in a northerly, southerly, easterly, or westerly direction and each step is assumed to be one unit, the estimate of the girl's distance from the post after any step is given by $\sqrt{n}$, where n is the number of steps. Moreover, the mathematical *law of disorder* predicts that as long as she keeps walking she will keep returning to the lamppost. This simple example illustrates the random-walk principle of modern physics—once brilliantly employed to describe the movement of tiny particles suspended in a liquid by a 26-year-old physicist named Albert Einstein.

A somewhat more sophisticated example illustrates the application of the random-walk technique to the solution of a partial differential equation of the form

$$\frac{\partial^2 u}{\partial x^2} + \frac{\partial^2 u}{\partial y^2} = 0 \tag{11.3.1}$$

Consider, for example, a thin rectangular plate with insulated faces

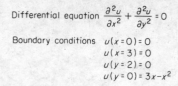

Differential equation $\dfrac{\partial^2 u}{\partial x^2} + \dfrac{\partial^2 u}{\partial y^2} = 0$

Boundary conditions $u(x=0)=0$
$u(x=3)=0$
$u(y=2)=0$
$u(y=0)=3x-x^2$

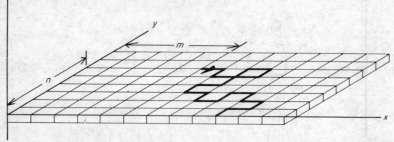

Fig. 11.4 Random-walk technique for finding the steady-state temperature at a point on a plate.

bounded by $x = 0$, $y = 0$, $x = 3$, and $y = 2$ (see Fig. 11.4). The edge $y = 0$ is held at temperature $u(x) = 3x - x^2$ and the other three edges, at $0°$. The steady-state temperature at any point (x,y) is desired using Monte Carlo techniques.

One approach is to cover the disk with a grid of spacing h and assign temperatures at the boundary according to the given conditions. In order to find the temperature at any point (m,n) start at that point and "take a random walk" until a boundary point is reached; note the temperature at the boundary point, return to the starting point, and repeat the process. The average of the temperatures at the boundary points reached in this manner provides an estimate of the desired temperature $u(m,n)$ at the starting point.

The theory behind this procedure involves changing the partial differential equation (11.3.1) to a difference equation

$$\frac{u(m+1,\,n) - 2u(m,n) + u(m-1,\,n)}{\Delta x^2}$$

$$+ \frac{u(m,\,n+1) - 2u(m,n) + u(m,\,n-1)}{\Delta y^2} = 0 \quad (11.3.2)$$

Since the grid interval is $\Delta x = \Delta y = h$, one obtains

$$u(m,n) = \tfrac{1}{4}[u(m+1,\,n) + u(m-1,\,n) + u(m,\,n+1)$$

$$+\, u(m,\,n-1)] \quad (11.3.3)$$

Equation (11.3.3) gives the desired temperature in terms of the temperatures of the four points surrounding it.   Stated another way, the temperature at (m,n) is equal to the probability of the first step being to $(m + 1, n)$ times the temperature $u(m + 1, n)$ plus the probability of the first step being to $(m - 1, n)$ times $u(m - 1, n)$, etc.   Continuing in this form, it can be shown that the expected temperature $u(m,n)$ is the sum of the products of the temperature at a boundary point (point of known temperature) times the probability that a random walk from (m,n) will lead to that boundary point.   Thus, in practice, one has only to program 1,000 or so random walks and calculate the average temperature at the boundary points reached to have a good estimate of the desired temperature.

Moore[3] shows how this procedure can be generalized to solve a much broader class of partial differential equations than the Laplace equation of (11.3.1).   Consider the partial differential equation

$$\frac{\partial^2 u}{\partial x^2} + \frac{\partial^2 u}{\partial y^2} + \frac{1}{x}\frac{\partial u}{\partial x} = 0 \tag{11.3.4}$$

existing on a square disk with known boundary conditions.   The associated difference equation is

$$u(m,n) = \frac{1}{4 - \dfrac{1}{m}}\left[\left(1 - \frac{1}{m}\right)u(m + 1, n) + u(m - 1, n)\right.$$

$$\left. + u(m, n + 1) + u(m, n - 1)\right] \tag{11.3.5}$$

This time the temperature at any point may be obtained by taking a *weighted* random walk instead of a random walk, the latter implying a weight of $\frac{1}{4}$ in each of the four grid directions.   It is apparent from (11.3.5) that in this case the weight depends on the x coordinate, m, which may be found from the identity $x = mh$, where h is the grid interval.   The sum of the coefficients for the four directions must of course always add up to unity.   Although for the purposes of illustration u has been associated with temperature, the technique is applicable to partial differential equations describing any phenomenon.

11.4   SIMULATED SAMPLING (SIMULATION)

There is no universally accepted terminology distinguishing Monte Carlo, random walks, and simulation.   However, the first two concepts are probabilistic in nature, whereas simulation may be either probabilistic (digital) or deterministic (analog).   In the study of vehicular traffic, use of the word *simulation* without any qualification will usually imply the use of a digital computer. A Monte Carlo method is often reserved for a procedure in which the process sampled has been modified to increase precision, whereas the term *simulation* is used when the process sampled is a close model of the real system. If the latter subtlety is subscribed to, one must concede that one advantage of simulation over a Monte Carlo method is that the detailed results give a direct qualitative impression of what the behavior of the system should look like under the conditions postulated.

Although Monte Carlo and the random-walks methods have contributed to the evolution of the art of simulation on a digital computer, the most significant origin lies in the theory of mathematical statistics. In its infancy, the subject of statistics consisted of the collection and display in numerical and graphical form of facts and figures from the fields of economics and science. One of the most useful forms of display was the histogram, or frequency chart, and the transformation of statistics began when it was realized that the occurrence of such diagrams could be explained by invoking the theory of probability.

Since a probability distribution is by its nature, in most cases, composed of an *infinite* number of items, whereas frequency charts by their nature are composed of a *finite* number of items, the latter had to be thought of as samples from an underlying theoretical probability distribution. Tocher[2] explains how the problem then arose of how to describe a probability distribution, given only a sample from it. Because of the seemingly immense mathematical difficulties associated with this, such steps as were taken required experimental verification to give early workers confidence. Thus was born the sampling experiment. A close approximation to a probability distribution was created; samples were taken, combined, and transformed in suitable ways; and the resulting frequency chart of sampled values compared with the predictions of theory.

It is not hard to visualize situations arising where some method of sampling is indicated but where the actual taking of a physical sample is either impossible or too expensive. In such situations, useful information can often be obtained from some type of simulated sampling. Typically, simulated sampling involves replacing the actual universe of items by its theoretical counterpart, a universe described by some assumed probability distribution, and then sampling from this theoretical population by means of random numbers. The advent of automatic digital computers to perform the tedious calculations associated with these sampling experiments has revitalized this as a possible approach to the solution of problems still beyond the reach of analysis. The method of taking such a sample is called *simulation;* the decision problems which rely heavily on such sampling methods are often referred to by the catchall label of *Monte Carlo methods.*

11.5 RANDOM NUMBERS

We have seen that random numbers are required in the sampling experiments associated with simulation. Humans are too full of associations to think up truly random numbers—no one would pick three 4s in a row, although such a sequence might be part of a random series.

The idea of using tables of random numbers was introduced by Tippett,[4] who constructed a table of 10,400 random digits by taking the terminal digits of entries in a census table. The RAND Corporation used an electronic roulette wheel to prepare the million-digit book of random-number tables (hence the name *Monte Carlo*). Actually a wide variety of natural phenomena have been used to produce randomness, although some controversy exists about the validity of such procedures. Even a few philosophers have argued whether any digits committed to tabular form can still be regarded as random, regardless of how they were obtained. For practical purposes, these arguments are irrelevant; one is forced to accept any phenomenon as random whose behavior is not predictable by any obvious deterministic laws and whose numbers satisfy several standard tests of randomness to ensure, for example, that each decimal digit occurs with equal frequency without any serial correlation.

From the standpoint of checking out computer programs, there are advantages in having a reproducible sequence of numbers instead of purely random numbers. Programs for digital computers can be written which will output a sequence of numbers which satisfy the various statistical tests of randomness that have been devised. Random numbers such as these, which are generated in a nonrandom fashion, are called *pseudorandom numbers*. In automatic computation, the storage of the large volume of random number used becomes a serious problem. It is this storage problem that has led to the abandonment of the use of tables of random numbers in computers. A satisfactory computer program for generating pseudorandom numbers should (1) require little storage space in the computer, (2) be relatively fast in operation, and (3) generate a sequence of numbers that satisfies the tests of randomness. The following section describes how such programs are written.

11.6 GENERATION OF PSEUDORANDOM NUMBERS

Pseudorandom-number generators may be divided into two groups: those in which the resulting sequences can be predicted theoretically and those in which theoretical prediction is impossible.

An example of the latter group is generated by the midsquare technique.[2] An n-digit number R_0 is squared, and from the resulting $2n$

digits the n middle digits are taken as the next random number R_1. The number R_1 is then squared and the process repeated. This technique, as are all practical techniques, is cyclic—that is, the sequence repeats itself. However, one disadvantage of this particular method lies in the possibility of obtaining a zero term during the cycle.

In order to appreciate the more sophisticated approach to random-number generation involving techniques in which the resulting sequences can be predicted theoretically, one should review a few fundamentals of number systems, computer operation, and number theory.

11.7 NUMBER SYSTEMS

In several aspects of large-scale systems—notably in digital computers—numbers are represented in the binary system rather than in the familar decimal system. The base of a number system (for example, 10, in the familiar system) is called the *radix* and may be represented symbolically by b. There must be exactly b different symbols, each of which is associated with one of the values from 0 to $b - 1$. A number is represented by writing a row of n digits, each being one of the b symbols. The meaning of a row of digits, ordered from right to left, is that it has a value R given by the equation

$$R = \sum_{i=-j}^{n-1} a_i b^i \tag{11.7.1}$$

in which a_i, must also be one of the b symbols.[5] If the lower limit starts at 0, the value is a whole number; if the lower limit is negative, R is an improper fraction with the position of a_0 indicated by a dot to its right called the *radix point* (decimal point in the decimal system). For example, the number 654.73 in the decimal system means

$$R = \sum_{-2}^{2} a_i (10)^i$$

where $a_{-2} = 3$, $a_{-1} = 7$, $a_0 = 4$, $a_1 = 5$, and $a_2 = 6$, or

$$R = 6 \times 10^2 + 5 \times 10^1 + 4 \times 10^0 + 7 \times 10^{-1} + 3 \times 10^{-2}$$

Using another example, the binary number 101101 as interpreted by Eq. (11.7.1) means

$$R = 1 \times 2^{101} + 0 \times 2^{100} + 1 \times 2^{011} + 1 \times 2^{010} + 0 \times 2^{001} + 1 \times 2^{000}$$

The decimal equivalent is

$$R = 1 \times 2^5 + 0 \times 2^4 + 1 \times 2^3 + 1 \times 2^2 + 0 \times 2^1 + 1 \times 2^0 = 45$$

The basis of arithmetic in any number system may be stated in terms of addition and multiplication tables. For the binary system the addition table becomes $0 + 0 = 0$, $0 + 1 = 1$, $1 + 1 = 10$, etc. For multiplication one obtains $0 \times 0 = 0$, $0 \times 1 = 0$, $1 \times 1 = 1$, $10 \times 10 = 100$, $11 \times 11 = 1001$, etc.

The property of division, as related to number systems, is particularly important in the generation of random numbers, as shall be seen when number theory is considered. For the present it suffices to show that in the decimal system if the number 65,473 is divided by 10^3, the remainder is 473. Similarly in the binary system' the number 101101 divided by 2^3 yields a remainder of 101. In both cases, the remainder is simply the three least significant digits.

Now the word length in a digital computer is the maximum number of digits n that can be stored in a normal storage location. For example, $n = 10$ decimal digits for the IBM 650; $n = 35$ binary digits for the IBM 709, 7090, and 7094; $n = 32$ for the IBM 360 and $n = 48$ for the CDC 6600. The product of two numbers is normally formed in a location having $2n$ digits (called the *upper* and *lower accumulators* in the 650, and the *accumulator* and the *MQ units* in the IBM 7094). Then the least significant n digits (the number in the lower accumulator in the 650, or the MQ unit in the 7094) of a product is exactly equal to the remainder after division of the entire product by b , where b is the number base of the computer. Thus, for the IBM 7094, $b^n = 2^{35}$.

11.8 POWER RESIDUE METHOD

Two numbers A and B which, when each is divided by a given number C (called the *modulus*), give the same remainders are called *congruent numbers*. The congruence between A and B with modulus C is written:

$$A = B \bmod C \tag{11.8.1}$$

and is read "A is congruent to B, modulus C." From (11.8.1) and what has been said it is evident that the difference in the two numbers must be divisible by the modulus

$$\frac{|A - B|}{C} = D \tag{11.8.2}$$

where D must be an integer.

This suggests a method for the generation of pseudorandom numbers by computer as follows: An assumed starting number R_0 is multiplied by an appropriate multiplier k. The less significant half of the product is taken as the random number R_1. The second random number R_2 is formed by using the first as a starting number and the same multiplier. This method, expressed in symbolic form, is

$$R_m = kR_{m-1} \bmod b^n \tag{11.8.3}$$

where R_m is the mth random number, b is the number base of the computer, and n is the word length in a normal storage location. The effect of mod b^n is to use only the low-order half of the full $2n$-digit product. For example, if $k = 5$, $b = 2$, $n = 4$ and 5 is used as a starting number, then

$$
\begin{aligned}
5 \cdot 5 \quad \bmod 2^4 &= 9 \\
5 \cdot 9 \quad \bmod 2^4 &= 13 \\
5 \cdot 13 \bmod 2^4 &= 1 \\
5 \cdot 1 \quad \bmod 2^4 &= 5 \\
5 \cdot 5 \quad \bmod 2^4 &= 9
\end{aligned} \tag{11.8.4}
$$

. . . .

It can be shown that the length of cycle is given by b^{n-2} or, for the above example, 4. Thus on the IBM 7094, one can generate 2^{33} distinct random numbers called *power residues*, before cycling. In practice, the starting number is chosen as 5^{15} and the multiplier as 5^{15} since 15 is the smallest power of 5 that fills out all 35 bits in storage. Using this *power-residue method*, as it is called, 8,589,934,592 random numbers will be generated before the sequence will start to repeat.

11.9 RANDOM SEQUENCES SATISFYING DESIRED DISTRIBUTIONS

The numbers resulting from the power-residue method form a uniformly distributed pseudorandom sequence of numbers, that is, one in which the probability of a number falling in a given interval is proportional to the width of the interval and does not depend on the location of the interval. Random numbers so generated may be interpreted as random integers or as random fractions. The latter is more appropriate since, as shall be seen in the following sections, the basic problem in simulation consists in sampling from statistical distributions for which any associated probability must, by definition, be a fraction of unity.

There are two principal methods employed for the conversion of random fractions to *random deviates* satisfying the desired frequency distribution: the inversion method and the point-distribution method. The theory behind these two methods and applications of each to common distributions used in traffic simulation will be discussed in some detail.

11.10 THE METHOD OF INVERSION

The distribution of a variable can be described by a density function $f(t)$ in the continuous case or by a set of frequencies in the discrete case, but both cases can be described by means of the cumulative distribution function $P(t < T)$ which specifies the probability of obtaining the given value or less from a distribution. Therefore, one possibility consists of the analytic inversion of this cumulative distribution function and the calculation of the value of this function for the value of a selected uniform random fraction.

In symbolical terms, the above means

$$P(t < T) = \int_0^T f(t) \, dt \qquad (11.10.1)$$

Equation (11.10.1) must be solved for the random deviate T by inversion. Since the left side of the equation is equivalent to a uniformly distributed random fraction between 0 and 1 (see Fig. 11.5), one obtains

$$R = \int_0^T f(t) \, dt \qquad (11.10.2)$$

where R is a pseudorandom fraction generated as explained in a preceding section. It should be evident that the success of this method depends on (1) being able to integrate the density function $f(t)$ and (2) being able

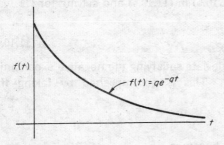

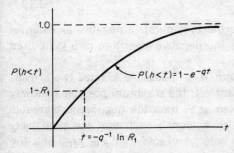

Fig. 11.5 Method of inversion illustrated for the negative exponential distribution.

to take the inverse of (11.10.2) after integration.   There are some simple, but quite important, cases in which this is possible.

Many phenomena characterized by random arrivals, as in traffic situations, may be described by use of the negative exponential distribution, which has a probability density function of the form

$$f(t) = qe^{-qt} \qquad (11.10.3)$$

where t is the time between arrivals or headway, usually expressed in seconds, and q is the rate of arrivals or flow, usually expressed in vehicles per second.   Substitution in (11.8.3) and integration yields

$$P(t < T) = 1 - e^{-qT}$$

or in more general terms

$$P(h < t) = 1 - e^{-qt} \qquad (11.10.4)$$

where h is the headway between successive arrivals and t is time.

Equations 11.10.3 and 11.10.4 are shown in Fig. 11.5.   If R_1 is a uniformly distributed random fraction, $0 \leq R_1 \leq 1$, generated on the computer as explained, then $1 - R_1$ must be a uniformly distributed random fraction, $0 \leq (1 - R_1) \leq 1$, and

$$1 - R_1 = P(h < t) \qquad (11.10.5)$$

since $P(h < t)$ is also a uniformly distributed random variable, $0 \leq P(h < t) \leq 1$.   Substituting (11.10.5) in (11.10.4) and solving for (11.10.4) gives

$$R_1 = e^{-qt} \qquad (11.10.6)$$

Solving (11.10.6) for t, a random deviate satisfying the negative exponential distribution, is called *inversion*. This is accomplished by taking the logarithm of each side

$$\ln R_1 = -qt$$

and the desired random deviate becomes

$$t = -q^{-1} \ln R_1 \qquad (11.10.7)$$

Equation (11.10.7) tells us that in order to generate a negative exponential random deviate, one simply takes the negative logarithm of a generated pseudorandom fraction.

Because vehicles possess length, it is impossible to have two arrivals in the same lane at the same instant.   If the minimum possible headway between successive arrivals is taken as τ, then the negative exponential distribution should be shifted by an amount τ such that the probability of a headway between successive vehicles of less than τ is zero and the cumulative distribution function becomes

$$P(h < \tau) = 0$$
$$P(\tau < h < t) = 1 - e^{-(t-\tau)/(\bar{l}-\tau)} \qquad (11.10.8)$$

Inversion of (11.10.8) gives

$$t = \tau - (\bar{l} - \tau) \ln R_1$$

where $\bar{l}$ is the average headway.

The probability density functions of many useful distributions such as the normal distribution, gamma distribution, and beta distribution are difficult to integrate. Moreover, there are cases in which it is possible to integrate the probability density function and represent the cumulative distribution function analytically, but the inversion is either impossible or impractical analytically. In such cases, the point-distribution method may be used.

11.11 POINT-DISTRIBUTION METHOD

Consider a bounded probability function (if it is not bounded, a point t_{max} may be chosen in the abscissa which is sufficiently large so that the probability of t falling to the right of t_{max} is negligible) such that

$$f(t) = \begin{cases} 0 & t < t_{min} \\ f(t) & t_{min} \le t \le t_{max} \\ 0 & t > t_{max} \end{cases} \qquad (11.11.1)$$

The steps in this method consist of (1) generating two random numbers, T_1 and T_2, meeting the conditions

$$T_1 = (t_{max} - t_{min})R_1 + t_{min} \qquad (11.11.2)$$
$$T_2 = f(t)_{max}R_2$$

and (2) checking to see if T_1 is the desired random deviate by the test

$$f(T_1) \ge T_2 \qquad (11.11.3)$$

If (11.11.3) is satisfied, then T_1 is accepted as conforming to the desired distribution; if (11.11.3) is not satisfied, T_1 is rejected and the two steps are repeated.

This method will be illustrated using the Erlang distribution, which has considerable conceptual appeal for describing the distribution of intervals between arrivals both in queueing situations in general and for

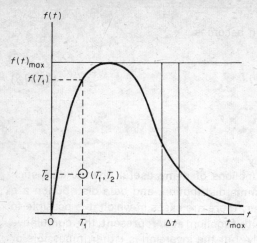

where $T_1 = t_{max} R_1$
$T_2 = f(t)_{max} R_2$

Fig. 11.6 Point-distribution method illustrated using the Erlang distribution.

the traffic phenomenon in particular.   Since the form of density function is

$$f(t) = \frac{(aq)^a}{(a-1)!} \, t^{a-1} e^{-aqt} \tag{11.11.4}$$

where a is a positive integer, the analytic form of the cumulative distribution can only be obtained by successive integration by parts, making use of the inversion technique impractical.   Equation (11.11.4) has been plotted in Fig. 11.6 for the case $a = 2$.   Two random fractions R_1 and R_2 are generated and T_1 and T_2 calculated using (11.11.2) in which $t_{min} = 0$ and $f(t)_{max}$ is the ordinate of the mode.   Then T_1 is substituted in (11.11.4), and the value of $f(T_1)$ obtained is compared to T_2 according to (11.11.3). It should be apparent from Fig. 11.6 that the number of T_1's accepted in an interval Δt in the abscissa is proportional to the area under the curve in that interval, which is precisely the desired distribution.

11.12   DISCRETE RANDOM DEVIATES

There are two main types of distributions from which simulated sampling is required.   We have already considered those in which the statistical variable can take any value giving rise to a continuous frequency function. The second type, those in which the statistical variable is limited to discrete numbers, will now be considered.

It has been shown in an earlier chapter that any convergent series of positive constants can be used as a basis for forming a discrete probability distribution.   This property of discrete distributions enables one to obtain the probability of $x + 1$ events in terms of the probability of x events. In analytical terms

$$P(x + 1) = k(x)P(x) \tag{11.12.1}$$

where $k(x)$ is some function of x and $P(x)$ and $P(x + 1)$ are the probabilities of x and $x + 1$ events, respectively.

The cumulative distribution function for a discrete distribution is given by

$$P(\leq x) = \sum_{i=0}^{x} P(i) \tag{11.12.2}$$

in which $P(i)$ is the probability of x or fewer events during some arbitrary interval.

Consider the computer generation of vehicular arrivals satisfying some known discrete distribution.   A random fraction R is generated and compared to $P(\leq x)$ for $x = 0$.   If $R \leq P(\leq 0)$ then there were no

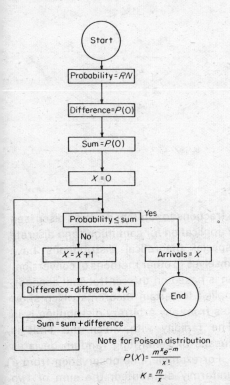

Fig. 11.7 Generation of discrete random deviates.

arrivals.   If $R > P(\leq 0)$, one knows only that there were more than zero arrivals.   Now x in (11.12.2) is increased by 1 and the test

$$R \leq P(\leq x) \qquad x = 1 \qquad (11.12.3)$$

is repeated.   If (11.12.3) is satisfied, it is apparent that since there was at most 1 arrival, and there was not zero arrivals, there must have been exactly 1 arrival.   And so the process of increasing x by 1 is repeated according to (11.12.2) until (11.12.3) is satisfied; then this value of x is chosen as the number of arrivals corresponding to the desired distribution.   The flow diagram of a computer program for accomplishing this generation is given in Fig. 11.7.

The generation of random deviates satisfying the Poisson distribution is accomplished by substituting the following forms in Eq. (11.12.1):

$$k(x) = \frac{m}{x + 1}$$

and

$$P(x) = \frac{m^x e^{-m}}{x!}$$

For the binomial distribution

$$k(x) = \frac{n - x}{x + 1}\frac{p}{1 - p}$$

and

$$P(x) = \binom{n}{x} p^x q^{n-x}$$

11.13   SPECIAL CONVERSION METHODS

The methods of converting random fractions to random deviates discussed in Secs. 11.10 to 11.12 are of general application for continuous and discrete distributions.   However, a few special theoretical distributions have certain characteristics that lend themselves to other methods of conversion.

One such method of conversion is based on the central-limit concept. The term *central-limit theorem* is applied to certain theorems which show that the sum of independent variates from any arbitrary distribution is in the limit normally distributed.   The rapidity with which this limit is approached makes it a practical means of generating random deviates satisfying the normal distribution.   For example, one observation from a uniform distribution is naturally uniformly distributed; the sum of two

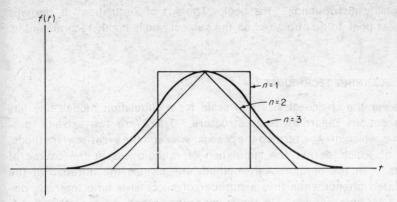

Fig. 11.8 Distribution of sums of n observations from a uniform distribution.

observations from a uniform distribution is distributed according to a triangular distribution; and the sum of three observations from the same uniform distribution is parabolic and very close to the desired normal (see Fig. 11.8). Thus if n random fractions R_i are summed the resulting variate t is normally distributed

$$t = \frac{\mu - \sum_{i=1}^{n} R_i}{\sigma} \tag{11.13.1}$$

with mean $\mu = n/2$ and standard deviation $\sigma = \sqrt{n/12}$. Half the range between $t_{\min}$ and $t_{\max}$ is obtained by substituting the forms for μ and σ in (11.13.1) and choosing $\sum_{i=1}^{n} R_i = 0$ or n

$$t = \sqrt{3n} \tag{11.13.2}$$

In practice, the number of random fractions n chosen will vary from 6 to 12 depending on the range of the phenomenon simulated and the precision desired.

A second special method of converting random fractions to random deviates is based on the theory of convolutions. It has been shown that the Erlang distribution, as expressed in (11.11.4), represents the a-fold convolution of the negative exponential distribution. This relationship suggests that a random deviate t belonging to the Erlang distribution can be generated by a generalization of (11.10.7),

$$t = -q^{-1} \ln (R_1 R_2 \cdots R_a) \tag{11.13.3}$$

Tocher[2] discusses several special sampling methods for many common probability distributions. His book "The Art of Simulation" is perhaps the best book to be published on the subject and is highly recommended.

11.14 SCANNING TECHNIQUES

Whatever the choice of size and scale for a simulation model, one has some options regarding basic structure. There are two extreme possibilities which may be termed the *periodic-scan* and the *event-scan* methods, both of which permit the simulation of randomness in the course of events.[6] In the periodic-scan method, one divides the duration of the simulated phenomenon into a number of successive time intervals displaced by time periods. In the event-scan method, on the other hand, after a given event has occurred, one determines and "stores" a set of "imminent" significant events and times at which they will occur, and selects the earliest. The occurrence of this next significant event may alter the possibility or timing of other events that had been listed, so that a new set of events and times may then be calculated. Thus an event-scan program is essentially asking, "What happens next?" whereas the periodic-scan program asks, "What will the situation be one time unit from now?"

The event-scan technique is much faster and can result in an increase of computing speed by a factor of about 10 but usually requires greater program complexity. The periodic-scan technique is very straightforward and is usually much easier to program. In practice, the periodic-scan and event-scan methods may be implemented in many ways and partially combined, in order to produce a program that is artfully suited to the problem.

An important aspect of the periodic scan concerns the generation of Poisson arrivals. A technique often used consists in comparing a random fraction R to the flow per second of the traffic stream for the scanning period Δt, written

$$R \leq q\,\Delta t \tag{11.14.1}$$

Thus for a traffic lane volume of 900 vph and a scanning period of 1 sec, a random fraction less than 0.25 will generate an arrival. The proof that arrivals generated in this fashion are Poisson follows from the derivation of the Poisson distribution, itself.

In considering the generation of vehicles in a traffic lane at some point on the road, one can immediately exclude the possibility of two or more vehicles arriving at the point at the same time. It is also reasonable to suppose that the probability of arrival of a vehicle is proportional to the length of interval, and hence the probability of generating an arrival is proportional to the scanning period Δt. Lastly, the number of arrivals generated in an interval of time t does not depend on the number of arrivals generated prior to t. When these three conditions are satisfied, we can prove that we have generated a Poisson process.

Mathematically, the conditions for randomness may be stated as follows:

$$P_0 \, \Delta t = 1 - q \, \Delta t \qquad \text{(Probability of 0 arrivals in } \Delta t)$$
$$1 - P_\iota \, \Delta t = q \, \Delta t \qquad \text{(Probability of 1 arrival in } \Delta t)$$
$$P_n \, \Delta t = (q \, \Delta t)^n = 0 \qquad n > 1 \qquad \text{(Probability of } > 1 \text{ arrival in } \Delta t)$$

where q is the rate of flow. Let $P_n(t)$ be the chance of generating n arrivals in a time interval t according to (11.14.1). In an interval of length $t + \Delta t$, n arrivals may be generated in two ways, either n arrivals in the interval t and none in Δt, or $n - 1$ arrivals in t and 1 arrival in Δt. Thus

$$P_n(t + \Delta t) = P_n(t)(1 - q \, \Delta t) + P_{n-1}(t)q \, \Delta t \qquad (11.14.2)$$

This may be written in the form

$$\frac{P_n(t + \Delta t) - P_n(t)}{\Delta t} = qP_{n-1}(t) - qP_n(t) \qquad (11.14.3)$$

and then taking the limit as $\Delta t \to 0$, one obtains

$$P'_n(t) = qP_{n-1}(t) - qP_n(t) \qquad n = 1, 2, 3, \ldots \qquad (11.14.4)$$

For the case $n = 0$

$$P_0(t + \Delta t) = P_0(t)(1 - q \, \Delta t) \qquad (11.14.5)$$

giving

$$P'_0(t) = -qP_0(t) \qquad (11.14.6)$$

The solution of this differential equation is

$$P_0(t) = e^{-qt} \qquad (11.14.7)$$

and by induction it can be shown that the solution of the general equation (11.14.4) is

$$P_n(t) = \frac{(qt)^n e^{-qt}}{n!} \qquad (11.14.8)$$

which is the desired Poisson distribution.

11.15 STEPS IN SIMULATION

In performing the simulation of a system, a normal sequence of events has evolved. It should be emphasized that the steps are neither sacred nor chronological. These steps indicate phases to be covered in an approximate order:

1. *Definition* of the problem specifically, in familiar terms and symbols, with placement of the necessary limitations
2. *Formulation* of the model, including the statement of assumptions, choice of criteria for optimization, and the selection of the operational procedure or rules of the road
3. *Construction* of the block diagram establishing the functional relationship between the components of the system to be simulated
4. *Determination* of inputs for the simulation program
5. *Preparation* of the computer simulation program
6. *Conducting experimental runs* of the simulated system including experimental design to determine the number of runs and the parameter values to be used and to establish confidence limits
7. *Evaluation* and testing of the simulated system

The most important step in a computer simulation of a traffic system is the formulation of the model. The computer is important in that it makes solution of the model practical, and the programming merely represents the means of communication between the investigator and the computer. However, it must always be remembered that neither the simulation model nor the computer program represents an end in itself but is merely a means to solving a complex problem regarding the operation of an existing traffic system or the design of a future one.

Important in the formulation of the model are simplifying assumptions. Although including the color of the vehicles would add realism to the simulation model, the effect of a vehicle's color on traffic operation has not been documented to the extent that such information would contribute anything to the solution of a practical problem. In simulating a high-type intersection in which vehicle-to-vehicle and vehicle-to-pedestrian conflicts have been eliminated, it might suffice to simulate a single approach. It is not uncommon for the novice to add extraneous components to his simulation in his zeal to "play God."

A second important aspect of the model formulation is the establishing of the basic rules by which the design or operational improvement to be simulated can be measured. This is best accomplished by formulating the simulation model in such a way that the figures of merit are expressed as functions of the variables of the system being studied. Several such measures of effectiveness worthy of consideration are (1) travel time and speed—their averages, variances, and distributions; (2) percentage of vehicles required to travel at some arbitrary fraction of their desired speeds; (3) acceleration noise in the analysis system; (4) number of lane changes per vehicle per second; (5) average platoon length; and (6) the level of service as described by the energy-momentum concept. Thus several criteria may be important, or it may be desirable to use different criteria at different times. Herein lies one of the advantages of simulation: In analysis one may use only those criteria which are mathematically tractable (for example, one may use least squares but not maximum absolute deviation; one may use the mean but not the entire distribution, etc.). In simulation, one may select any criterion, measuring it continually if necessary.

Inherent to the formulation of the model are the determination of the significant input and output variables. Inputs may be divided into four categories: geometrics, traffic characteristics, driver policy, and vehicle performance.

Important geometric variables are curvature, grade, number of lanes, angle of convergence of ramps, location of ramps, sight distances, length of acceleration lane, etc. Such system considerations as interchange type, ramp configuration, frontage roads, and upstream or downstream bottlenecks may be important, as well as the presence of signs, traffic signals, and other control devices.

The three fundamental characteristics of traffic movement—speed, volume, and density—define the operational requirements of the traffic stream. The speed of a maneuvering vehicle has a significant effect on that particular maneuver—crossing, merging, weaving, or lane changing. On the one hand, the size of gap required for the maneuver varies directly with the relative speed between the maneuvering vehicle and the stream across which, into which, or through which it desires to maneuver.

The transverse distribution and longitudinal distribution of vehicles on the simulated system may be described in terms of either volume or density. The longitudinal placement of vehicles in the traffic stream affects the driver's choice of speed. The reciprocal of this longitudinal placement yields volume if the measurement is made in time, and density if distance is the measurement parameter. Equally important to the ability of an individual vehicle to maneuver through the simulated system is the transverse or lane use distribution of vehicles where there are more than one traffic lane in a given direction.

The principal characteristics of traffic concern the abilities, requirements, and performance of the driver and vehicle, who together form the discrete unit of the simulation model. The objectives of the driver must be incorporated into the model. It may be to minimize his delay or maximize his safety. To complicate things a driver's policy may not be consistent with the capabilities of his vehicle. A vehicle which gets trapped behind a slow-moving vehicle is forced to reduce its speed. When a passing opportunity presents itself, the accelerating potential of the vehicle becomes yet another significant input to the model.

The power of simulation as a tool for the study of traffic flow lies in the ability to include the effect of the random nature of traffic. We have seen that the number of variables—some associated with the characteristics of the roadway, some with the characteristics of the driver, and some with the characteristics of the vehicle—is very large for most traffic systems. These variables are expressed as frequency distributions and input into the simulation model, using the techniques of Secs. 11.9 to 11.13.

11.16 THE SIMULATION PROGRAM

The simulation logic for stepping the vehicles through the system, once the inputs are known, may be divided into three classifications: (1) flow logic for unimpeded vehicles, (2) car-following logic for platooned vehicles, and (3) maneuvering logic for vehicles executing maneuvers involving more than a single stream of traffic. In actuality, all drivers of vehicles within the roadway system are continually and simultaneously making decisions and modifying their behavior. In the course of a simulation, the classification of most vehicles—unimpeded, following, or maneuvering—will change many times. The computer, however, can make only one simple logical choice at a time. To control all the occurrences at any given instant, it must process all decisions sequentially. In other words, it must process each decision for every vehicle, for each vehicle in every lane, and for each lane within the system. It must do this in accordance with a prescribed sequence for each instant of time to be considered.

The success of the digital computer lies in the fact that extensive computations can be described as repetitions of cycles. Thus, the computer program consists of blocks of instructions which effect the computations required, each terminated by tests which direct the computer to the next appropriate block. The functional diagram used for describing the simulation problem for the computer programmer is called the *block diagram*. Figure 11.9 depicts a block diagram for illustrating the relationship between the input generation, unimpeded flow logic, car-following logic, and maneuvering logic.

The simulation of a complex traffic system demands a good understanding of computer programming. After the various traffic manifestations of the problem have been built into the model, a program must be written to enable the computer to carry out the simulation. Ideally, a

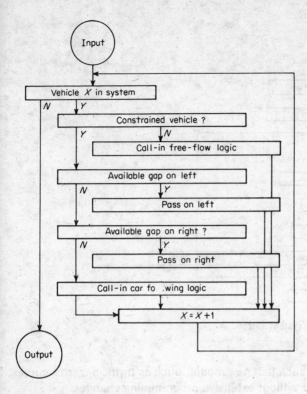

Fig. 11.9 Traffic stream logic.

traffic simulation should represent a cooperative effort between a traffic theorist and a computer technologist.   Based on this, a good simulation program should fulfill the following requirements:

1. It must provide an easy, inexpensive method of highway simulation.
2. It must be general enough that any highway configuration can be simulated by the input of the proper geometric parameters.
3. The input to such a system must be easily understood and capable of execution by non-computer-oriented personnel.
4. It must furnish output which is easily readable and which contains all parameters needed by the traffic engineer for application in the design or modification of highway systems.
5. It must be written in a modular fashion, such that any of the moduli can be changed without affecting the rest of the program.   (That is, the car-following process should be completely independent of the input generation process, etc.)

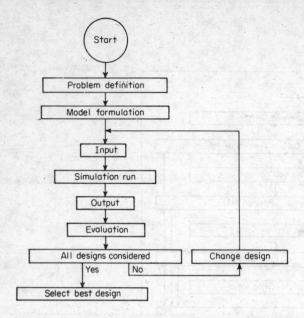

Fig. 11.10 Simulation for system design.

6. It must be written such that new moduli, such as traffic hazards, curves, etc., can be added without extensive programming changes.
7. It must be machine independent, written in one of the higher-level languages such as Fortran IV in such a manner that a novice programmer can modify it.

The major components of a simulation program are illustrated in Fig. 11.10. The input data, input generation, and program logic have already been discussed. Moving on, there are two basic outputs. One of these normally consists of a one- or two-page report giving the input geometry and overall performance of the system under study. A second form of output is a time-space diagram which is desirable in "debugging" and evaluating the program visually. Of course any type of plot, such as percentage of gap acceptance, "contour" maps, "profile" maps, can be incorporated.

The internal bookkeeping procedures have not been discussed. These are vitally important to the efficient, and one might even say the successful, operation of the program. To implement any practical simulation program the method of bookkeeping needed must keep core storage at a minimum and lend itself to fast sequential processing.

Several procedures can and have been used to represent the flow of traffic by the computer. The memorandum notation[6] uses an entire word to represent a vehicle. Various parts of the word are used for such individual characteristics as its time of entry into the system and its desired velocity. Each vehicle's characteristics are identifiable as it moves through the system, making it possible to compute the delays associated with an individual vehicle. Distance along the roadway, using this method, is quantified using a unit block which is one lane wide and has a length equivalent to some fractional part of the length of an average vehicle. Thus, a vehicle may occupy only a limited number of discrete positions. Each vehicle can be advanced by changing the record to show its position one time unit later (periodic scan). This is done by multiplying the vehicle's speed by the time increment and adding the product to the present position. Thus, vehicles move through the system in much the same manner that players move their pieces through a monopoly game.

In another procedure,[7] the entire roadway system is represented by a three-dimensional mathematical array (Fig. 11.11). The length dimension corresponds to relative position along the roadway; that is, vehicle data is stored in adjoining array elements in the same order as the vehicles occupied by a particular lane. The vertical dimension of the array accommodates all the information characteristics of each particular vehicle, and the width dimension represents the several traffic lanes. In contrast with the memorandum notation, this mathematical notation procedure allows each vehicle to be associated with its own position indicator. Therefore, a vehicle's position is essentially continuous, and speed and acceleration are no longer step functions. In addition to the circular array, this procedure utilizes two special registers for each traffic lane—the index position of the lead vehicle and the number of vehicles in the lane.

Sandefur[8] has developed a method of simulation bookkeeping known as *list processing* or *chaining*. The assumption made in chaining is that every information set is a matrix $\bar{C}$ made up of information vectors $\bar{X}_i$. In each of the information vectors there are *pointers* indicating the preceding vector $\bar{X}_{i-1}$, and the following vector $\bar{X}_{i+1}$. Take for example a (4,8) information matrix $\bar{C}$, made up of the four vectors $\bar{X}_i$ where $x_{i,1}$ and $x_{i,2}$ are the chain pointers and $x_{i,j}$, where $j = 3, 4, \ldots, 8$, is the information associated with the vector $\bar{X}_i$. The matrix would appear in the following manner:

	Machine loc	Last	Next		
$\bar{X}_1$	loc A		loc C	$I_{1,j}$	$j = 3, 4, \ldots, 8$
$\bar{X}_2$	loc C	loc A	loc Z	$I_{2,j}$	$j = 3, 4, \ldots, 8$
$\bar{X}_3$	loc Z	loc C	loc B	$I_{3,j}$	$j = 3, 4, \ldots, 8$
$\bar{X}_4$	loc B	loc Z		$I_{4,j}$	$j = 3, 4, \ldots, 8$

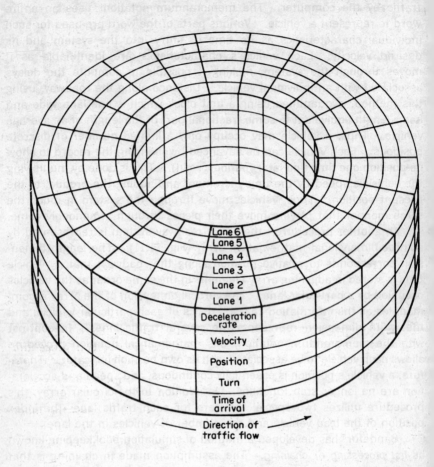

Lane 6
Lane 5
Lane 4
Lane 3
Lane 2
Lane 1
Deceleration rate
Velocity
Position
Turn
Time of arrival

⟶ Direction of traffic flow ⟶

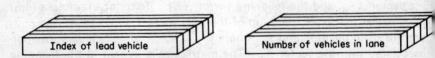

Index of lead vehicle   Number of vehicles in lane

Fig. 11.11 Computer representation of traffic.

where loc (A) is the computer location (address) of the vector $\bar{X}_i$. In the above manner, information vectors can be added to, or deleted from, a chain without data movement, resulting in a decrease in machine time. It also decreases storage requirements because many chains can share the same location pool.

11.17 MODEL CALIBRATION

If programmed properly, the realism of the computer output is a function only of the realism of the system model and the inputs to the model. A simulation model is essentially a hypothesis and therefore must be tested before it can be accepted as fact. Such tests include its feasibility, realism, and validity.

Since the analyst is generally interested in steady-state conditions, some attention must be given to the initial state of the system. This may be handled by allowing the simulation to start in an easily identified state and run for a period long enough to effect steady-state conditions. Other approaches include "warm-up periods," or "warm-up sections," to allow the system to become loaded with normal traffic before the desired production runs are made or sampled.

Simulation is of course a sampling technique (we sometimes even refer to it as *simulated sampling*). Although simulation on a digital computer is fast, it is impractical to attempt to utilize the whole range of each variable because several thousand simulation runs would be required. A statistical design should be adopted to handle some finite number of *treatment combinations*. Thus, for an intersection simulation three levels might be chosen for each input variable, as illustrated in Table 11.1.[9]

Table 11.1 Three levels used for each independent
variable in seven-factor second-order rotatable design

Independent variable		Design level	
No. Name	−1	0	1
Approach flow (lanes 1 + 2), vehicles/15 min	75	150	225
Opposing flow (lanes 3 + 4), vehicles/15 min	75	150	225
Right turns in lane 1, percent	5	15	25
Left turns in lane 2, percent	0	10	20
Left turns in lane 3, percent	0	10	20
Right turns in lane 4, percent	5	15	25
Amount of time signal is red, percent	40	50	60

Replication is extremely desirable once the experimental design is devised. Statistical studies show that the accuracy of simulation increases as the square root of the number of samples. It is better to have the results of four 15-min runs than a single 1-hr run. There are a number of techniques available to reduce the sample size and attain the same accuracy.

Finally it is essential that the results of simulation be compared with known real world responses to the same input in order to satisfy the analyst that modeling has been satisfactory. Validation is the process whereby the simulation model is evaluated to determine whether it satisfactorily duplicates real traffic behavior. Since, as mentioned before, it is not the intent in simulation to reproduce all minute details of a real system, it is necessary to establish in the beginning those characteristics of real traffic the model should duplicate to be useful, in other words, what criteria are to be used in the validation.

Theoretically, the model must duplicate the characteristics that the highway engineer uses as design criteria or the characteristics that the traffic engineer uses as operational criteria. As is the usual case, however, universally useful design and operational criteria cannot be precisely defined, and each design application requires the selection of suitable criteria based on engineering judgment. One might adopt a "microscopic" philosophy in which attempts would be made to duplicate, in the computer, the specific details of field samples of moderate time length. Otherwise, a "macroscopic" approach might be utilized in which computer runs seek to reproduce gross statistical properties of field samples accumulated over long periods of time. Inasmuch as traffic is a stochastic process, valid arguments can be raised in support of either approach.

11.18 WHY SIMULATE[10]

Simulation is resorted to when the systems under consideration cannot be analyzed using direct or formal analytical methods. There are a few additional reasons for simulation, most of which have been found to pertain to the simulation of traffic systems.

1. The task of laying out and operating a simulation is a good way to systematically gather pertinent data. It makes for a broad education in traffic characteristics and operation.
2. Simulation of complex traffic operations may provide an indication of which variables are important and how they relate. This may lead eventually to successful analytic formulations.
3. In some problems information on the probability distribution of the outcome of a process is desired, rather than only means and variances such as obtained in queueing. Where traffic interaction is involved, the Monte Carlo method is probably the only tool which can give the complete distribution.
4. A simulation can be performed to check an uncertain analytic solution.
5. Simulation is cheaper than many forms of experiment. Imagine the

cost saving in simulating to find the optimum spacing of freeway interchanges.

6. Simulation gives an intuitive feeling for the traffic system being studied and is therefore instructive.

7. Simulation gives a control over time. Real time can be compressed, and the results of a long amber phase can be observed in a few minutes of computer time. On the other hand, machine time can be expanded and run slower than real time so that all the manifestations of the complex interactions of freeway movement can be comprehended.

8. Simulation is safe. It provides a means for studying the effect of traffic control measures on existing highways. The effect of signals, speed limits, signs, and access control all can be studied in detail without confusing or alarming drivers. Simulation offers the ability to determine in advance the effect of increased traffic flow on existing facilities. Probable congestion points and accident locations can be anticipated and changes in the physical design of the highway can be effected before the need is demonstrated through accident and congestion experience.

The above summarizes some advantages of the simulation model. As a form of model, it should be compared (1) with analysis, which involves using analytical, rigid, and probabilistic models, and (2) with trial and error, which involves devising some kind of trial solution and then taking it into actual traffic and trying it out. It will be useful to refer to Table 11.2, prepared by Goode,[11] which compares the relative merits of analysis, simulation, and trial and error.

Referring to the table, one sees that the traffic problem has been attacked in the past with the tools of both analysis and trial. Simulation is actually a combination of both methods but allows attack on the most complicated of processes (which analysis does not) and does not affect traffic until the solution has been reached (which trial methods do). Simulation, it will be noted in Table 11.2, is almost always midway between

Table 11.2

Criterion	Analysis	Simulation	Trial
Cost	Least	Medium	Most
Time	Least	Medium	Most
Reproducibility	Most	Medium	Least
Realism	Least	Medium	Most
Generality of results	Most	Medium	Least

analysis and trial.   But as the situation which is being studied becomes more complex (as in the traffic problem), the differences between methods in terms of cost, time, etc., become more pronounced, until finally neither of the extremes can be tolerated and simulation becomes the only feasible method.

Simulation is a powerful tool, and like all powerful tools it can be dangerous in the wrong hands.   The increased emphasis on simulation studies and the corresponding lack of experience on the part of some people who attempt to apply the method can lead to a type of pseudosimulation. Pitfalls exist in simulation as in every human attempt to abstract and idealize.   Some rules to follow in avoiding these pitfalls are (1) no assumption should be made before its effects are clearly defined, (2) no variables should be combined into a working system unless each one is properly explained and its relationships to the other variables are set and understood, and (3) it must be remembered that simplification is desirable, but oversimplification can be fatal.

For the most part it can be said that the goals achievable by simulation in the traffic process are clear-cut and offer a rewarding payoff.   Simulation is an ideal technique for traffic research.   The simulation model is not just another means for accomplishing what we can do today but is a tool for solving problems which cannot be solved today.

11.19   AN EXAMPLE—FREEWAY MERGING[12]

The simulation program developed to study the freeway merging process in this section can be represented physically by the configuration diagramed in Fig. 11.12.   The study was made as general as possible, with the ability to make changes in the input data to change the physical representation

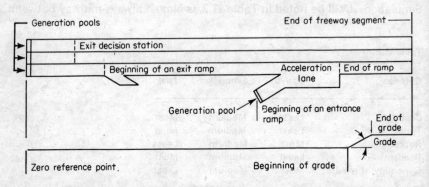

Fig. 11.12   Physical representation of the freeway segment.

of the freeway, allowing the study to simulate a number of different free-way designs.   Some of the variable design factors are:

1. Number of lanes, from one to six
2. Length of simulated freeway segment
3. Number of on-ramps and off-ramps, from none to two
4. Location and lengths of on-ramps and off-ramps
5. Length of each acceleration lane
6. Overall length of each on-ramp
7. Location of the start of a grade relative to the beginning of the freeway segment
8. The grade itself

The model performs all logical operations to accomplish safe travel of a vehicle through the freeway segment.   The general logical organiza-tion is shown in Fig. 11.13.

Vehicles are initially placed in a random array on the freeway segment, ensuring that each vehicle on the freeway segment is a safe distance from neighboring vehicles.   Then the vehicles in the system are processed, starting with the vehicle which is most distant from the zero reference point regardless of its lane.   Vehicles are treated successively, in order of decreasing distance, until all vehicles in the system have been updated.

Each vehicle is assigned a number of characteristics: current speed, desired speed, and distance from the zero reference point.   These are the characteristics necessary to determine what the vehicle is doing and what it desires to do in the future.

This simulation study employs a periodic-scan technique for the processing of vehicles in the system.   This technique updates information on each vehicle in the system during each time interval, repeating for each time increment, until the study is completed.   The vehicle data are updated by determining which characteristics are to be altered during the time interval and then performing the required alterations.   The vehicle information consists of all the characteristics of each vehicle in the system after processing for a time interval.

The normal logic allows a vehicle to be processed along the freeway segment under certain restrictions.   If the vehicle being processed can maintain its desired speed, it is simply allowed to proceed down the free-way.   If it encounters a slower vehicle in its lane, an attempt is made to weave.   The weaving logic will allow a vehicle to attempt to weave left, then right, whenever possible.   If the weaving attempt is unsuccessful, the vehicle will decelerate.   The normal logic also allows a freeway vehicle to exit on an exit ramp from the shoulder lane.   If a vehicle is not in the shoulder lane when it approaches its desired exit ramp, it attempts to

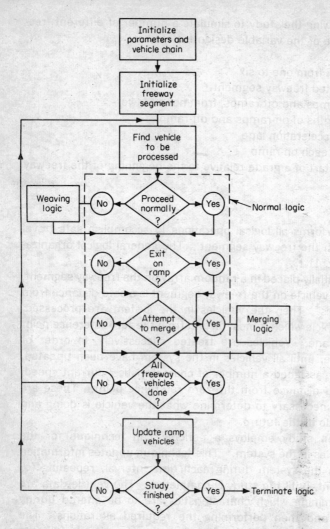

Fig. 11.13 Logic flow diagram of the system.

weave right. If it is unable to weave right, it will decelerate and try to weave during the next time period. If it passes the desired exit ramp before it enters the shoulder lane, it will keep trying to weave right and will exit on any of the following exit ramps. The normal logic also initiates the merging logic to allow a ramp vehicle to enter the shoulder lane from the acceleration lane of an entrance ramp. If a ramp vehicle reaches the end of the acceleration lane without merging, it must stop before entering the freeway segment. After the normal logic has processed all freeway vehicles in the system, the ramp operation logic updates the ramp vehicles.

In order to process vehicles through the system, each vehicle is identified by a unique subscript, and its relative position to all other vehicles in the system is logically organized by a chaining technique.   By this technique, each vehicle is assigned a value in each of two characteristic arrays which contain the subscripts of the vehicles immediately ahead of and behind it in the same lane.   The arrays of these characteristics are named LAST and NEXT, respectively.   This gives the processing program access to the characteristics of the vehicles between which a vehicle being processed is situated and allows determination of those characteristics of the vehicle which must be changed.

Since each vehicle has characteristics in arrays LAST and NEXT, the processing program has an overall picture of the placement of vehicles in the system.   These two arrays can be thought of as a chain, with each link one subscript of a vehicle.   For example a chain array might appear as shown in Fig. 11.14.

A chain from the arrays NEXT and LAST containing the subscripts of vehicles shows the organization of one lane of traffic.   The lane represented by this chain would physically look like Fig. 11.15.

This chaining logic permits an organized handling of vehicle characteristics while using a method which is easily adapted for digital computation.   In the digital computer this process reduces the number of internal storage locations required.   If this process were not used and specific locations were required for the characteristics of each vehicle, either on or off the system, the number of vehicles passing through the system would be more restricted.   Using the chaining technique, the restriction imposed on the system is that not more than 500 vehicles can be on the freeway segment at one time.

The program is organized into independent logical divisions with one monitor division to direct the control among the other divisions.   By using separate divisions, experimentation can be carried out in one division with-

Subscript	Last	Next
2	1	4
4	2	8
8	4	10
10	8	23
23	10	49
49	23	75
75	49	76

Fig. 11.14   Array organization of a chain.

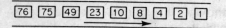

Fig. 11.15 Physical representation of a chain.

out changing any other logic. The monitor division incorporates the tasks of initializing all parameters, including the chains, initializing the freeway segment for the first time period, and handling for the normal or through flow of traffic. This section determines which of the vehicles on the freeway segment is the most distant nonprocessed vehicle and performs a series of tests on it. The tests are in the form of the following questions:

1. Is this vehicle going to travel past the end of the freeway segment during the next time period?

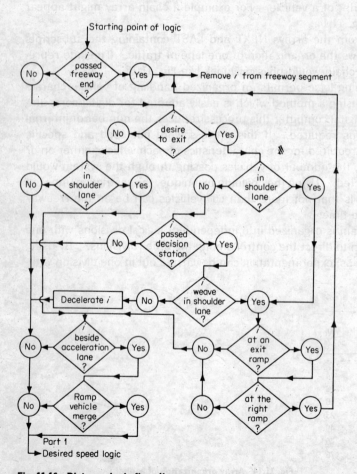

Fig. 11.16 Distance-logic flow diagram.

2. Does this vehicle exit on a ramp?
3. Does a vehicle merge in front of this vehicle from the acceleration lane of an entrance ramp?
4. If this vehicle is not in the shoulder lane and desires to exit soon, can it weave into the shoulder lane?
5. Is this vehicle traveling as fast as it desires?

Once the monitor division has operated on all freeway vehicles and has updated them, the ramp vehicles are updated.

The normal flow section determines the behavior of vehicles in the same lane and is divided into two segments: the distance section and the desired-speed section. A flow diagram of the distance section is shown in Fig. 11.16. The distance logic stems from the fact that certain points on the freeway segment are of particular importance in the processing of vehicles. These are the end of the freeway segment, the beginning of the exit ramps, the exit decision stations, and the area adjacent to the acceleration lane. The number of these points and their distances from the zero reference point can be varied to give the scheme flexibility. In this section it is determined whether a vehicle has passed one of the above positions during a given time period by calculating a tentative displacement. If this tentative displacement indicates that the vehicle is about to pass one of these positions, different algorithms are employed, depending on which position is involved.

When the vehicle has passed the end of the freeway segment, a sequence is initiated to remove it from the chain of that lane and place its subscript into a free links pool. The exit decision stations are points upstream of the exit ramps, which give a vehicle that desires to exit time to weave into the shoulder lane. After a vehicle has passed such a decision station, it will attempt to weave right into the shoulder lane and continue to attempt to weave until it is successful, even after it has passed the exit ramp. Once such a vehicle reaches the shoulder lane, it will be processed onto the first exit ramp it reaches. The distance logic is based on the assumption that there is no change in the speed of a vehicle during a time interval. This is, of course, not necessarily true. After the position of a vehicle on the freeway segment has been compared with all the above mentioned positions and it does not leave the freeway, a closer look is taken to see if its speed must be changed.

The desired-speed logic is broken down into three parts, based on the relationship between the vehicle being processed and the vehicle ahead of it in the same lane. The following distance and an acceptable following distance are calculated for a pair of vehicles, such as i and j where i is the vehicle under consideration and j is directly in front of it. The acceptable following distance depends not only on the speed but also on the relative speed of the two vehicles, being smaller if the leading vehicle is the faster of the two.

First, the following distance between vehicle i and vehicle j is compared with the calculated value of the acceptable following distance. If the following distance is less than the acceptable distance and vehicle i is moving faster than vehicle j, i will decelerate at a rate that depends on how close i and j are. However, if i is slower than j, no speed changes occur, since the distance between them is increasing and thus becoming safer as time progresses.

If the following distance is equal to the acceptable following distance and the speed of vehcle i is greater than the speed of vehicle j, vehicle i will either slow down, in one time increment, to the speed of vehicle j or, if i cannot decelerate fast enough, it will decelerate by a maximum amount. On the other hand, if i is slower than j, it is accelerated; but if its speed is still slower than it desires, an attempt to weave is made.

The last part of the speed logic is concerned with the following distance being greater than the acceptable following distance. This is the most desirable condition, since vehicle i is a safe distance away from vehicle j. A look is now taken at the present situation to see if i will come too close to j during the present time period if its speed remains the same. A new tentative following distance is calculated, which will be the following distance at the end of the scan interval. When this distance is less than acceptable, a hazardous condition will exist if the speed of i is not changed; therefore vehicle i is decelerated. If the acceptable distance remains less than the following distance, the distance between i and j is still safe, but now consideration is given to processing the vehicle to eliminate erratic movement.

When it is determined that a vehicle desires to weave into another lane, the possibility of weaving first left and then right is explored. If the vehicle is on either the shoulder lane or the innermost lane, only one weave is attempted. If all weave attempts are unsuccessful, the vehicle remains in its present lane. If a weave is successful, the weaving vehicle is removed from its old lane to the new lane by rearranging the appropriate chains. After a lane has been chosen into which an attempted weave will be made, the current speed of the weaving vehicle is increased to a new speed, called the *weaving speed*. The vehicle will weave if it will not come hazardously close to the vehicle ahead or behind it in the new lane. If the gap is inadequate, the weave attempt is unsuccessful.

The first criterion for a successful weave concerns the following distance and the acceptable following distance to the vehicle j, behind which it attempts to weave. If the following distance is acceptable, the vehicle will weave, provided that the speed of the lane it weaves into is higher than the speed of the lane it weaves out of and provided that the second criterion, discussed below, is satisfied.   If the following distance is less than acceptable, an attempt to change speed, such that a safe distance will exist between the two vehicles, is investigated.   If this is possible, the vehicle will weave, provided all other conditions are satisfied. If the required speed change is not possible or some of the other conditions are not satisfied, the weave is unsuccessful.

The second criterion concerns the relationships between the weaving vehicle i and the vehicle k in front of which it attempts to weave.   If the speed of i is greater than the speed of k, a certain minimum distance is required between them for a successful weave.   On the other hand, if i is slower than k, it will weave only if the *lag time*—defined as the time required by vehicle k to come within the minimum safe distance of i—is greater than a certain minimum acceptable lag time.

The behavior of vehicles on an entrance ramp is treated in two parts: the movement of vehicles on the ramp and the merging of vehicles onto the freeway segment from the ramp.

The movement of vehicles on a ramp is handled similarly to the vehicles on the freeway segment, but the ramp vehicles are not allowed to weave while on the ramp.   The ramp vehicle closest to the end of the ramp is processed as a special case, because this vehicle must stop at the end if it is unable to merge.   This vehicle is gradually decelerated as it approaches the end of the ramp.   Once this vehicle reaches the end, its speed is set to zero.   The remaining ramp vehicles are processed similarly to the freeway vehicles by calculating an acceptable following distance which is shorter than those for the through vehicles, since it is known that ramp vehicles crowd closer together when attempting to merge.   The procedure determines whether a vehicle will accelerate or decelerate by comparing the acceptable following distance with its following distance during the last time period, the present time period, and the next time period.   Again smooth processing is the key criterion, and when a vehicle approaches another vehicle, it will do so gradually.

The merging subroutine processes vehicles from the acceleration lane onto the shoulder lane of the freeway segment.   A successful merge will occur if the following conditions are met: a gap is present on the shoulder lane into which a ramp vehicle can merge, and merging vehicles will not come hazardously close to the vehicles ahead of and behind it on the freeway segment.

After a ramp vehicle is chosen to attempt to merge into a given gap, the merging logic examines the relationship between the merging vehicle i and the vehicle j ahead of it on the freeway segment and calculates a following distance and an acceptable following distance.   If the following distance is greater than the acceptable following distance, the relationships between vehicle i and the following freeway vehicle k are investigated to determine if i will merge.   If the following distance is less than acceptable and the speed of i is less than the speed of j, the merge does not take place.   However, if the speed of i is greater than that of j, i will attempt to decelerate to a safe distance.   If this is not possible, no merge takes place, but if it is, the relationships between i and k are investigated.

Once it has been determined that a gap is available and that the minimum requirements between the merging vehicle i and the lead vehicle j are satisfied, the relative positions of i and the following freeway vehicle k are investigated.   If vehicle i is faster than vehicle k and at least a certain distance away from it, the merge is successful.   If vehicle k is faster, a gap time is calculated which is the time it takes vehicle k to crowd vehicle i excessively.   If this is less than the maximum crowding gap time, the merge is successful.   If a merge is successful, the merging vehicle is updated, taken off the ramp, and placed onto the shoulder lane by manipulating the NEXT and LAST arrays, as was done for the weaving logic.

Each vehicle on the acceleration lane is given a chance to merge into the gap presented to the merging logic.   The logic will attempt to merge each vehicle on the acceleration lane into a gap in succession, starting with the vehicle closest to the end of the acceleration lane.   Even if one does merge into a gap, the remaining ramp vehicles behind it still have an opportunity to merge into the gap.

The normal output of a simulation system is a table of average values of quantities such as travel times and volumes processed.   This does give an overall view of the operation of a system, but it does not give any information about individual vehicle movement.   A number of small mistakes can be present in the logic without being detected.   A more complete picture of the operation can be developed, displaying the movement of vehicles on the freeway segment in a graphical form by means of time-space diagrams.

The abscissa of the time-space diagram represents time, the ordinate represents distance from the zero reference point, and one continuous line represents the movement of an individual vehicle along the freeway segment.   This plot gives the position of each vehicle on the freeway segment for the entire length of the study plus a picture of relative speeds and positions.   Different lanes, and the ramp movement, can be plotted using continuous lines and dashed lines or different-colored lines to distinguish among them.   Each curve of a plot is made up of a number of points with a line drawn between them, where each point represents the position and time of a vehicle in the system.   Since the digital computer program makes a periodic scan, it decides when data are to be saved and writes these items on magnetic tape, one record per block, indicating the name of the vehicle, its position on the freeway segment, and the time.

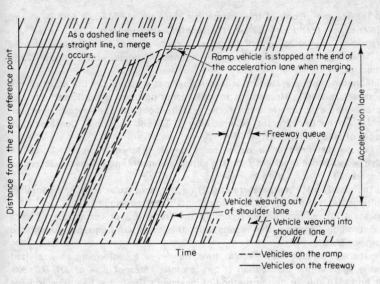

As a dashed line meets a straight line, a merge occurs.

Ramp vehicle is stopped at the end of the acceleration lane when merging.

Freeway queue

Vehicle weaving out of shoulder lane

Vehicle weaving into shoulder lane

Distance from the zero reference point

Acceleration lane

Time

– – – Vehicles on the ramp
——— Vehicles on the freeway

Fig. 11.17   Time-space diagram.

After the study is complete, this tape is sorted so that the data on the tape are in a proper form to plot the time-space diagrams.

An example of the time-space output is shown in Fig. 11.17 and serves to demonstate the realism of the simulation model. Note that vehicles are clustered in queues moving down the freeway segment at the same speed. Note also that vehicles are weaving into and out of the shoulder lane. Vehicles move down the entrance ramp in an orderly fashion at a slower speed than the through-lane vehicles. The ramp vehicles start to merge as soon as they reach the beginning of the acceleration lane, merging anywhere along the acceleration lane. If there is no opportunity for a vehicle to merge before it reaches the end of the ramp, it stops, waiting for an acceptable gap. All these features of Fig. 11.17 show that the merging model and the overall model present a fairly true representation of the movement of vehicles on a freeway.

PROBLEMS

11.1. What is the equitable payment for playing in the "gambler's ruin" game described in Sec. 11.1 if the bank's wealth is limited to $1 million?

11.2. Evaluate the following integrals using the *point-distribution method* and check your results using the *mean-value method*.

$$\int_0^1 (x - x^{3/2})\, dx \qquad \int_0^1 x(1 - x^2)^{1/2}\, dx$$

11.3. Find the volume of one of the wedges cut from the cylinder $x^2 + y^2 + a^2$ and the planes $z = 0$ and $z = y$, using the Monte Carlo approach.

11.4. Explain how you would obtain the water surface area of an irregularly shaped lake which contains several irregularly shaped islands, using Monte Carlo techniques.

11.5. An inebriated graduate student is leaning against a column in the lobby of the Sheraton-Park Hotel. He decides to walk, going nowhere in particular. He can take one step at a time in a northerly, easterly, southerly, or westerly direction. How far will he be from the column after 5, 10, and 25 steps? Consider each number of steps separately.

11.6. Lay out parallel lines on a paper some equal distance W apart. Throw an unsharpened pencil of length L on the paper repeatedly and estimate π according to Eq. (11.2.6).

11.7. Find the steady-state temperature at (1,1) for the rectangular plate and boundary conditions described in Sec. 11.3, using Monte Carlo techniques.

11.8. Generate 1,000 random digits, recording the frequency of occurrence of each decimal digit. Compare the results with the expected frequency, using the chi-square test.

11.9. Generate and output 300 vehicular time headways t conforming to a shifted negative exponential distribution ($t > 1$ sec) for rates of flow of 300, 600, 900, 1,200, and 1,500 vph, using the method of inversion. Group the headways in 1-sec classes and plot the five graphs.

11.10. Generate and output 1,000 spot speeds approximating a normal distribution with a mean of 40 mph and a standard deviation of 10 mph, using the central-limit concept. Perform the problem using 3, 6, 9, and 12 random numbers. Compare the observed distribution in each case with the normal distribution, using the chi-square test.

11.11. Generate and output 300 vehicular time headways conforming to an Erlang distribution for $a = 1$ and $a = 2$ for a volume of 900 vph, using the point-distribution method.

11.12. Repeat Prob. 11.11 for $a = 1$, 2, 3, and 4, using a method based on convolution theory. Group the hideways in 1-sec classes and plot the four curves.

11.13. A signalized intersection equipped with a 3-phase pretimed controller is to be operated such that the left-turn phase has one-third of the cycle. Studies show that a 10-sec phase will handle four vehicles 20 percent of the time, three vehicles 75 percent of the time, and two vehicles the remainder of the time. After 10 sec, starting headways tend to become constant and equal to 2.5 sec per vehicle. Assuming Poisson arrivals for left-turn volumes ranging from 100 to 400 vph in 100 vph increments over cycle lengths ranging from 30 to 90 sec in 15-sec increments, plot the number of failing cycles and the maximum lengths of queues in each case. Program for 1 hr in each case, generating the arrivals according to Sec. 11.12.

11.14. Repeat Prob. 11.13, assuming that the arrivals conform to a binomial distribution with a variance equal to three-fourths of the mean.

11.15. Generate freeway arrivals for 1 hr, using a periodic scan (1-sec scan). Program the lane volumes ranging from 300 to 1,500 vph in increments of 300 vph. Compare the frequency of 1-min arrivals in each case with the expected frequency assuming a Poisson distribution, using the chi-square test.

11.16. What factors should be considered in establishing the scanning intervals for the periodic-scan technique?

11.17. A researcher using three random numbers and making use of the central-limit theorem generates a random deviate. Deciding that four random numbers would give him a better approximation to the normal distribution, he draws one more random number and adds to it the other three. If the fourth random number is equal to one-half the sum of the first three and the new sum generates the same random deviate as before, what is the fourth random number?

11.18. If a random number equaling 0.386 will generate a headway of 7.2 sec conforming to a shifted negative exponential distribution, what is the hourly volume of traffic?

11.19. Using a periodic scan, a researcher draws 1,000 random numbers and generates 20 arrivals. If the minimum headway that can be generated between arrivals is 4 sec, what is the probability of generating a headway greater than 1 min?

11.20. Assume that the distribution of critical gaps for ramp drivers conforms to an Erlang $a = 6$ distribution and that the mean critical gap is 3.0 sec. Find the mean delay to a ramp vehicle in position to merge for outside freeway lane volumes of 600, 1,200, and 1,800 vph, where the distribution of freeway gaps conforms to an Erlang $a = 2$ distribution. Simulate 1,000 ramp vehicles for each case.

REFERENCES

1. Holstein, W. K., and W. R. Soukup: Monte Carlo Simulation, Institute Paper 23, Institute for Quantitative Research and Economics and Management, Graduate School of Industrial Administration, Purdue University, Lafayette, Ind., p. 1, 1962.

1a. Bergami, D.: "Mathematics," Life Science Library, Time Inc., New York, 1963.

2. Tocher, K. D.: "The Art of Simulation," D. Van Nostrand Company, Inc., Princeton, N.J., 1963.

3. Moore, B. C.: "Notes on Monte Carlo Techniques," unpublished.

4. Tippett, L. H.: Random Sampling Number, "Tracts for Computers," Cambridge University Press, New York, 1960.

5. Goode, H. H., and R. E. Machol: "System Engineering," p. 207, McGraw-Hill Book Company, New York, 1957.

6. Gerlough, D. L.: Simulation of Traffic Flow: An Introduction to Traffic Flow Theory, *Highway Res. Board Spec. Rept.* 79, p. 97, 1964.

7. Lewis, R. M., and H. L. Michael: Simulation of Traffic Flow to Obtain Volume Warrants for Intersection Control, *Purdue Univ. Eng. Reprint* CE 205, West Lafayette, Ind., 1964.

8. Sandefur, G. G.: "A Method of Sequential Processing for Traffic Simulation," unpublished.

9. Dart, O. K.: "Development of Factual Warrants for Left-turn Channelization through Digital Computer Simulation," doctoral dissertation, Texas A&M University, College Station, 1966.

10. Morgenthaler, G. W.: The Theory and Application of Simulation in Operations Research, "Publications in Operations Research No. 5," p. 364, John Wiley & Sons, Inc., New York, 1961.

11. Goode, H. H.: "Computers, Simulation, and Traffic," unpublished.

12. Drew, D. R., T. C. Meserole, and J. H. Buhr: Digital Simulation of Freeway Merging Operation, *Texas Transportation Inst. Rept.* 430-6, Texas A&M University, College Station, 1967.

CHAPTER TWELVE
DETERMINISTIC
RELATIONSHIPS

Logical consequences are the scarecrows of fools and the
beacons of wise men.

Thomas Hardy

12.1 INTRODUCTION

Optimization is a term used to indicate the selection of the best operating conditions, subject to the abilities of the traffic system and to the constraints of the driver and environment, in respect to certain objectives. We have seen in the previous chapters on queueing and simulation that the first step is to look for possible simplifying assumptions without loss of generality. These assumptions are subject to revision; yet on them must be based the block of mathematical equations to be built.

In order to make the statement that a traffic system has been optimized, a criterion must have been established to determine the quality of the operation in this particular respect. It is imperative that the measure of effectiveness inherent in the criterion of optimization be one which can be expressed as a function of the traffic variables entering into the problem. The problem, then, is to choose those values of the control variables that optimize the effectiveness function.

In the deterministic approach to traffic problems, the subject of the next three chapters, it is assumed that the functional relationships between the input variables and the parameters measuring effectiveness are constant. In other words, one and only one value of the effectiveness function will occur for any given set of values for the input variables. This assumption was not invoked in the various stochastic models discussed in Chaps. 7 to 11.

Although it is evident that the ultimate objective of the traffic engineer is to optimize the operation of existing traffic systems and the design of future facilities, his immediate problem is arriving at a rational description of the traffic phenomena. Streams of traffic are actually successions of discrete occurrences of vehicles, although many theories exist which treat traffic flow as though it were a continuous process. A dimensional analysis suggests that a relationship exists between vehicular velocity, space gap, and time headway

$$\text{Velocity (ft/sec)} = \frac{\text{space gap}}{\text{time headway}} \frac{\text{ft}}{\text{sec}}$$

and between velocity, concentration (density), and flow (volume)

$$\text{Velocity (ft/sec)} = \frac{\text{flow}}{\text{concentration}} \frac{\text{sec}^{-1}}{\text{ft}^{-1}}$$

The general equation of the traffic stream is usually expressed in the form

$$q = ku \tag{12.1.1}$$

where q = mean rate of flow
      u = mean speed
      k = mean concentration

If any two of these three stream variables are known, the third is uniquely determined. In Chap. 4, we saw that flow or volume is the easiest of the traffic variables to measure and that speed probably ranks next. It is not surprising that concentration or density has generally been considered as the dependent variable, because the other two have been the measured or independent variables. Actually, there is no single dependent variable, any more than length could be considered the dependent variable and width and depth the independent variables in describing a point on a cube. It is helpful, then, in visualizing the basic traffic-stream equation, to consider the surface which represents the equation when plotted on mutually perpendicular axes as shown in Fig. 12.1.[1] Efforts to relate various pairs of the three fundamental elements have been based on (1) simple curve fitting, (2) deduction from the boundary conditions, and (3) physical analogies.

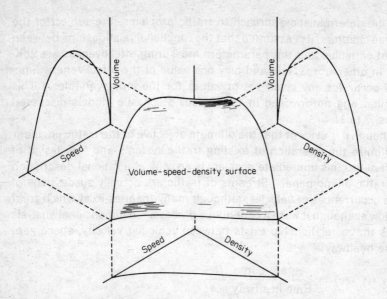

Fig. 12.1 Fundamental volume-speed-density surface.

12.2 CURVE FITTING

The deterministic approach began with a suitable algebraic equation to explain the flow-concentration relationship. Greenshields[2] chose to plot speed u against concentration k for one lane of traffic because it represented the only one of the three planes in the fundamental surface with a single-valued relationship.

The problem of fitting a curve to a set of points objectively is essentially the problem of estimating the true parameters of the curve. The best known method of performing the estimation of such parameters is known as the *method of least squares*. In applying this procedure to speed-concentration graphs, one is seeking a function $u = f(k)$ which fits the point such that the sum of the squares of the deviations between the ordinates of the points and the ordinates of the curve are minimized.

The additional problem of determining the accuracy of the least-squares line is a statistical problem. Thus, the least-squares line $y = a + bx$ is called a regression line, where a and b are the intercept and slope of the regression line. The test of significance of regression is performed on the regression coefficient b, using the t test:

$$t = \frac{b}{s_b} \qquad \mathrm{df} = n - 2 \tag{12.2.1}$$

where

$$b = \frac{\Sigma xy - n^{-1}\Sigma x \Sigma y}{\Sigma x^2 - n^{-1}(\Sigma x)^2} \qquad (12.2.2)$$

and

$$s_b = \left[\frac{1}{D} \frac{\Sigma y^2 - n^{-1}(\Sigma y)^2 - bN}{n-2}\right]^{\frac{1}{2}} \qquad (12.2.3)$$

[N and D refer to the numerator and denominator of the expression for b in Eq. (12.2.2).]   Inherent in the t test is the hypothesis that B, the true regression coefficient estimated by b, is zero and that the error in the estimate of B is normally distributed.   If the 5 percent level of confidence is chosen and $t > t_{0.05}$, the null hypothesis concerning B may be rejected. It then may be concluded that a cause-and-effect relation does exist between the variables x and y.

Although the preceding method for finding the slope of a regression line can be generalized to find confidence limits for the regression coefficients in curvilinear and exponential regression, it is also possible by simple transformations of the variables to represent the relationships as straight lines in the transformed variates.   Thus, a *parabolic* model may be written

$$y_2 = a_2 + b_2 x_2 \qquad (12.2.4)$$

where $y_2 = u$
  $a_2 = u_f$
  $b_2 = -u_f/k_j^{\frac{1}{2}}$
and $x_2 = k^{\frac{1}{2}}$

In the case of an *exponential* model, taking logarithms of both sides, one obtains a linear relation of the form

$$y_3 = a_3 + b_3 x_3 \qquad (12.2.5)$$

where $y_3 = \ln k$
  $a_3 = \ln k_j$
  $b_3 = -u_m^{-1}$
  $x_3 = u$

The linear model is simply

$$y_1 = a_1 + b_1 x_1 \qquad (12.2.6)$$

where $y_1 = u$
  $a_1 = u_f$
  $b_1 = -u_f/k_j$
  $x_1 = k$

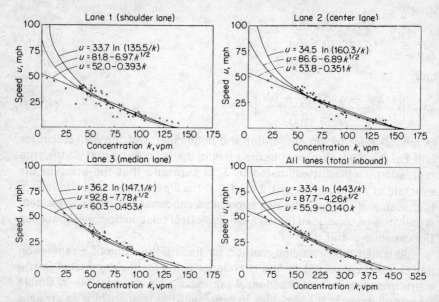

Fig. 12.2 Regression curves of macroscopic equations of state.

The various parameters b_1, b_2, b_3, and a_1, a_2, a_3 are obtainable by solving Eq. (12.2.2) and the following expression

$$a = n^{-1}(\Sigma y - b\Sigma x) \tag{12.2.7}$$

using the speed-concentration data (see Fig. 12.2).

12.3 THE BOUNDARY–CONDITION APPROACH

If the fundamental equation $q = ku$ is differentiated with respect to k, the result is

$$\frac{dq}{dk} = k\frac{du}{dk} + u \tag{12.3.1}$$

Since a general relationship exists between q and u, the k corresponding to $dq/dk = 0$ can be set equal to k_m. Thus

$$0 = k_m\frac{du}{dk} + u \tag{12.3.2}$$

Separating the variables and integrating both sides results in:

$$\ln u = -\frac{k}{k_m} + C \tag{12.3.3}$$

where C is a constant of integration. Application of the boundary condition at $k = 0$ and $u = u_f$ enables one to equate C to $\ln u_f$, yielding, after terms are rearranged,

$$k = k_m \ln \frac{u_f}{u} \qquad (12.3.4)$$

Substitution into the general equation gives

$$q = u k_m \ln \frac{u_f}{u} \qquad (12.3.5)$$

Edie[3] obtained the same result by differentiating (12.3.1) with respect to u. Edie's equation (12.3.5) represents an effort to make the macroscopic q-u-k relations more accurate for traffic densities less than optimum. He suggests that the conventional forms still be used for densities greater than optimum, giving rise to a "discontinuous" form of the steady-state surface. This theory as modified for noncongested traffic may help to describe quantitatively the sudden change of state occurring in a traffic stream going from a relatively free-flowing condition to a crawling stop-and-go condition and back again.

In support of the discontinuous approach, May[4] defines three zones which may be described as constant speed, constant volume, and constant rate of change of volume with density. In zone 1, the speed of the vehicle is determined by the facility itself and the volume matches the demand. Zone 2 represents impending poor operations; average speed drops but the flow rates may be sustained at a high level. In zone 3 both speed and volume rates decrease, which in itself may serve as a definition of congestion.

The studies cited have placed some limits on the shapes of the fundamental traffic surface. Differentiation of the fundamental equation of state $q = f(k)$ with respect to k yields the wave velocity equation. Setting the wave velocity equation equal to zero yields the optimum concentration k_m, which is that value of k for which flow is a maximum.

Morrison[5] obtains the capacity parameter in a slightly different manner. He assumes that the average space headway depends on the average length of vehicle L, the average driver's reaction time C_1, and some function of the vehicle's braking capabilities C_2; then the safe headway becomes

$$s = L + C_1 u + C_2 u^2 \qquad (12.3.6)$$

Substituting Eq. (12.3.6) in $q = ku$ yields

$$q = \frac{u}{L + C_1 u + C_2 u^2} \qquad (12.3.7)$$

Setting $dq/du = 0$ gives the optimum speed u_m or that speed for which flow is a maximum:

$$u_m = \left(\frac{L}{C_2}\right)^{\frac{1}{2}}$$

(12.3.8)

12.4 THE HEAT–FLOW ANALOGY

Most situations in nature are so complicated that they cannot be dealt with exactly by mathematics. The regular procedure is to apply mathematics to an ideal situation having only important features of the actual one. The results are approximations having a practical importance which depends on the closeness of the approximation as verified by reasoning and experiment. The deterministic approach to traffic theory involves analyzing the pertinent traffic characteristics, devising a theory, and then applying methods in the development of which differential equations usually play a prominent role. In this treatment the relationships obtained by observation, experimentation, and reasoning are given; the researcher is required to express them in mathematical symbols, solve the resulting differential equations, and interpret the solutions.

Consider the problem of one-dimensional heat flow in a long, slender insulated rod satisfying the differential equation

$$a^2 \frac{\partial p}{\partial t} + \frac{\partial^2 p}{\partial x^2} = 0$$

(12.4.1)

where p = temperature
$\quad x$ = distance
$\quad t$ = time
$\quad a^2 = sk/c$
$\quad$ where c = conductivity of material
$\quad\quad\quad s$ = specific heat
$\quad\quad\quad k$ = density

The boundary conditions are statements of the initial conditions at the ends of the rod at any time and the initial conditions throughout the rod at time zero. What is usually desired is a description of the temperature at any time and place along the rod or, in other words, a solution of the differential equation expressed in the form

$$p(x,t) = f(L,a,x,t)$$

(12.4.2)

On the other hand if a continuous record in time were kept of the temperature at various points along the rod, one could solve for a and thus determine some property of the material such as its conductivity c.

In many traffic situations, a single traffic lane acts as a long, slender insulated rod (controlled access and no opportunity for lane change). If p in (12.4.1) is allowed to assume the role of some parameter related to such conventional traffic variables as speed, concentration, and flow, the solution, Eq. (12.4.2), provides the means for evaluating some property of the highway—call it the *trafficability*. All that need be done is to measure the traffic characteristics along the stretch of highway defined by L.

The heat equation (12.4.1) is, in the study of partial differential equations of physics, called the *equation of diffusion*. It is satisfied by the concentration p of any substance which penetrates a porous solid by diffusion. Thus, Harr and Leonards[6] postulated that the movement of traffic resulted from a motivating "pressure potential" analogous to p. In their solution the parameter was then eliminated (since it has no physical significance in the traffic phenomenon and therefore cannot be measured), leaving vehicular speed as a function of what we have defined above as trafficability and thus affording a means of rating such various geometric features as a curve or grade.

It is interesting to note that several variations of (12.4.1) have been studied in classical physics. For a rod in which heat is being generated at a constant rate, while the lateral surface is insulated, the heat equation takes the form

$$a^2 \frac{\partial p}{\partial t} + \frac{\partial^2 p}{\partial x^2} = c \qquad (12.4.3)$$

where c is a positive constant. Reddy[7] has suggested Eq. (12.4.3) as a basis for evaluating geometric features of the roadway rather than Eq. (12.4.1), which is of course a special case of (12.4.3). In this manner the theory was generalized to account for the effects of vehicles entering or leaving a section of roadways such as in the vicinity of entrance and exit ramps.

12.5 THE FLUID–FLOW ANALOGY

It should be pointed out that the difficulty in utilizing the deterministic approach is not in solving a differential equation but in finding one that expresses the physical condition realistically. For example, Harr and Leonards[6] define p so that

$$\frac{\partial p}{\partial x} = -c_1 u \qquad (12.5.1)$$

and

$$\frac{\partial k}{\partial p} = c_2 k \tag{12.5.2}$$

where u and k are speed and concentration. Solving (12.5.1) and (12.5.2) simultaneously would suggest the following speed-concentration relationship:

$$u = -(c_1 c_2 k)^{-1} \frac{\partial k}{\partial x} \tag{12.5.3}$$

This implies that speed might be negative if the change in vehicular density with respect to distance is increasing. This is not realistic and therefore reduces the conceptual appeal of the model.

It is more logical to express an equation of motion such as given in (12.5.3) in terms of acceleration rather than velocity or speed, since the sign (positive or negative) would not specify forward or backward movement but merely speeding up or slowing down. Consider the following equation of motion,[8] which expresses the acceleration of the traffic stream at a given place and time as

$$\frac{du}{dt} = \frac{-c^2}{k} \frac{\partial k}{\partial x} \tag{12.5.4}$$

This states that a driver adjusts his velocity at any instance in accordance with the traffic conditions about him as expressed by $k^{-1}(\partial k/\partial x)$. If traffic is thinning out (negative $\partial k/\partial x$) the driver accelerates (positive du/dt) and vice versa.

Equation (12.5.4) has the form of the equation of motion of a one-dimensional compressible fluid with a concentration k and a fluid velocity u. This is illustrative of several new theories of traffic flow described in terms of fluid or hydrodynamic flows. Their analyses are based on an assumed relation between flow and concentration, such as Eqs. (12.5.3) and (12.5.4), and a partial differential equation expressing the conservation of matter.

Fluid mechanics is based on, among other things, the principle of conservation of mass. Imagine a volume in space: If the outflow of mass is greater than the inflow, the principle of conservation of mass requires that there is an equal decrease of mass stored in the volume. When this principle is stated mathematically as an equation, it is called the *equation of continuity*.

Considering traffic flow as a conserved system, the change in the number of vehicles on a length of road dx in an interval of time dt (see Fig. 12.3) must equal the difference between the number of vehicles entering the section at x and the number of vehicles leaving the section at

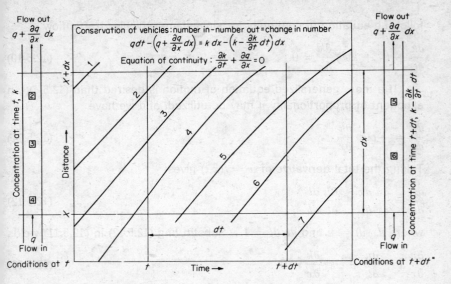

Fig. 12.3 Derivation of the equation of continuity for a traffic stream.

$x + dx$. If the number of vehicles on the length dx at time t is $k\,dx$ and the number of vehicles entering in time dt at x is expressed as $q\,dt$, the conservation of vehicles can be expressed symbolically as

$$k\,dx - \left(k - \frac{\partial k}{\partial t}\,dt\right)dx = q\,dt - \left(q + \frac{\partial q}{\partial x}\,dx\right)dt \qquad (12.5.5)$$

Making use of $q = ku$ yields the equation of continuity for traffic flow

$$\frac{\partial k}{\partial t} + \frac{\partial(ku)}{\partial x} = 0 \qquad (12.5.6)$$

Taking the derivative of the product in the second term yields

$$\frac{\partial k}{\partial t} + \frac{u\,\partial k}{\partial x} + \frac{k\,\partial u}{\partial x} = 0 \qquad (12.5.7)$$

It is well established in the theory of traffic flow that vehicular velocity varies inversely with the concentration of vehicles

$$u = f(k) \qquad (12.5.8)$$

As a consequence of (12.5.8)

$$\frac{\partial u}{\partial k} = \frac{\partial u}{\partial x}\frac{\partial x}{\partial k} = \frac{du}{dk} = u' \qquad (12.5.9)$$

Solving for $\partial u/\partial x$ from (12.5.9) and substituting in (12.5.7), one obtains the following equation of continuity for single-lane vehicular traffic flow

$$\frac{\partial k}{\partial t} + (u + ku') \frac{\partial k}{\partial x} = 0 \tag{12.5.10}$$

If a more generalized equation of motion is desired than (12.5.4), an exponent of proportionality n may be utilized, and we have

$$\frac{du}{dt} = -c^2 k^n \frac{\partial k}{\partial x} \tag{12.5.11}$$

Taking the total derivative of $u = f(x,t)$ gives

$$\frac{du}{dt} = \frac{\partial u}{\partial x} \frac{dx}{dt} + \frac{\partial u}{\partial t} \frac{dt}{dt} \tag{12.5.12}$$

where $dx/dt = u$ and $dt/dt = 1$. Substituting (12.5.12) in (12.5.11) yields

$$\frac{\partial u}{\partial x} u + \frac{\partial u}{\partial t} + c^2 k^n \frac{\partial k}{\partial x} = 0 \tag{12.5.13}$$

From (12.5.9), it is equally apparent that

$$\frac{\partial u}{\partial t} = u' \frac{\partial k}{\partial t} \tag{12.5.14}$$

Solving for $\partial u/\partial x$ from (12.5.9) and substituting in (12.5.13), substituting for (12.5.14) in (12.5.13), then dividing through by u', we see that Eq. (12.5.13) becomes

$$\frac{\partial k}{\partial t} + \left(u + \frac{c^2 k^n}{u'} \right) \frac{\partial k}{\partial x} = 0 \tag{12.5.15}$$

which is the generalized equation of motion. The nontrivial solution of Eqs. (12.5.10) and (12.5.15) is obtained by equating the quantities within the brackets

$$(u')^2 = c^2 k^{(n-1)} \tag{12.5.16}$$

Finally, because of the inverse relation between velocity and concentration,

$$u' = -ck^{(n-1)/2} \tag{12.5.17}$$

Greenberg[8] has solved (12.5.17) for $n = -1$ obtaining

$$u = c \ln \left(\frac{k_j}{k} \right) \tag{12.5.18}$$

The solution of (12.5.17) for $n > -1$ is as follows:

$$u = \frac{-2c}{n+1} k^{(n+1)/2} + C_1 \qquad n > -1 \tag{12.5.19}$$

where the constant of integration is to be evaluated by the boundary conditions inherent in the vehicular velocity-concentration relationship. Thus, since no movement is possible at jam concentration k_j,

$$C_1 = \frac{2c}{n+1} k_j^{(n+1)/2} \qquad n > -1 \qquad (12.5.20)$$

and

$$u = \frac{2c}{n+1} [k_j^{(n+1)/2} - k^{(n+1)/2}] \qquad n > -1 \qquad (12.5.21)$$

Similarly, the implication exists that a driver is permitted his free speed, u_f, only when there are no other vehicles on the highway ($k = 0$). Therefore,

$$u_f = \frac{2c}{n+1} k_j^{(n+1)/2} \qquad n > -1 \qquad (12.5.22)$$

and the constant of proportionality takes on the following physical significance:

$$c = \frac{(n+1)u_f}{2k_j^{(n+1)/2}} \qquad n > -1 \qquad (12.5.23)$$

Substitution of (12.5.23) in (12.5.21) yields the generalized equations of state

$$u = u_f \left[1 - \left(\frac{k}{k_j} \right)^{(n+1)/2} \right] \qquad n > -1 \qquad (12.5.24)$$

$$q = ku = ku_f \left[1 - \left(\frac{k}{k_j} \right)^{(n+1)/2} \right] \qquad n > -1 \qquad (12.5.25)$$

Differentiation of (12.5.25) with respect to k equated to zero gives the optimum concentration k_m, which is that concentration yielding the maximum flow of vehicles:

$$\frac{dq}{dk} = \left[1 - \frac{(n+3)}{2} \left(\frac{k}{k_j} \right)^{(n+1)/2} \right] u_f = 0$$

$$k_m = \left(\frac{n+3}{2} \right)^{-2/(n+1)} k_j \qquad n > -1 \qquad (12.5.26)$$

Substituting (12.5.26) in (12.5.24), one obtains the optimum velocity

$$u_m = \frac{n+1}{n+3} u_f \qquad n > -1 \qquad (12.5.27)$$

The maximum flow of vehicles of which the highway lane is capable

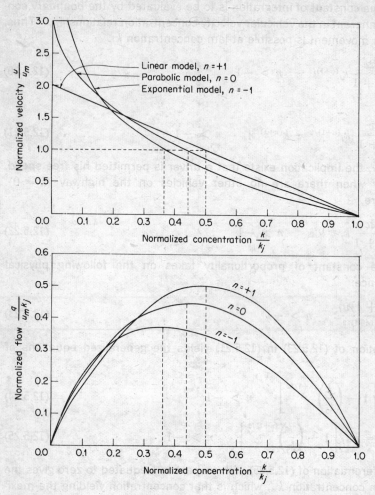

Fig. 12.4 Solution of generalized equation of traffic motion. $du/dt + c^2 k^n (\partial k/\partial x) = 0$, $n = -1, 0, +1$.

(capacity) is obtained from the product of (12.5.26) and (12.5.27)

$$q_m = \left[\frac{n+1}{(1/2)^{2/(n+1)}(n+3)^{[2/(n+1)]+1}} \right] u_f k_j \qquad n > -1 \qquad (12.5.28)$$

Some special cases of (12.5.24) through (12.5.28) have proven to be of significance. Greenshields'[2] linear model is obtainable by setting $n = 1$, and Drew[9] has discussed the case for $n = 0$. These cases, as well as Greenberg's model, are summarized in Fig. 12.4 and Table 12.1.

Table 12.1 Comparison of macroscopic models of traffic flow

Element	General $n > -1$	Exponential $n = -1$	Parabolic $n = 0$	Linear $n = 1$
Equation of motion	$\dfrac{du}{dt} + c^2 k^n \dfrac{\partial k}{\partial x} = 0$	$\dfrac{du}{dt} + \dfrac{c^2}{k}\dfrac{\partial k}{\partial x} = 0$	$\dfrac{du}{dt} + c^2 \dfrac{\partial k}{\partial x} = 0$	$\dfrac{du}{dt} + c^2 k \dfrac{\partial k}{\partial x} = 0$
Constant of proportionality	$c = \dfrac{(n+1)u_f}{2k_j^{(n+1)/2}}$	u_m	$\dfrac{u_f}{2k_j^{1/2}}$	$\dfrac{u_f}{k_j}$
Equation of state	$q = ku_f\left[1 - \left(\dfrac{k}{k_j}\right)^{(n+1)/2}\right]$	$kv_m \ln\dfrac{k_i}{k}$	$ku_f\left[1 - \left(\dfrac{k}{k_j}\right)^{1/2}\right]$	$ku_f\left(1 - \dfrac{k}{k_j}\right)$
Optimum concentration	$k_m = \left(\dfrac{n+3}{2}\right)^{-2/(n+1)} k_j$	$\dfrac{k_j}{e}$	$\dfrac{4k_j}{9}$	$\dfrac{k_j}{2}$
Optimum speed	$u_m = \dfrac{n+1}{n+3}u_f$	c	$\dfrac{u_f}{3}$	$\dfrac{u_f}{2}$
Capacity	$q_m = \dfrac{(n+1)u_f k_j}{(\frac{1}{2})^{2/(n+1)}(n+3)^{[2/(n+1)]+1}}$	$\dfrac{1}{e}u_m k_j$	$\dfrac{4}{27}u_f k_j$	$\dfrac{1}{4}u_f k_j$
Wave velocity	$\dfrac{dq}{dk} = u_f\left[1 - \dfrac{n+3}{2}\left(\dfrac{k}{k_j}\right)^{(n+1)/2}\right]$	$u_m\left[\ln\left(\dfrac{k_i}{k}\right) - 1\right]$	$u_f\left[1 - \dfrac{3}{2}\left(\dfrac{k}{k_j}\right)^{1/2}\right]$	$u_f\left(1 - \dfrac{2k}{k_j}\right)$

12.6 THE MOVING–VEHICLE METHOD

A method of estimating the speed and flow of traffic from a moving vehicle was developed by the Road Research Laboratory, Traffic and Safety Division, and was described by Wardrop and Charlesworth in a paper[10] given to the Institution of Civil Engineers in London in 1954.   The method consists of making a series of runs in a test vehicle.   The observers in the test vehicle measure and record (1) the number of vehicles passed by the test vehicle, (2) the number of vehicles which overtake the test vehicle, and (3) the travel time for the test vehicle.

Some insight into how this information can be used to estimate the speed and flow of a traffic stream is obtained by considering the two special cases illustrated in the time-space diagrams of Fig. 12.5.   In case 1 when the test vehicle is stopped, the flow q is obviously just the number overtaking the test vehicle n_o divided by the duration of the study T:

$$q = \frac{n_o}{T} \tag{12.6.1}$$

In case 2 when all vehicles are stopped except the test vehicle, the density of traffic k is the number of vehicles passed by the test vehicle n_s divided by the length of the test section L:

$$k = \frac{n_s}{L} \tag{12.6.2}$$

Since

$$L = vT \tag{12.6.3}$$

where v is the speed of the test vehicle and T is the travel time over section L, substitution of (12.6.3) in (12.6.2) gives

$$k = \frac{n_s}{vT} \tag{12.6.4}$$

Solving (12.6.1) for n_o and (12.6.4) for n_s and then subtracting (12.6.4) from (12.6.1) gives

$$n_o - n_s = qT - kvT \tag{12.6.5}$$

Letting $n = n_o - n_s$, one obtains the fundamental equation of the moving-vehicle method,

$$\frac{n}{T} = q - kv \tag{12.6.6}$$

It is apparent that since n, T, and v are measured variables, q and k can be solved simultaneously if the test is replicated so that n, T, and v

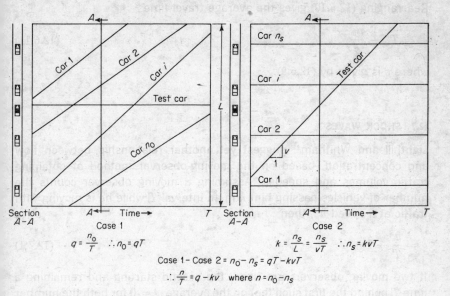

$$q = \frac{n_0}{T} \quad \therefore n_0 = qT$$

$$k = \frac{n_s}{L} = \frac{n_s}{vT} \quad \therefore n_s = kvT$$

Case 1 – Case 2 $= n_0 - n_s = qT - kvT$

$$\therefore \frac{n}{T} = q - kv \quad \text{where } n = n_0 - n_s$$

Fig. 12.5 Derivation of moving vehicle equation.

are different for the two runs.   It is convenient to base this replication on a test run with traffic and a test run against traffic.   Using the subscripts of w for *with traffic* and a for *against traffic* gives

$$\frac{n_w}{T_w} = q - kv_w \tag{12.6.7}$$

and

$$\frac{n_a}{T_a} = q + kv_a \tag{12.6.8}$$

Solving (12.6.7) and (12.6.8) simultaneously for q yields

$$q = \frac{n_w + n_a}{T_w + T_a} \tag{12.6.9}$$

To get the average travel time of the traffic stream $\bar{T}$, one uses (12.6.7), replacing k by $q/\bar{v}$ and v_w by L/T_w:

$$\frac{n_w}{T_w} = q - \frac{q}{\bar{v}} \frac{L}{T_w}$$

$$= q - \frac{q\bar{T}}{T_w} \tag{12.6.10}$$

Rearranging (12.6.10) gives the average travel time

$$\bar{T} = T_w - \frac{n_w}{q} \tag{12.6.11}$$

where q is given by (12.6.9).

12.7 SHOCK WAVES

Lighthill and Whitham[11] suggest yet another relationship between flow and concentration, based on the moving-observer method of obtaining traffic volumes and speeds. Assuming a moving observer counts the number of vehicles passing him n in an interval T while he is moving with traffic at a speed u_w, then

$$q = \frac{n}{T} + k u_w \tag{12.7.1}$$

If two moving observers are used, the second starting and remaining a time T behind the first such that on the average $n = 0$ for both the number of vehicles between observers is constant, then

$$q_1 - q_2 = u_w(k_1 - k_2) \tag{12.7.2}$$

In other words, when changes in flow are occurring, the changes through the stream of vehicles travel at a velocity, called the *wave velocity*, given by

$$u_w = \frac{dq}{dk} \tag{12.7.3}$$

If the equation of continuity (12.5.6) is multiplied by the wave velocity $u_w = dq/dk$, the characteristic equation of one-dimensional wave motion is obtained as

$$\frac{\partial q}{\partial t} + u_w \frac{\partial q}{\partial x} = 0 \tag{12.7.4}$$

Consider the lines in the space-time plane given by

$$u_w = \frac{dx}{dt}$$

Substitution in (12.7.4) gives

$$\frac{\partial q}{\partial t} + \frac{dx}{dt} \frac{\partial q}{\partial x} = 0$$

$$\frac{\partial q}{\partial t} dt + \frac{\partial q}{\partial x} dx = 0 \tag{12.7.5}$$

The left side of (12.7.5) is of course the total derivative of flow

$$dq = \frac{\partial q}{\partial t} dt + \frac{\partial q}{\partial x} dx = 0$$

Since $dq = 0$ along the lines u_w, q must be constant along the path of u_w in the space-time plane.

Differentiating $q = ku$ with respect to k gives

$$\frac{dq}{dk} = u_w = u + k \frac{\partial u}{\partial k} \tag{12.7.6}$$

Equation (12.7.6) requires that the wave velocity always be less than the space mean speed since $\partial u / \partial k$ is negative. Finally, just as it was shown that q is constant along u_w, it may be shown that k and u are also constant along these waves.

12.8 THE BOTTLENECK–CONTROL APPROACH

Closely related to the theory of traffic waves is the concept of the bottleneck—a stretch of roadway with a flow capacity less than the road ahead. The upper q-k curve in Fig. 12.6 is for the roadway ahead of the bottleneck, and the lower one refers to the bottleneck itself. When the traffic volume

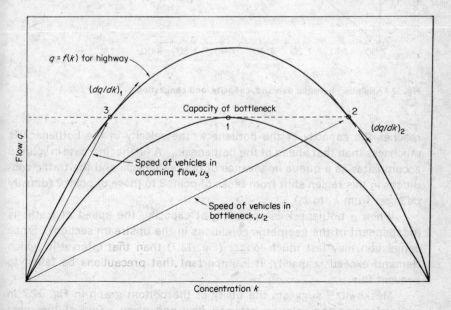

Fig. 12.6 Traffic flow in a bottleneck.

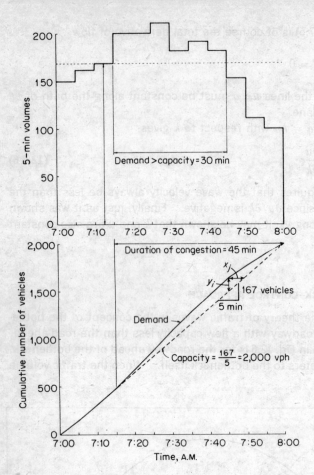

Fig. 12.7   Relation between demand, capacity, and congestion.

reaches the capacity of the bottleneck, the velocity in the bottleneck is much less than that ahead of the bottleneck.   A further increase in volume accumulates as a queue in advance of the bottleneck, and the traffic conditions in this region shift from those of point 3 to those of point 2 (density changes from k_3 to k_2).

When a bottleneck is operating at capacity, the speed of traffic is independent of the geometric conditions in the upstream section.   Since congestion may last much longer (Fig. 12.7) than that interval in which demand exceeds capacity, it is important that precautions be taken to prevent this.

Moskowitz[12] suggests the utility of the bottom graph in Fig. 12.7 in understanding the relation between flow and delay.   One of the most useful things about the graph is that it shows flow as a rate (i.e., a mathematical slope) which is time related, instead of as a point which is not time related.

The figure represents cars going through a bottleneck as a function of time during a peak period.   The slope of the upper curve at any point is the arrival rate, and the slope of the lower curve is the service rate, or the actual flow.   It can be seen that at 7:15 the arrival rate begins to exceed the service rate.   At 7:45 the arrival rate is equal to the service rate, and from 7:45 until 8:00 the arrival rate is less than the service rate.   At 8:00 the congestion is over.

It can be seen that theoretically the delay to an individual vehicle, x_j in the figure, is much greater at 7:49 (when vehicle j goes through the bottleneck) than it is at 7:15 or at any other time of day between 7:15 and 7:45, although the predicted demand is *less* at 7:49 than it is at any time between 7:15 and 7:45.   Using a speed-volume curve, a person might infer that the speed was close to 0 at 7:15 $+ \Delta t$, just after the queue begins to form, because the volume/capacity ratio at 7:15 $+ \Delta t$ is very high and is considerably greater than 1.0, whereas in fact the delay is very small, as can be seen on the graph.   The same person might infer that subsequent to 7:45, speed would be very high because the demand/capacity ratio is less than 1, and yet a glance at the figure shows that the jth vehicle, arriving at 7:45, suffers the greatest delay of any vehicle.

The actual flow rate stays nearly constant from 7:15 to 7:45, but the journey time of any individual vehicle varies considerably, and the queue length y_j also varies.   The spot speed immediately upstream of the bottleneck will be variable, from 0 up, on a moment-to-moment basis, and this is what many investigators call *forced flow*.   The spot speed in the bottleneck or downstream of it will be more uniform during time and will vary, depending on how far downstream the observation is made, because cars will generally be accelerating.   If the bottleneck is a merging area on a freeway, the volume or flow upstream of the bottleneck is less than the volume or flow in the bottleneck by the amount added at the ramp; i.e., the ramp volume plus the freeway volume is equal to the downstream volume.

The type of input output graph illustrated in Fig. 12.7 to explain congested freeway operation is also useful in depicting operation at a signalized intersection.   In Fig. 12.8, the conditions at a fixed-time signalized intersection approach with a constant arrival rate q and constant departure rate during the green interval Q are shown.   It is apparent that the arrivals per signal cycle are less than the capacity for departures or

$$(R + G)q < GQ \qquad (12.8.1)$$

The maximum queue n on the approach exists at the end of the red phase

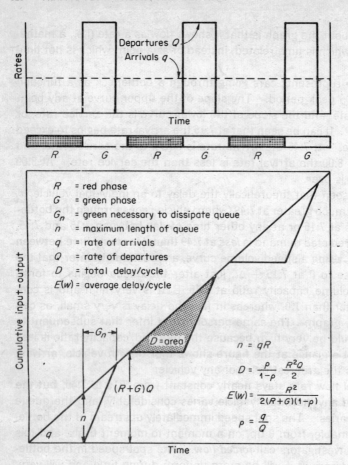

Fig. 12.8 Demand-capacity relations at a signalized intersection.

and is given by the expression

$$n = qR \tag{12.8.2}$$

The green time G_n necessary to dissipate the queue of n vehicles is obviously less than the total green phase G and may be expressed symbolically as

$$G_n = \frac{\rho}{1 - \rho} R \tag{12.8.3}$$

where ρ is the queueing utilization factor q/Q, also called the *degree of saturation*. The total delay D is represented by the area between the cumulative input and cumulative output curves, or

$$D = \frac{\rho}{1 - \rho} \frac{R^2 Q}{2} \tag{12.8.4}$$

The average delay $E(w)$ to all vehicles using the approach is the total delay divided by the total number of vehicles:

$$E(w) = \frac{R^2}{2(R + G)(1 - \rho)}$$
(12.8.5)

The delay-flow relationship for intersection flow has been well established by Webster,[13] Miller,[14] Newell,[15] and others for fixed time signals. The first term of their forms is the same as (12.8.5), with corrections added to account for the simplifying assumption regarding arrivals.

12.9 STREAM MEASUREMENTS

In the study of traffic flow on a busy facility, it is necessary to describe the motion of a great number of vehicles. Continuing with the hydrodynamic analogy developed in this chapter, we see this task is similar to that of describing the motion of an infinite number of fluid particles. A quantity such as velocity must be measured relative to some convenient coordinate system. The two methods commonly used in fluid mechanics are the Euler and Lagrangian methods of analysis.

One may choose to remain fixed in space and observe the fluid pass by a given point. This method, whereby a fixed coordinate system is established, is referred to as the Euler method of analysis. The Lagrangian method of analysis involves establishing a coordinate system relative to a moving fluid particle as it flows through the continuum and measuring all quantities relative to the moving particle.

Similar methods of analysis may be utilized to describe a traffic system. They are summarized in Fig. 12.9. By careful design of a freeway study, point, instant, or floating studies can be used to give essentially continuous coverage in both time and space. The point and instant methods are more applicable to the measurement of the macroscopic properties of the traffic stream. Moving-vehicle methods are microscopic in nature and analogous to the Lagrangian method of analysis used in fluid mechanics.

In studying road traffic as a stream or fluid continuum, one is primarily interested in the three principal characteristics of the stream—flow (quantity per unit time), concentration (quantity per unit space), and speed (space per unit time)—and how they vary in time and space. In pursuit of this interest, several methods of observation have been described above.

Since road traffic is actually made up of discrete vehicles, rather than being in fact a continuous fluid, measurements are made on individual vehicles. The need to convert these discrete measurements into the desired continuum characteristics has led to precise definitions of the latter

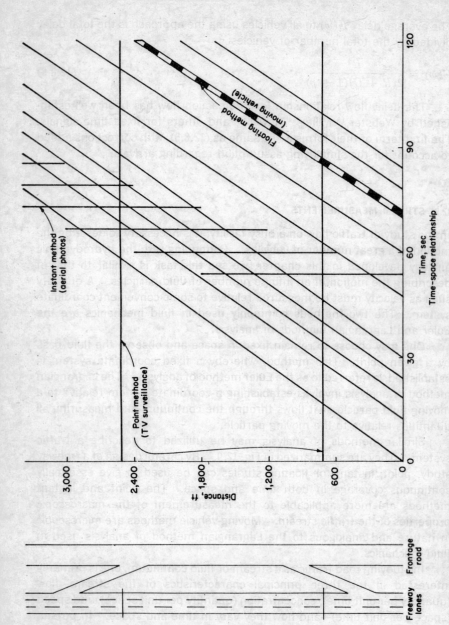

Fig. 12.9 Illustration of study procedures.

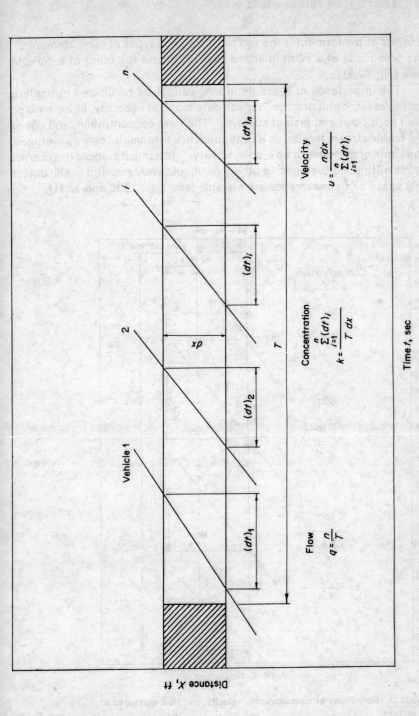

Flow
$$q = \frac{n}{T}$$

Concentration
$$k = \frac{\sum_{i=1}^{n}(dt)_i}{T \, dx}$$

Velocity
$$u = \frac{n \, dx}{\sum_{i=1}^{n}(dt)_i}$$

Time t, sec

Fig. 12.10 Definitions of flow, concentration, and velocity measured at a point.

in terms of the former for the two most common types of measurements,[16] one type made at a point in space (Fig. 12.10) and the other at a point in time (Fig. 12.11).

The importance of these defintions cannot be minimized in modern traffic research and practice in which data must, of necessity, be assembled from both point and instant surveys. The flow, concentration, and speed of a traffic stream treated as a continuum are meaningful only as averages. The kinds of averages to be employed vary. In some instances it is correct to use arithmetic averages, in others harmonic averages, and in still others only space or time averages are suitable (see Figs. 12.10 and 12.11).

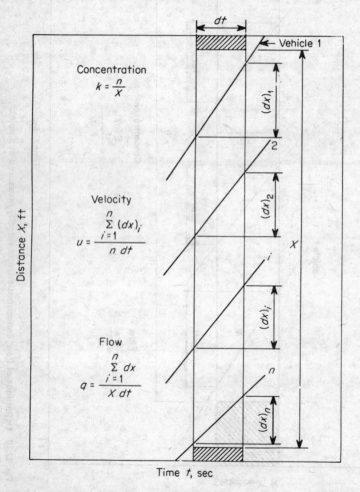

Fig. 12.11 Definitions of concentration, velocity, and flow measured at an instant.

The primary characteristics of traffic movement are concerned with speed, density, and volume. These three fundamendal characteristics are dependent on the geometric design of the roadway, the composition of the traffic stream, and combinations of both. Interest in these characteristics is manifest in the need for specific indications of impending traffic congestion, which can then be utilized in control processes intended to maintain maximum efficiency.

Contour maps provide an ideal means of documenting the variations in these characteristics in time and space. Figure 12.12 illustrates volume, speed, and density contours for the three outbound lanes of a 6-mile area of the Gulf Freeway throughout the 2-hr afternoon peak. These maps were obtained by using time as the ordinate and distance along the freeway under the plan view as the abscissa. Ten aerial flight strips were utilized over the 2-hr peak, and the speed-density characteristics were averaged over each of some 42 photos per strip; each of the contour maps was therefore drawn from about 420 points. Just as conventional land contours connect points of equal elevation, speed contours connect points of equal speed, and density contours connect points of equal density.

It is evident from the contour maps in Fig. 12.12 that the facility south of the Griggs Street overpass (station 220) operates very efficiently throughout the entire 2-hr peak. However, operation between stations 20 and 220 begins to deteriorate at about 4:50 P.M. and does not recover at the north extremity until almost 6 P.M. It is noted that almost without exception increases in density are accompanied by decreases in speed. The range in average speed and the range in average density plainly decrease as one travels away from the CBD during the afternoon peak.

The analysis of highway traffic is more than making measurements and collecting facts. Although this exploration of the true nature and characteristics of freeway traffic is the necessary beginning in providing new ways of improving performance, the freeway traffic phenomenon is so complex that a collection of bare data tells little more than is already known.

From the contour maps, one can conclude that congestion is the severest at the north end of the freeway (station 20) and least severe at the south extremity in the vicinity of the Reveille Interchange. However, the exact degree of congestion or level of operation is not known. Some control parameters, calculated by the regression techniques, are plotted on the profiles: (1) the speed at capacity (critical speed), (2) the speed at the optimum service volume, (3) the optimum service volume, etc. These parameters afford a rational quantitative means for interpreting the level of operation on the facility according to the energy-momentum definitions to be discussed in Chap. 14.

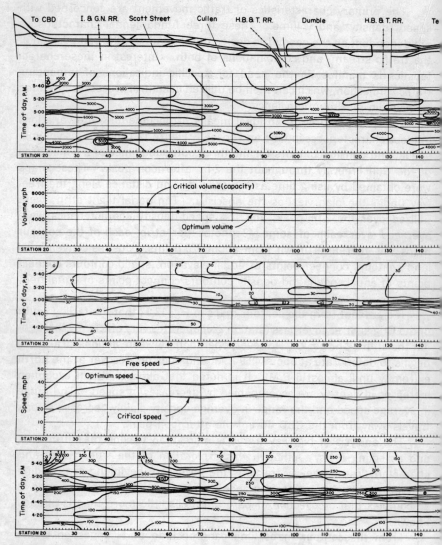

Fig. 12.12 Outbound operational characteristics, all lanes.

12.10 SYSTEM MODELS

The models discussed so far deal primarily with events at a point in space or time. In developing plans for optimizing the operation of an entire freeway system (one direction of freeway several miles in length), two types of models have proved useful.[17] They are (1) a model for the estimation of demand at freeway bottlenecks and (2) linear programming models. Both models would use total freeway capacity as the control parameter

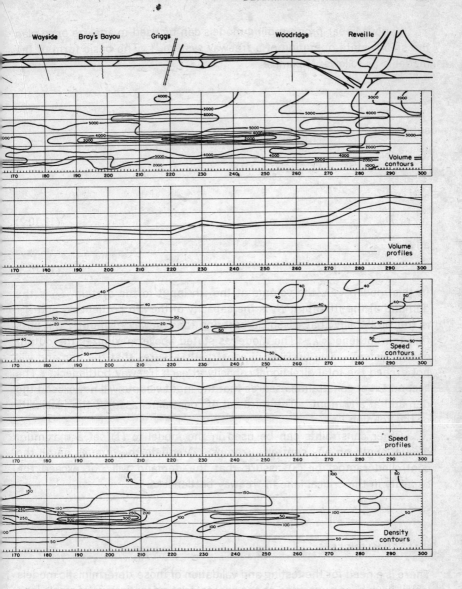

and attempt to keep the demand less than the capacity (or service volume) at each bottleneck.

The demand at a freeway bottleneck consists of vehicles which entered at upstream entrance ramps. If the entering volume is known at each input upstream of the bottleneck and if for each input the percent of vehicles which pass through the bottleneck is known, the demand on the bottleneck can be estimated. This technique has proved valuable in developing plans for controlling the inputs to a freeway system.[18]

Various linear programming models can be used to analyze or to plan the steady-state operation of a freeway system.[19]   The basic form of the model is as follows: Maximize

$$\sum_{j=1}^{n} X_j$$

subject to

$$\sum_{j=1}^{n} A_{jk}X_j \leq B_k \qquad k = 1, \ldots, n \tag{12.10.1}$$

and

$$X_j \leq D_j \qquad j = 1, \ldots, n \tag{12.10.2}$$

and

$$X_j \geq 0 \qquad j = 1, \ldots, n \tag{12.10.3}$$

In this model X_j is the volume at the jth input to the system, D_j is the demand at the jth input, B_k is the capacity of the kth bottleneck, and A_{jk} is the probability that a vehicle entering the freeway at input j will pass through bottleneck k. The model as stated above maximizes the output of the system. The first set of constraints assures that the demand will not exceed the total directional capacity at each freeway bottleneck. The second set of constraints assures that the input volume will not exceed the demand at any input. The third set of constraints assures the feasibility of the solution.

Other constraints can be used in the model to (1) set a maximum queue length on any controlled ramps, (2) equate the queues on all the controlled ramps, or (3) assure that the demand will not exceed the capacity in the critical lane in freeway bottlenecks.

12.11 CHALLENGES

There is a need for the testing and validation of those deterministic models which have been suggested as the characteristic descriptions of single-lane traffic flow. This accomplished, there is a necessity in determining which, of the many models that have been advanced, most accurately describes traffic flow on different types of unidirectional highways, tunnels, bridges, and elevated freeways.

In short, research should be undertaken to discover which of the mathematical descriptions of traffic flow describe the "real world" phenomena.

PROBLEMS

12.1. Fit a curve of the form $y = \alpha + \beta x$ to the points (0,6), (1,7), (2,9), (4,8). First plot the data and fit a straight line curve by estimation, then fit by a *least-squares* line.

12.2. Evaluate the following macroscopic equation of motion for realism, limiting discussion to 50 words or less:

$$\frac{dx}{dt} = -\frac{c^2}{k}\frac{\partial k}{\partial x}$$

12.3. Starting with $q = ku$, develop the following equation of state through application of boundary conditions:

$$q = uk_m \ln \frac{u_f}{u}$$

12.4. Test vehicles A and B traverse a 1-mile section of one-way street. Vehicle A makes the run in two-thirds of the time it takes vehicle B. Vehicle A passes 15 more vehicles than pass it, whereas vehicle B is overtaken by 10 more vehicles than it passes. Find the average number of vehicles on the test section during the runs.

12.5. Utilizing the following data taken from a time-lapse aerial-photography flight over a freeway, fit the traffic flow model

$$u = u_c \ln \frac{k_j}{k}$$

to the data by *least-squares* regression. Plot the speed-vs.-density, speed-vs.-flow, and flow-vs.-density curves, showing the data points on each plot.

Speed-concentration measurements

Speed, mph	Density, vpm
11.5	125.5
28.1	70.2
31.3	41.8
39.8	36.6
39.9	39.6

12.6. Given a roadway where section A is approaching a $\frac{1}{4}$-mile school zone, section B is the $\frac{1}{4}$-mile school zone (operates only 15 min), and section C is the road downstream of the school zone. Stream measurements are as follows:

Section A $q_a = 1{,}250$ vph $u_a = 50$ mph

Section B $q_b = 1{,}000$ vph $u_b = 20$ mph

Section C $q_c = 1{,}200$ vph $u_c = 30$ mph

Determine the speed of the shock wave at $A\text{-}B$ (the start of the school zone) and at $B\text{-}C$ (the end of the school zone).

Note: Use the linear flow model where necessary.

12.7. Observations at a signalized intersection of 2 one-way streets near a university located in a city of 20,000 persons yield the following input output data on the university-bound approach. The number of vehicles counted in by 1-min intervals starting at 7:30 A.M. are 15, 19, 17, 25, 21, 32, 31, 33, 34, 36, 28, 31, 30, 29, 26, 21, 20, 15, 10, 10, 16, 14, 15, 17, 13, 18, 12, 14, 10, 16. The number of vehicles counted out by 1-min intervals starting at 7:30 A.M. are 15, 19, 17, 25, 21, 25, 25, 25, 25, 25, 25, 25, 25, 25, 25, 25, 25, 25, 25, 25, 16, 15, 17, 13, 18, 12, 14, 10, 16. Utilizing this information (a) plot input vs. time and output vs. time on the same set of coordinate axes, (b) plot the cumulative value of input output vs. time, (c) plot the estimated 1-min queue length vs. time, (d) determine the maximum estimated queue length, (e) estimate the capacity of the approach, (f) estimate the duration of congestion on the approach, and (g) estimate how long the input exceeds the capacity of the approach.

12.8. An energetic graduate student would rather count calories than cars. On his way home along a one-way street he alternately runs awhile and stops to rest, etc., counting cars all the time. He gets home 10 min later and tells his wife that he has counted 10 vehicles while running with traffic and 20 vehicles while resting and that his average running speed was $2\frac{1}{2}$ times his overall speed. Before filing for separate maintenance, she determines the travel time of the traffic stream. Can you?

12.9. The following vehicle displacements were measured on a traffic circle of 210-ft radius from two aerial photos taken 1 sec apart: 44, 44, 44, 66, 66, 33, 88, and 33 ft. Based on this information, how much traffic do you estimate uses the traffic circle in 1 hr?

12.10. Fit Greenshields' linear traffic flow model expressed as

$$q = ku_f\left(1 - \frac{k}{k_j}\right)$$

to the following freeway data. Plot the speed-vs.-density, speed-vs.-flow, and flow-vs.-density curves with data points on each plot. What is the significance of the various parameters that can be obtained from the model?

Speed-concentration measurements

Speed, mph	Density, vpm
46.0	22.2
56.0	29.3
48.0	30.1
42.0	40.7
22.9	90.0
56.6	29.2
9.2	113.3
17.9	88.2
14.5	103.1
29.5	60.2
25.0	67.4
32.0	70.6
35.5	41.1
36.5	49.6

12.11. Fit a curve of the form $u = u_f e^{-k/k_m}$ to the following speed-density data, minimizing the sum of the squares of the deviations; then determine the lane capacity of the facility from which the data were taken:

$$k = 20,\ 65,\ 120 \text{ vpm}$$

$$u = 44,\ 26,\ 6 \text{ mph}$$

REFERENCES

1. Haynes, J. J.: Some Considerations of Vehicular Density on Urban Freeways, *Freeway Surveillance Control Rept.* 24-6, Texas Transportation Institute, Texas A&M University, College Station, 1965.
2. Greenshields, B. D.: A Study in Highway Capacity, *Highway Res. Board Proc.* 14, pp. 448–477, 1934.
3. Edie, L. C.: Car-following and Steady-state Theory for Non-congested Traffic, *Operations Res.*, vol. 9, no. 1, pp. 66–67, January–February, 1961.
4. May, A. D., P. Athol, W. Parker, and J. B. Rudden: Development and Evaluation of Congress Street Expressway Pilot Detection System, *Highway Res. Board Record* 21, 1963.
5. Morrison, R. B.: The Traffic Flow Analogy to Compressible Fluid Flow, *Advanced Res. Eng. Bull.* M 4056-1, Ann Arbor, Mich., 1964.
6. Harr, M. E., and G. E. Leonards: A Theory of Traffic Flow for Evaluation of the Geometric Aspects of Highways, *Highway Res. Board Bull.* 308, 1961.
7. Reddy, M. S.: "Quantitative Evaluation of the Effect of Merging Vehicles on Freeway Operation," doctoral dissertation, Texas A&M University, College Station, 1966.
8. Greenberg, H.: An Analysis of Traffic Flow, *Operations Res.*, vol. 7, no. 1, pp. 79–85, January–February, 1959.
9. Drew, D. R.: "A Study of Freeway Traffic Congestion," doctoral dissertation, Texas A&M University, College Station, 1966.
10. Wardrop, J. G., and G. Charlesworth: A Method of Estimating Speed and Flow of Traffic from a Moving Vehicle, *Proc. Inst. Civil Engrs. (London)*, pt. 2, vol. 3, 1954.
11. Lighthill, M. J., and G. G. Whitham: On Kinematic Waves, II: A Theory of Traffic Flow on Long Crowded Roads, *Proc. Roy. Soc. (London)*, ser. A, vol. 229, pp. 317–345, May 10, 1955.
12. Moskowitz, Karl, and Leonard Newman: Notes on Freeway Capacity, *Highway Res. Board Record* 27, 1963.
13. Webster, F. V.: Traffic Signal Settings, *Dept. Sci. Ind. Res. Road Res. Tech. Paper* 39, London, 1958.
14. Miller, A. J.: "Settings for Fixed-cycle Traffic Signals," *Australian Road Res. Board Proc.*, Melbourne, 1964.
15. Newell, G. F.: Queues for a Fixed Cycle Traffic Light, *Ann. Math. Statist.*, vol. 31, no. 3, 1960.
16. Edie, L. C.: "Discussion of Traffic Stream Measurements and Definitions," The Port of New York Authority, N.Y., 1964.
17. Wattleworth, J. A.: "Peak-period Control of a Freeway System: Some Theoretical Considerations," doctoral dissertation, Northwestern University, Evanston, Ill., 1963. (See also *Rept.* 9., Chicage Expressway Surveillance Project, 1963.)
18. Pinnell, C., D. R. Drew, W. R. McCasland, and J. A. Wattleworth: Inbound Gulf Freeway Ramp Control Study I, *Texas Transportation Inst. Res. Rept.* 24-10, Texas A&M University, College Station, 1965.
19. Wattleworth, J. A.: Estimation of Demand at Freeway Bottlenecks, *Traffic Eng.*, vol. 35, no. 5, pp. 21–26, 1965.

CHAPTER THIRTEEN
THE MAN - MACHINE SYSTEM

Man is the measure of all things.

Protagoras

13.1 INTRODUCTION

As is well known, the automobile driving situation consists of three major factors—driver (man), car (machine), and environment—all interacting in a complicated manner in both time and space. In view of this, a control-systems approach to describe the relationships between stimulus and response or between input and output is a logical one.

For the purpose of organization, it is convenient to classify control systems into two broad types. An *open-loop system*, such as a rocket-firing system, once started is no longer controlled. The other type of system actually measures the quantity to be controlled, compares the actual value to the desired value, and if they are not the same, institutes corrective action. This type of system is called a *closed-loop* or *feedback control system*. The differences between these two basic types are worth emphasizing. There is no unified theory which is common to all open-loop systems; the differences among these open-loop systems are more signifi-cant than the similarities. On the other hand, analysis techniques for closed-loop systems constitute a unified theory independent of the particular applications and as such are useful in traffic flow theory.[1]

13.2 FEEDBACK CONTROL

Systems which utilize feedback for control purposes have become essential elements in modern technology.   They range from simple toys to complex automatic equipment.   In a simple feedback system, the output of the system is compared with the input to the system.   Such a system is self-correcting in the sense that any deviations from the desired performance are used to produce corrective action.   A good example of such a regulating system is a home heating system.   The input to the system is the setting of the thermostat.   Whenever the difference between the room temperature and the thermostat setting is large enough, a sensing device either opens or closes an electric circuit which controls the furnace.   Thus, the system regulates the temperature of the home within some predetermined range.

A system of the type described above can be represented by a general block diagram, as is shown in Fig. 13.1.   The most important elements of closed-loop notation are shown in the figure.   From the functional standpoint the two most fundamental variables in the system are the input and output, denoted by r (for reference input) and c (for controlled variable). When the actual functions are being indicated, lowercase letters are used; specific dependence on time is indicated by (t).   Uppercase letters are generally used if the Laplace transform of a time function is implied, with the dependence on the complex variables indicated by (s).   Thus $C(s)$ indicates the transform of the time function $c(t)$.   Two other variables of considerable importance in control-system analysis are the feedback signal $b(t)$ and the error $e(t)$, which is obtained by subtracting the feedback signal from the output.   The symbol G is used for the *transfer function* of elements in the forward branch of the loop and the symbol H is used for elements in the feedback path.[2]

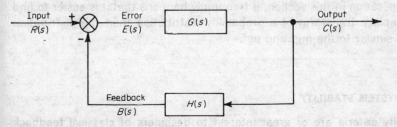

Fig. 13.1 Block diagram showing nomenclature for feedback system.

A transfer function is essentially a mathematical description of the ratio of the output of a component or a system to its input. The characteristics of feedback control may be summarized in the following equations:

$$E(s) = R(s) - H(s)C(s) \tag{13.2.1}$$
$$C(s) = E(s)G(s) \tag{13.2.2}$$

The solution of these equations for the ratio between the output and input yields the transfer function T:

$$T = \frac{G(s)}{1 + G(s)H(s)} \tag{13.2.3}$$

13.3 THE LAPLACE TRANSFORM

The theory of Laplace transforms, also referred to as operational calculus, has in recent years become an essential part of the mathematical background required of engineers, physicists, mathematicians, and other scientists.
The notation associated with the Laplace transform was mentioned in the previous section; if a function of t is indicated in terms of a lowercase letter such as $f(t)$, the Laplace transform of the function is denoted by the corresponding uppercase letter $F(s)$. By definition then the Laplace transform of $f(t)$, denoted by $\mathcal{L}[f(t)]$, is

$$\mathcal{L}[f(t)] = F(s) = \int_0^\infty e^{-st} f(t)\, dt \tag{13.3.1}$$

Just as logarithms facilitate arithmetic operations, the utility of the Laplace transformation lies in the fact that it may be applied to a differential equation in t; the resulting equation in s may then be manipulated in a purely algebraic (and therefore simpler) manner to yield an expression which may be transformed back (analogous to getting the antilog) into an expression in t which is the solution of the differential equation. Just as there are tables of logarithms, there are tables of Laplace transforms which eliminate most of the difficulty in applying the operator. In addition to its application in this section, it frequently happens that it is easier to find the Laplace transform of a probability distribution than the distribution itself, similar to the mgf and pgf.

13.4 SYSTEM STABILITY

Stability criteria are of great interest to designers of classical feedback systems since the possibility of instability is inherent in all such systems.

For any system, stability must ultimately be defined in terms of the response or time variation of some system variable. This variable, which may be a displacement, temperature, voltage, or other physical quantity, depends upon the input as well as upon the initial conditions of the system.

Theoretically, in absolutely unstable systems the controlled variable approaches infinity (Fig. 13.2). This is of course physically impossible because it would require a source of infinite power to drive the controlled variable. What happens is that amplifiers and motors saturate and valves go wide open, and then pressure vessels may explode or safety valves may pop. All these occurrences will lead the system either to a dead stop or into a sustained nonlinear oscillation.[1]

In systems work we must content ourselves with a mathematical model which fits the physical situation to a reasonable degree of approximation. As in all applied mathematics, we only expect the mathematical model to fit over a finite range of variables involved. From the above discussion the following practical criterion may be arrived at: A physically

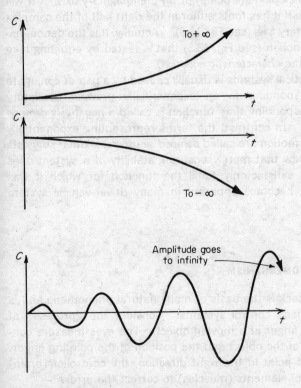

Fig. 13.2 Theoretical behavior of an unstable system.

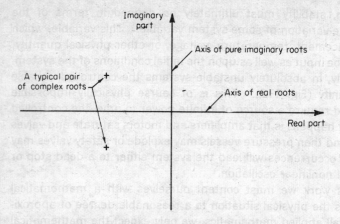

Fig. 13.3 Root location in complex plane.

realizable transfer function will be said to be stable if it has no roots in the right half of the complex plane (s plane) or on the imaginary axis.[3]   It will be classed as unstable if it has roots either in the right half of the complex plane or on the imaginary axis (see Fig. 13.3).   Actually, it is the denominator of the transfer function (see Fig. 13.3) that is tested by equating it to zero.   This is called the *characteristic equation.*

Instability in practical systems is usually caused by a pair of conjugate complex roots corresponding to an exponentially growing oscillatory response.   The corresponding time function is called a *negatively damped* or *undamped sinusoid*.   In contrast, the terms representing exponentially decreasing oscillatory motion are called *damped sinusoids*.   Wilts[2] suggests several reasons to show that merely ensuring stability of a system does not guarantee it will satisfactorily fulfill the function for which it was intended, a fact that becomes apparent in many driver-vehicle system situations.

13.5 THE HUMAN SERVOMECHANISM

The principle of feedback is the basis of many natural phenomena and is not limited to man-made control systems.   Consider the "process" of a person pointing his finger at a moving object.   The eyes measure continuously the position of the object and the position of the pointing finger. If the finger does not point in the right direction, the controller (brain) signals the final control elements (muscles) to correct the error.

The current development of complex man-machine systems such as missiles, space ships, surveillance systems, and automated production systems has dramatized the role of man in the closed-loop system. In contemplating the human being within this system and attempting to describe him in transfer-function terms, it would seem that the human operator serves in a manner analogous to a servomechanism. The peculiarities of the input/output ratio involved in describing the transfer functions of people are those relating to sensory input (usually visual signals) and physical response (typically the operation of controls). Of course, such variables as perception, decision, reaction time, and physical control should be included within the framework of this "human transfer function."[4]

Driving an automobile is an example of a man-machine feedback control system.[5] The system input is the panorama of the winding road and stream of vehicular traffic as seen through the outline of the driver's own windshield. The display includes such instruments as his speedometer. The controls are the steering wheel, accelerator, and brake. The output of these controls acts upon the vehicle to yield the movements of the auto, or system output.

To illustrate use of feedback in the driver-vehicle system, Fig. 13.4 contains a block diagram describing the control of automobile speed by a human operator. The driver decides upon an appropriate speed (input) for the highway on which he is driving. He observes the actual speed (output) by looking at the speedometer. He makes a mental evaluation of the relationship between the observed speed and his desired speed (error). If he is traveling too slowly, he depresses the accelerator in proportion to the deficiency in speed. This action supplies more fuel to the engine, and the car speeds up. If a later observation reveals he has overcompensated, the driver releases his pressure on the accelerator, and the car slows down.

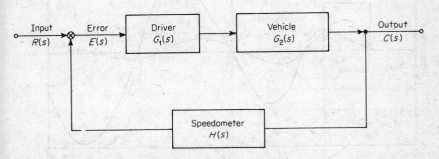

Fig. 13.4   Block diagram for automobile driving control.

13.6   TIME LAGS

Many real-world systems such as fractionating towers, production lines, producer-to-wholesaler-to-retailer schedules, and open flume flows are difficult to control because of dead time or a transportation lag.[6]   Since mechanical, chemical, electrical, and human reactions are never instantaneous, perfect control is impossible.   One end of a river can be full or empty for a while before the other end "sees" it; the composition or temperature can be changed at one end of a pipe before the other end "knows" it.

Figure 13.5 serves to illustrate the concept of system time lag and its relation to stability.[5]   The ordinates are shower temperature and the abscissas are time.   At $t = 0$ sec, the temperature is the desired controlled temperature, and at $t = 1$ sec, the temperature begins to rise. In the idealized situation given in Fig. 13.5a, the man taking the shower feels this increase in temperature with a time delay of 1 sec, and, instituting corrective action, he returns the temperature to desired value.

In 13.5b considerable realism has been added.   First, the temperature is shown rising gradually; second, corrective action is applied gradually (turning on the "cold" faucet).   This drop in temperature continues until at $t = 2.5$ sec the temperature drops below the desired value.   When the bather notices this, he institutes a correction proportional to the observed "error."   The dampening of the time-temperature curve in Fig. 13.5b illustrates a stable system with a time lag, as opposed to instability in Fig. 13.5c.   Here, reaction to the first disturbance led to an over-correction—that is, a larger error in the opposite direction.

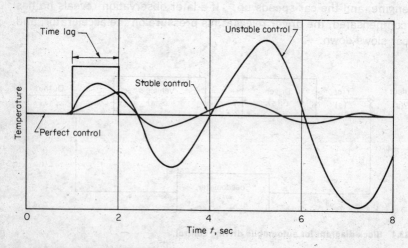

Fig. 13.5   Time-lag and stability concepts in system control.

The essence of the time-lag problem is that a correction that was right at the time of the observation may be wrong at the time that it is actually carried out in the system.[7]   Extreme sensitivity may cause a rapid change that continues after there is no need for it, and an error is caused in the opposite direction.   When we walk carrying a tray of water, we may be tempted to watch the water in the tray and try to tilt the tray to keep the water from spilling.   This is often disastrous.   The more we tilt the tray to avoid spilling the water, the more wildly the water may slosh about. When we apply feedback to change a process on the basis of its observed state, the overall situation may be unstable.   That is, instead of reducing small deviations from the desired goal, the control we exert may make them larger.

13.7 LINEAR CAR–FOLLOWING

We have seen in Fig. 13.4 how driving an auto on the open road is an example of man-machine feedback control systems.   On a busy highway, the average driver is motivated less by time-distance economy than he is by safety.   The driver is faced with a difficult situation; he must adjust to other vehicles in the traffic stream.   It may be observed that whenever a driver approaches fairly close to the vehicle ahead of him and is unable to pass, he slows down.   The purpose of this present discussion is to develop a model of the dynamics of a stream of traffic which results on assuming that the drivers of the various vehicles strive to maintain a safe headway.   This safe-headway concept will be explained first.

It is convenient to show the progress of a stream of vehicles, designated as $i = 1, 2, \ldots , n$, in terms of a time-space diagram (Fig. 13.6).   The ordinate represents distance x and the abscissa represents time t.   The slope of the line gives the rate of change of distance with respect to time, dx/dt, or vehicular velocity, $\dot{x}_i$ or v_i.   A curved line means a changing slope or changing velocity designated as vehicular acceleration, $\ddot{x}_i$ or $\dot{v}_i$.   When following other vehicles, a driver often adjusts his speed to maintain a safe headway.   The concept of safe headway s_{i+1} is illustrated in Fig. 13.6. It is that headway which a vehicle must maintain at an instant $(t - T)$ to avoid a rear-end collision with a preceding vehicle which is in the process of braking to a stop:

$$s_{i+1}(t - T) = C_{i+1}v_{i+1}^2(t) + L_i + Tv_{i+1}(t) - C_i v_i^2(t - T)$$

where $C_i = (2f_i g)^{-1}$ and T is a time lag.   If it is assumed that the braking capabilities and speeds of the vehicles are equal, and then writing the average safe headway in terms of the position x_{i+1} and x_i, one obtains

$$x_i(t - T) - x_{i+1}(t - T) = Tv_{i+1}(t) + L_i \tag{13.7.1}$$

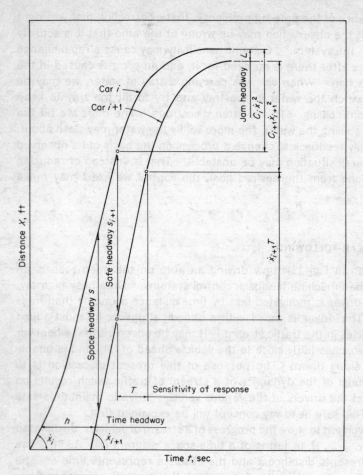

Fig. 13.6   Fundamental relations of traffic in time and space.

It is interesting to note that (13.7.1) provides the basis for the popular "rule of thumb" for following another vehicle, namely, to allow one car length for every 10 mph of speed.

Differentiating Eq. (13.7.1) with respect to time gives

$$v_i(t - T) - v_{i+1}(t - T) = T\dot{v}_{i+1}(t) \qquad i = 1, 2, 3, 4 \tag{13.7.2}$$

From Eq. (13.7.2) the motion of any vehicle in the line can be found if the motion of the lead car is known.[8,9] To accomplish this, it is convenient to introduce the Laplace transform of the velocity $v_i(t)$, utilizing the following notation:

$$\mathcal{L}[v_i(t)] = V_i(s) \tag{13.7.3}$$

The expressions for the vehicular motion at time $t - T$ may be obtained by employing the translation theorem of the transformation, giving

$$\mathcal{L}[v_i(t - T)] = V_i(s)e^{-Ts} \qquad (13.7.4)$$

By the transformation formula for the first derivative of a function, we obtain

$$\mathcal{L}[\dot{v}_{i+1}(t)] = sV_{i+1}(s) - v_{i+1}(0) \qquad (13.7.5)$$

where $v_{k+1}(0)$ are the initial velocities of the vehicles at $t = 0$.

Substitution of the transforms (13.7.4) and (13.7.5) in (13.7.2) yields

$$V_{i+1}(s) = \frac{e^{-Ts}}{Ts + e^{-Ts}} V_i(s) + \frac{T}{Ts + e^{-Ts}} v_{i+1}(0) \qquad (13.7.6)$$

Equation (13.7.6) can be used to find the motion of any vehicle in a line of traffic if the motion of the lead car is given. This will be illustrated by considering the simple case of all cars at rest when $t < 0$ and the lead car starts to move at $t = 0$, with its initial velocity v_0.

Initial conditions are

$$v_1(t) = v_0 \qquad t \geq 0$$

or

$$V_1(s) = \frac{v_0}{s}$$

and

$$v_{i+1}(0) = 0 \qquad i = 1, 2, 3, \ldots$$

Applying the initial conditions to (13.7.6), one obtains the transform for the speed of any vehicle by induction

$$V_{i+1}(s) = \left(\frac{e^{-Ts}}{Ts + e^{-Ts}}\right)^i \frac{v_0}{s} \qquad (13.7.7)$$

Equation (13.7.7) can be rewritten in the form of the sum of the terms of a geometric series, which facilitates taking the inverses:

$$\mathcal{L}^{-1}\left(\frac{e^{-kTs}}{s^u}\right) \qquad u > 0 = \frac{(t - kT)^{u-1}}{u - 1} \qquad t > kT \qquad (13.7.8)$$

Utilization of (13.7.8) in the terms of the series expansion yields the normalized velocities of each succeeding vehicle in the line

$$\frac{v_2(t)}{v_0} = \frac{t - T}{T} - \frac{(t - 2T)^2}{2T^2} + \frac{(t - 3T)^3}{6T^3} - \cdots$$

$$\frac{v_3(t)}{v_0} = \frac{(t - 2T)^2}{2T^2} - \frac{(t - 3T)^3}{3T^3} + \frac{(t - 4T)^4}{8T^4} - \cdots \qquad (13.7.9)$$

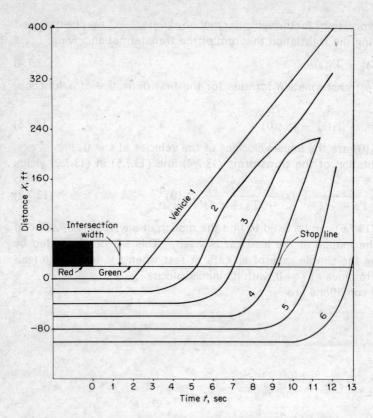

Fig. 13.7 Time-space relationship for vehicles obeying the car-following law of Eq. (13.7.2).

The distances traveled by the vehicles at a time t after the first vehicle has begun to move are obtained by integrating the terms of (13.7.9) from 0 to t. In Fig. 13.7 the simple case considered in the derivation is applied to the performance of a queue of vehicles at a signalized intersection.[10] If it is assumed that the queued vehicles are standing still during the red interval and the lead car suddenly acquires a constant velocity of 40 ft/sec 2 sec after the light turns green, the response of the following vehicles, based on (13.7.2), would be as shown. The starting headways for all vehicles after the lead car, based on a reaction time $T = 1$ sec, are realistic, but it can be observed that the fourth car collides with the third some 230 ft downstream from the intersection stop line. The amplitude of the response to the deceleration of vehicle 2 becomes larger as it is propagated through the queue, and the system of vehicles exhibits a response which could be considered unstable even though the velocities of each individual vehicle are bounded.

13.8 NONLINEAR CAR–FOLLOWING MODELS[11,12]

Car-following theory assumes that the traffic stream is a superposition of vehicle pairs where each vehicle follows the vehicle ahead according to a specific stimulus-response equation that approximates the behavioral and mechanical aspects of the driver-vehicle-road system. Put another way, the equations says: A driver faced with a situation (stimulus) does something about it (response).   How much he does represents his sensitivity coefficient; how quickly he does it determines the response lag of the man-machine system.

A general equation for vehicle following is given by

$$\ddot{x}_{i+1}(t) = a_m \frac{\dot{x}_i(t-T) - \dot{x}_{i+1}(t-T)}{[x_i(t-T) - x_{i+1}(t-T)]^m} \tag{13.8.1}$$

where T is the time lag of response to the stimulus.   The sensitivity is given by

$$\frac{a_m}{[x_i(t-T) - x_{i+1}(t-T)]^m}$$

where a is a constant of proportionality. In the previous section, the stability condition was established for a simple situation—that of vehicles starting from a stop.   This was referred to as linear car-following since the sensitivity factor [see Eq. (13.7.2)] was constant and, for the elementary illustration given, equal to T^{-1}.

The choice of one form of the general equation given in (13.8.1) over another form depends on how well they describe the traffic phenomena. There are two major classes of these phenomena that have received primary consideration.   They are stability, which has already been discussed, and steady-state flow characteristics. Unfortunately the mathematical investigation of stability is so formidable that it is impractical for any equations other than the linear case obtained by assuming a constant sensitivity.

On the other hand, fairly complete steady-state flow characteristics have been formed for many sensitivities, given by Eq. (13.8.1).   Equation (13.8.1) states that the acceleration of a car $\ddot{x}_{i+1}$ at a delayed time T is directly proportional to the relative speed of the car $\dot{x}_{i+1}$ with respect to the one ahead $\dot{x}_i$, and it is inversely proportional to the headway of the car $x_i - x_{i+1}$.   Since the right side of (13.8.1) is of the form dy/y^m, integration of (13.8.1) yields[13]

$$\dot{x}_{i+1}(t) = a \ln [x_i(t-T) - x_{i+1}(t-T)] + C_1 \qquad m = 1 \tag{13.8.2}$$

and

$$\dot{x}_{i+1}(t) = (-m+1)^{-1} a[x_i(t-T) - x_{i+1}(t-T)]^{-m+1} + C_2$$
$$m > 1 \tag{13.8.3}$$

The constants of integration are evaluated by observing that the velocity of a car approaches zero as its headway approaches the effective length of each car L:

$$C_1 = -a \ln L \tag{13.8.4}$$

$$C_2 = -(-m + 1)^{-1} a L^{-m+1} \qquad m > 1 \tag{13.8.5}$$

Substituting for C_1 and C_2, we see that Eqs. (13.8.2) and (13.8.3) become

$$\dot{x}_{i+1}(t) = a \ln \{L^{-1}[x_i(t - T) - x_{i+1}(t - T)]\} \qquad m = 1 \tag{13.8.6}$$

$$\dot{x}_{i+1}(t) = (m - 1)^{-1} a \{L^{-(m-1)} - [x_i(t - T) - x_{i+1}(t - T)]^{-(m-1)}\}$$
$$m > 1 \tag{13.8.7}$$

Equation (13.8.6) is credited to Gazis, Herman, and Potts,[13] who showed that the traffic equation of state could be derived from the microscopic car-following law, just as the gas equation of state can be derived from the microscopic law of molecular interaction. Since the space headway is the reciprocal of concentration k, Eqs. (13.8.6) and (13.8.7) become

$$u = a \ln \frac{k_j}{k} \tag{13.8.8}$$

and

$$u = (m - 1)^{-1} a (k_j{}^{m-1} - k^{m-1}) \qquad m > 1 \tag{13.8.9}$$

The constant of proportionality is evaluated at $u = u_f$ and $k = 0$, giving

$$a = \frac{m - 1}{k_j{}^{m-1}} u_f \qquad m > 1 \tag{13.8.10}$$

Some special cases of the generalized microscopic equation of motion have been discussed. Pipes's model and the modification proposed by Chandler, Herman, and Montroll[14] are obtainable from (13.8.1) by setting $m = 0$. The *reciprocal-spacing* model of Gazis, Herman, and Potts comes from the generalized equation $m = 1$. It is apparent that just as Greenberg's exponential continuous flow model and the reciprocal-spacing car-following model are related, comparable relationships exist for the linear and parabolic continuous flow models. These relationships are summarized in Table 13.1.

13.9 CAR MANEUVERING—A NEW CONCEPT

The similarities in the macroscopic and microscopic approaches have been emphasized. The former solves a differential equation of stream motion and a differential equation of continuity, both expressed in terms of speed

Table 13.1 Comparison of microscopic models of traffic flow

Element	General $m > 1$	$m = 1$	$m = \tfrac{3}{2}$	$m = 2$
Equation of motion	$\ddot{x}_i = \dfrac{a(\dot{x}_{i-1} - \dot{x}_1)}{(x_{i-1} - \dot{x}_i)^m}$	$\ddot{x}_i = \dfrac{a(\dot{x}_{i-1} - \dot{x}_i)}{x_{i-1} - x_i}$	$\ddot{x}_i = \dfrac{a(\dot{x}_{i-1} - \dot{x}_i)}{(x_{i-1} - x_i)^{3/2}}$	$\ddot{x}_i = \dfrac{a(\dot{x}_{i-1} - \dot{x}_i)}{(x_{i-1} - x_i)^2}$
Constant of proportionality	$a = (m - 1)u_f k_j^{-(m-1)}$	u_m	$\dfrac{u_f}{2k_j^{3/2}}$	$\dfrac{u_f}{k_j}$
Equation of state	$q = ku_f\left[1 - \left(\dfrac{k}{k_j}\right)^{m-1}\right]$	$kv_m \ln\left(\dfrac{k_i}{k}\right)$	$ku_f\left[1 - \left(\dfrac{k}{k_j}\right)^{1/2}\right]$	$ku_f\left[1 - \left(\dfrac{k}{k_j}\right)\right]$
Macroscopic counterpart*	$n = 2m - 3$	$n = -1$	$n = 0$	$n = +1$

* See Table 12.1.

u and density k to obtain an equation of state (the equation of the fundamental q-u-k traffic surface). The latter combines a differential equation of motion for an individual vehicle together with the appropriate boundary conditions to obtain an equation of state.

The car-following laws are simplified descriptions of a very complicated response to the world of stimuli that confronts a driver. In fact, only two stimuli are considered: the relative speed between a vehicle and the one ahead and the spacing between the two vehicles. Obviously a driver considers more. Considerable realism can be achieved by including several vehicles ahead of and perhaps the vehicle immediately behind the driver. Most drivers are continually evaluating several gaps ahead and are extremely conscious of the proximity of the car behind.

It can be seen that the verbal statement of the equations of motion in both the fluid-flow and car-following models applies equally well to the following formulation:

$$\ddot{x}_i = -c^2 k_i{}^n k_i'(x) \tag{13.9.1}$$

where the subscript i refers to a portion of the traffic stream. If i is assumed to be an individual vehicle and only three headways are considered, then the geometric significance of (13.9.1) is given by Fig. 13.8.

It is evident that before the slope $f'(x)$ of the $k = f(x)$ curve can be determined, the function $k = f(x)$ must be known. This can be resolved by fitting a polynomial to the points $(x_{i-1}, k_{i-1}; x_i, k_i; x_{i+1}, k_{i+1})$ and then

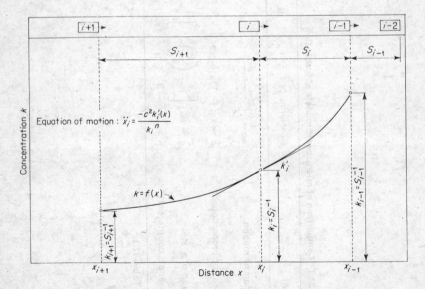

Fig. 13.8 Determination of variables for car maneuvering theory.

evaluating the derivative at x_i, k_i.   Using Lagrange's interpolation formula to fit the polynomial to the points,

$$k(x) = \sum_{v=i-1}^{i+1} \frac{\displaystyle\prod_{j=i-1}^{i+1} (x - x_j)}{\displaystyle\prod_{j=i-1}^{i+1} (x_v - x_j)} \, k_r \qquad j \neq v \tag{13.9.2}$$

Expanding under the summation yields

$$k(x) = \frac{(x - x_i)(x - x_{i+1})k_{i-1}}{(x_{i-1} - x_i)(x_{i-1} - x_{i+1})} + \cdots + \frac{(x - x_{i-1})(x - x_i)k_{i+1}}{(x_{i+1} - x_{i-1})(x_{i+1} - x_i)} \tag{13.9.3}$$

Evaluating the derivative at $x = x_i$, one obtains

$$k_i' = \frac{(x_i - x_{i+1})k_{i-1}}{(x_{i-1} - x_i)(x_{i-1} - x_{i+1})} + \frac{(2x_i - x_{i+1} - x_{i-1})k_i}{(x_{i+1} - x_{i-1})(x_{i+1} - x_i)}$$
$$+ \frac{(x_i - x_{i-1})k_{i+1}}{(x_{i+1} - x_{i-1})(x_{i+1} - x_i)} \tag{13.9.4}$$

Substituting for k_i' in (13.9.5) and solving for c^2 gives

$$c^2 = \frac{-\ddot{x}_i k_i^n}{k_i'} \tag{13.9.5}$$

where the acceleration $\ddot{x}_i$ and the position and reciprocal headway variables x and s^{-1}, which are used in finding k_i' and k_i^n, are measurable variables.

The above derivation covers the driver, two vehicles in front, and the driver behind.   Referring again to Eq. (13.9.1), however, it is seen that the microscopic view can be broadened by letting the subscript i refer to any combination of vehicles or the average of several vehicles.   Averaging vehicles over distances of Δx, intervals between x_{i-1} to x_i and x_i to x_{i+1} become equal, and k_i' in Eq. (13.9.4) reduces to

$$k_i' = \frac{k_{i+1} - k_{i-1}}{2 \, \Delta x} \tag{13.9.6}$$

The average acceleration of the vehicles in Δx is obtained from

$$\ddot{x}_i = \frac{u_{i+1}^2 - u_{i-1}^2}{4 \, \Delta x} \tag{13.9.7}$$

Substituting for k_i from (13.9.6) and for $\ddot{x}_i$ from (13.9.7) in (13.9.5) gives the means of evaluating the sensitivity coefficient from the data:

$$c^2 = - \frac{(u_{i+1}^2 - u_{i-1}^2)k_i^n}{2(k_{i+1} - k_{i-1})} \tag{13.9.8}$$

13.10 THE ANALOG SIMULATOR[15]

It has been illustrated how a line of traffic can be studied mathematically for certain special cases. However, if a complete study of linear car-following is desired based on complicated velocity functions of the leading vehicle, the analytical work may soon become formidable. In order to solve problems involving realistic speed changes of the leading vehicle and realistic car-following models, the problem can be attacked by use of an electrical analog computer. Electric circuits in which the potential distribution satisfies the same differential equations as the dynamical equations of the line of traffic may be constructed. The computer can be mechanized based on a deterministic model of a vehicle operator, described in the previous sections, that will simulate a line of vehicles in various traffic situations. The principal use of the computer is to study the stability of traffic as it relates to the individual vehicle operator and as it depends upon group dynamics. A second use to which the proposed computer may be applied is those of generating traffic flow data to supply intermediate data points to that actually observed. By programming the computer in a manner which yields traffic data actually observed for known conditions, other data can be generated for variations in such conditions.

A linear car-following model of traffic flow developed in Sec. 13.7 is described by the differential equation

$$\dot{v}_i(t) = k_i[v_{i-1}(t - T_i) - v_i(t - T_i)]$$

where v_i represents the velocity of the ith vehicle in a line, and T_i represents the reaction time delay of the operator of that vehicle. Laplace transforming, assuming initial conditions zero,

$$sV_i(s) = k_iV_{i-1}(s)e^{-T_is} - k_iV_i(s)e^{-T_is}$$

$$(s + k_ie^{-T_is})V_i(s) = k_iV_{i-1}(s)e^{-T_is}$$

$$\frac{V_i(s)}{V_{i-1}(s)} = \frac{k_ie^{-T_is}}{s + k_ie^{-T_is}} = \frac{(k_i/s)e^{-T_is}}{1 + (k_i/s)e^{-T_is}}$$

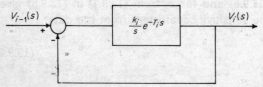

Fig. 13.9 Single-sense block diagram.

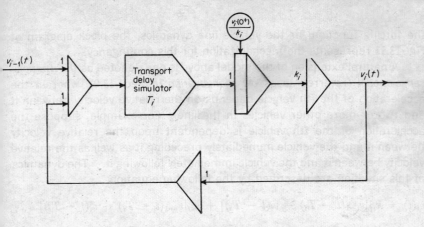

Fig. 13.10   Program for simulating the single-sense model.

This is of the form

$$\frac{C}{R} = \frac{G(s)}{1 + G(s)}.$$

which is representable in block diagram form, as illustrated in Fig. 13.9.

Note that the initial conditions were assumed to be zero. When developing transfer functions for simulation on an analog computer, it is convenient to neglect initial conditions since it simplifies the algebraic manipulations. The introduction of initial conditions into the analog computer program is a very simple matter once the transfer functions have been established.

This model will be referred to herein as the *single-sense model* because the acceleration of the ith vehicle is presumed to be dependent only on the sensing of the relative velocity between the ith vehicle and the vehicle immediately preceding it in the line. The weighting factor k_i will depend upon the individual operator of vehicle i. This block diagram can be simulated on an analog computer by means of the program of Fig. 13.10. If a line of vehicles is to be simulated by a collection of the single-sense models, where the velocity of the leading or first vehicle is presumed to be

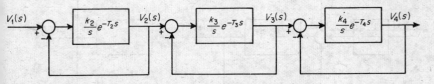

Fig. 13.11   Block diagram for single-sense four-vehicle simulation.

the forcing function for the vehicle line dynamics, the block diagram of Fig. 13.11 represents the mechanization for this contingency.

A natural extension of the model above is one denoted as the *multiple-sense model* (referred to as *car maneuvering* in Sec. 13.9), wherein the acceleration of the ith vehicle depends on the relative velocity between it and two or more other vehicles in the line. For example, suppose the acceleration of the ith vehicle is dependent upon the relative velocity between it and the vehicle immediately preceding it as well as the relative velocity between it and the vehicle immediately following it. The dynamics of this situation are described by the following equation:

$$\dot{v}_i(t) = k_{ip}[v_{i-1}(t - T_i) - v_i(t - T_i)] + k_{if}[v_{i+1}(t - T_i) - v_i(t - T_i)]$$

Laplace transforming, assuming zero initial conditions,

$$sV_i(s) = k_{ip}V_{i-1}(s)e^{-T_is} - k_{ip}V_i(s)e^{-T_is} + k_{if}V_{i+1}(s)e^{-T_is} - k_{if}V_i(s)e^{-T_is}$$

$$[s + (k_{ip} + k_{if})e^{-T_is}]V_i(s) = k_{ip}V_{i-1}(s)e^{-T_is} + k_{if}V_{i+1}(s)e^{-T_is}$$

Let $k_{ip} + k_{if} = k_i$

$$V_i(s) = \frac{k_{ip}e^{-T_is}}{s + k_ie^{-T_is}} V_{i-1}(s) + \frac{k_{if}e^{-T_is}}{s + k_ie^{-T_is}} V_{i+1}(s)$$

$$V_i(s) = \frac{\frac{k_i}{s}e^{-T_is}}{1 + \frac{k_i}{s}e^{-T_is}} \left(\frac{k_{ip}}{k_i}\right) V_{i-1}(s) + \frac{\frac{k_i}{s}e^{-T_is}}{1 + \frac{k_i}{s}e^{-T_is}} \left(\frac{k_{if}}{k_i}\right) V_{i+1}(s)$$

Since the unbracketed portion of the transfer functions above are identical to the single-sense case, one interprets the terms in the brackets as constant multiplier modifiers of the signals $V_{i-1}(s)$ and $V_{i+1}(s)$. This case results in the block diagram of Fig. 13.12, where $k_i = k_{ip} + k_{if}$. The

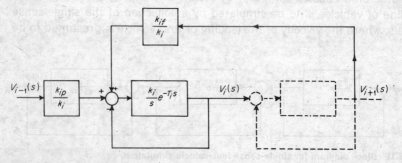

Fig. 13.12 Multiple-sense block diagram.

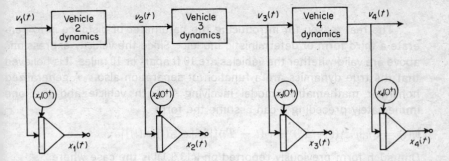

Fig. 13.13 Simulation for determining absolute distance.

multiple-sense model can, of course, derive the input of the ith vehicle from any other vehicle or vehicles in the line in a similar manner.

Since the deterministic models described above are described in terms of velocity, an additional calculation will be required to obtain displacement. The calculation is integration, which is an arithmetic operation easily implemented on the analog computer. There are two ways of specifying *distance* in this simulation.

One may integrate each velocity directly and determine absolute distance from some specified starting point. The program for doing this is given in Fig. 13.13. Again the introduction of initial displacements is possible, allowing a specification of initial vehicle positions.

The second way of specifying *displacements* is on the basis of vehicle separation, rather than the absolute displacement from an initial point. This form is implemented by determining the relative velocity between each vehicle and then integrating to ascertain relative displacement. Figure 13.14 illustrates the concept of this mechanization. $\Delta x_{ij}(0)$ represents the initial displacements between the ith and jth vehicles, which are assumed to be adjacent.

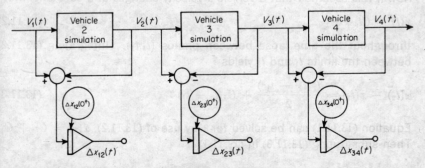

Fig. 13.14 Simulation for determining relative distance.

The real reason for introducing the parameter of distance is to generate a third form of deterministic model. Since the velocity expressions above are valid whether the vehicles are 10 ft apart or 10 miles, it is believed that the true dynamics are a function of separation also. A generalized nonlinear mathematical model involving the ith vehicle and the one immediately preceding it can assume the form

$$\dot{v}_i(t) = \{k_i[v_{i-1}(t - T_i) - v_i(t - T_i)]\}\{f_i[x_{i-1}(t),x_i(t)]\}$$

One such form previously reported on [(13.8.1)] is the case where

$$f_i[x_{i-1}(t).x_i(t)] = \frac{1}{[x_{i-1}(t) - x_i(t)]^m}$$

With the use of diode-function generators and multipliers available in electronic analog computers, a great many nonlinear functional relationships can be generated.

13.11 DETERMINATION OF CAR–FOLLOWING VARIABLES

The solution of the various microscopic equations of motion for the constant of proportionality depends on the evaluation of several variables in real traffic.   It can be seen that an overlap in time-lapse aerial photographic studies provides the means of obtaining vehicle speeds, averaged over one time-lapse interval, and vehicle acceleration, averaged over two time-lapse intervals (see Fig. 13.15).   Since the position trajectories x are continuous and differentiable in the interval between t_1 and t_2, there exists a point τ in (t_1,t_2) such that

$$\dot{x}(\tau) = \frac{x(t_2) - x(t_1)}{t_2 - t_1} \qquad t_1 < \tau < t_2 \tag{13.11.1}$$

Now if it is assumed that a vehicle's velocity changes at a constant rate,

$$x(t) = Ct + C_1 \tag{13.11.2}$$

throughout the time lapse between photos (t_1,t_2).   Integrating (13.11.2) between the limits t_1 and t_2 yields

$$x(t_2) - x(t_1) = \frac{C(t_2^2 - t_1^2)}{2} + C_1(t_2 - t_1) \tag{13.11.3}$$

Equation (13.11.1) can be solved for τ by use of (13.11.2), $\dot{x}(\tau) = C\tau + C_1$. Then substituting (13.11.3) in (13.11.1)

$$\tau = \frac{t_2^2 - t_1^2}{2(t_2 - t_1)} \tag{13.11.4}$$

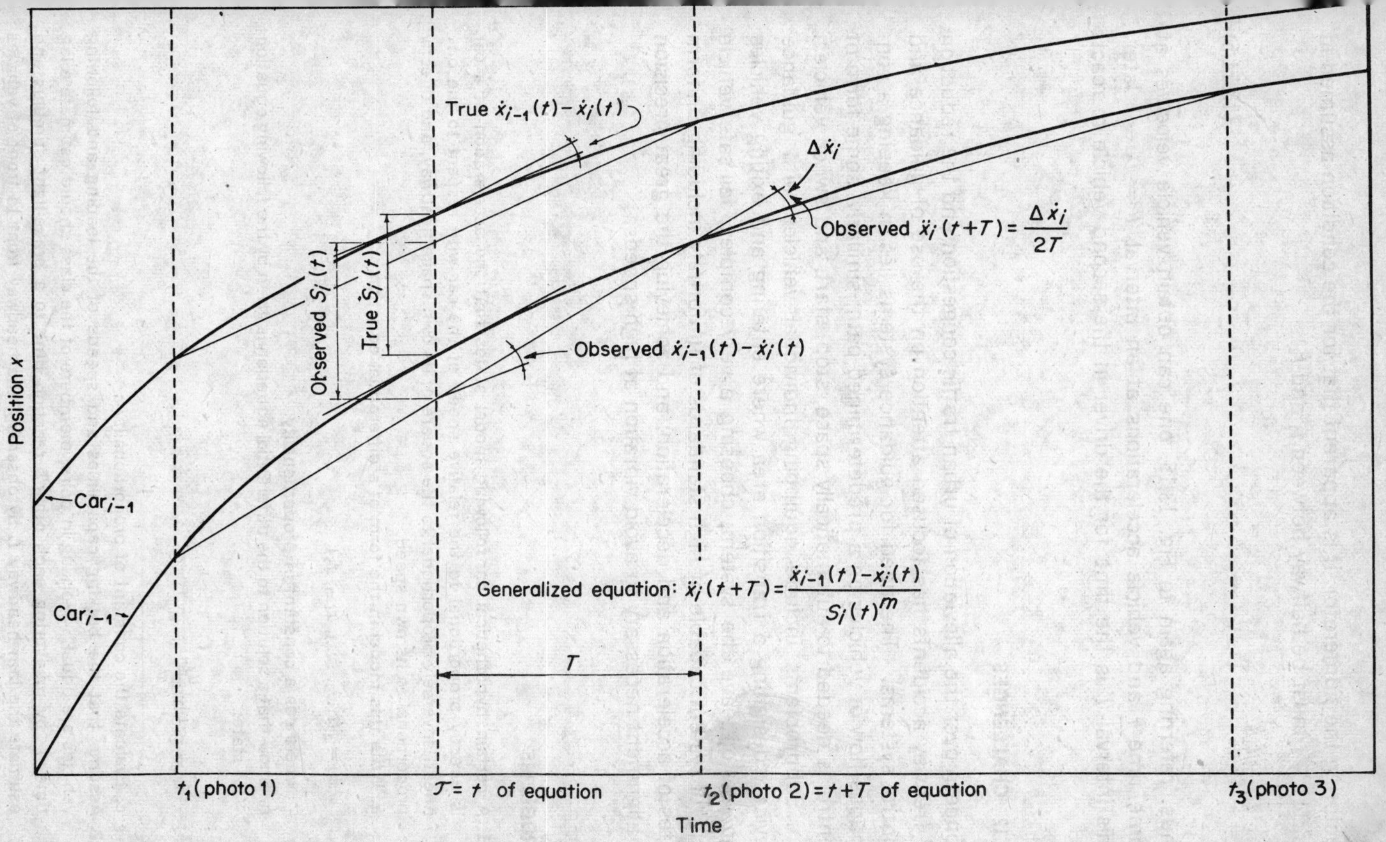

True $\dot{x}_{i-1}(t)-\dot{x}_i(t)$
$\Delta\dot{x}_i$
Observed $\ddot{x}_i(t+T)=\frac{\Delta\dot{x}_i}{2T}$
Observed $S_i(t)$
True $\dot{S}_i(t)$
Observed $\dot{x}_{i-1}(t)-\dot{x}_i(t)$
Position x
Car$_{i-1}$
Car$_{i-1}$
Generalized equation: $\ddot{x}_i(t+T)=\frac{\dot{x}_{i-1}(t)-\dot{x}_i(t)}{S_i(t)^m}$
T
t_1(photo 1)
$\mathcal{T}=t$ of equation
t_2(photo 2) $=t+T$ of equation
t_3(photo 3)
Time

Fig. 13.15 Comparison of true and observed car-following variables using time-lapse aerial photography.

Factoring the numerator, it is apparent that for the condition assumed in (13.11.2), τ must be halfway between t_1 and t_2

$$\tau = \frac{t_2 + t_1}{2} \tag{13.11.5}$$

Thus, referring again to Fig. 13.15, one can obtain vehicle velocities at some time τ and vehicle accelerations at an interval $T = t_2 - \tau$ later. This interval T is the time for the driver in the second vehicle to react.

13.12 CHALLENGES

Concern over the alleviation of urban traffic congestion and the reduction of freeway accidents has focused attention on the use of guidance and control systems. The need for guidance systems lies in keeping a continuous flow of vehicles on a predetermined path; similarly, some form of control is needed to automatically space, stop, start, and switch vehicles.

Subproblems include equipping individual vehicles with guidance devices; designing a transition area where entering and exiting vehicles can join or leave the system; choosing a very complex, fail-safe vehicle control system capable of high standards of uniformity including uniform rates of acceleration and deceleration; and achieving the great precision in alignment necessary to avoid vibration at high speeds.

PROBLEMS

13.1. A certain hypothetical car-following model states that the acceleration of a car is directly proportional to the relative speed of the car with respect to the one ahead, inversely proportional to the square of its own space headway, and directly proportional to its own speed.

 a. Write this model in the form of a differential equation

$$\ddot{x}_i = f(c, \dot{x}_1, \dot{x}_{i-1}, x_{i-1}, x_i)$$

where c is a constant of proportionality.

 b. Show that a solution to the differential equation is given by the following equation of state:

$$k = c^{-1} \ln \frac{u_f}{u}$$

 c. Evaluate the constant of proportionality c.

13.2. Assume that the braking capabilities and speeds of the leading and following vehicles are equal. Working in pairs, members of the class should estimate a lag time T by measuring each other's reaction time to a brake-light stimulus and estimate the jam headway L by observing the headway front to front of vehicles queued at a traffic signal or stop sign. Each individual should plot average safe headway vs. velocity for (1) himself, (2) the class average and (3) the rule of thumb for following. Compare the three plots and discuss any differences.

13.3. Discuss the differences between a *linear* car-following model and a *nonlinear* car-following model, including any advantages or disadvantages.

13.4. What is the basic difference between *car following* and *car maneuvering*?

13.5. Given the following measurements from aerial time-lapse photography:

Photo time	Car A at station	Car B at station
7:00:00.00 A.M.	238 + 01	238 + 00
7:00:02.00 A.M.	237 + 50	236 + 67
7:00:04.00 A.M.	236 + 15	235 + 25

Utilize the microscopic models of traffic flow to determine the jam density, optimum speed, free speed, and capacity of the traffic stream, based on this sample.

13.6. Discuss briefly the applicability of macroscopic vs. microscopic traffic flow models in each of the given conditions:
   a. A three-lane freeway with lane flows inside to out of 2,000 vph, 2,200 vph, 1,600 vph.
   b. A two-lane freeway section with lane volumes of 1,100 vph inside and 800 vph outside with stream speed at 50 mph.

REFERENCES

1. Doebelin, E. O.: "Dynamic Analysis and Feedback Control," McGraw-Hill Book Company, New York, 1962.
2. Wilts, C. H.: "Principles of Feedback Control," Addison-Wesley Publishing Company, Inc., Reading, Mass., 1960.
3. Brown, R. G., and J. W. Nilsson: "Introduction to Linear Systems Analysis," John Wiley & Sons, Inc., New York, 1962.
4. Goode, H. H., and R. E. Machol: "System Engineering," McGraw-Hill Book Company, New York, 1957.
5. Wilson, W. E.: "Concepts of Engineering System Design," McGraw-Hill Book Company, New York, 1965.
6. Truxal, J. G.: "Automatic Feedback Control System Synthesis," McGraw-Hill Book Company, New York, 1955.
7. Raven, F. H.: "Automatic Control Engineering," McGraw-Hill Book Company, New York, 1961.
8. Pipes, L. A.: An Operational Analysis of Traffic Dynamics, *J. Appl. Phys.*, vol. 24, pp. 274–281, 1953.
9. Kometani, E., and T. Sasaki: A Safety Index for Traffic with Linear Spacings, *Mem. Fac. Eng. Kyoto Univ.*, vol. 21, pt. 3, pp. 229–246, July, 1959.
10. Drew, D. R.: Microscopic Models of Traffic Flow, *Traffic Eng.*, vol. 36, no. 8, May, 1966.
11. Gazis, D. C., R. Herman, and R. W. Rothery: Nonlinear Follow-the-leader Models of Traffic Flow, *Operations Res.*, vol. 7, no. 4, pp. 499–505, July–August, 1959.
12. Newell, G. F.: Nonlinear Effects in the Dynamics of Car Following, *Operations Res.*, vol. 9, no. 2, pp. 209–229, March–April, 1961.

13. Gazis, D. C., R. Herman, and R. B. Potts: Car-following Theory of Steady-state Traffic Flow, *Operations Res.*, vol. 7, no. 4, pp. 499–505, July–August, 1959.
14. Chandler, R. E., R. Herman, and E. W. Montroll: Traffic Dynamics: Studies in Car Following, *Operations Res.*, vol. 6, pp. 165–184, 1958.
15. Drew, D. R., and M. G. Rekoff: "Testing and Evaluation of Deterministic Models of Traffic Flow Derived through Theoretical Deductions and Experimentation," Texas A&M Research Foundation, Texas A&M University, College Station, 1965.

CHAPTER FOURTEEN
ENERGY - MOMENTUM APPROACH TO LEVEL OF SERVICE[1-4]

Concepts, considered logically, never originate in experience;
i.e. they are not to be derived from experience alone.

Albert Einstein

14.1 INTRODUCTION

An evaluation of any change in operation or design requires a thorough study of those traffic parameters which will reflect changes in traffic flow quantitatively. The traffic parameters of volume, density, speed, and queueing of vehicles have been utilized to present a complete description of the quantity of flow on a facility. The measurement of other similar parameters, such as travel time and delay, to complete the quantitative analysis presents no immediate problem. The need exists, however, for methods of comparing the effects of various designs and controls on the quality of flow, or level of service.

Level of service, as applied to the traffic operation on a particular roadway, refers to the quality of driving conditions afforded a motorist by a particular facility. Factors which are involved in the level of service are (1) speed and travel time, (2) traffic interruption, (3) freedom to maneuver, (4) safety, (5) driving comfort and convenience, and (6) vehicular operating costs.

Many approaches to the problem of measuring level of service provided the motorist have been introduced in an attempt to obtain a qualitative measurement of traffic flow. A universal standard has not yet been adopted, however, perhaps because of the uncertainty of what traffic parameters should be measured to truly indicate the quality of flow.

The need for an objective evaluation of the level of service on a facility has long been recognized. Travel time on a roadway has often been used as a criterion for this evaluation.[5] However, travel time may not always reflect the true conditions on parts of a facility and therefore may not always indicate the congestion and discomfort experienced by the motorist.

For example, Fig. 14.1 represents a travel time and speed pattern between successive 500-ft sections of the inbound Gulf Freeway for one trip made near the end of the peak period. The figure helps to illustrate the delay in one section of the freeway, which is somewhat concealed by the value of the average travel time. The average speed for the entire trip was 41.5 mph, which indicates good flow. However, it is noticed that a speed differential of more than 30 mph existed between some of the sections. Also, the speed in one section of the freeway averaged only 24 mph and dropped below 20 mph. Therefore, the congestion in this section is dampened by the smooth flow in the remaining portions of the freeway when the flow is evaluated on the basis of travel time.

The fact that the discomfort experienced by the motorist is not neces-

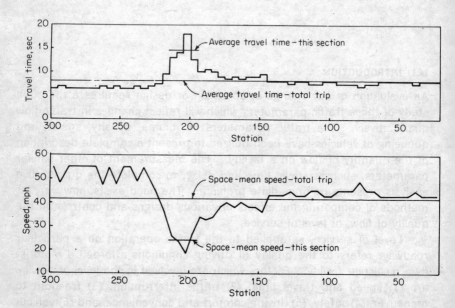

Fig. 14.1 Comparison of travel time and speeds within sections to total trip.

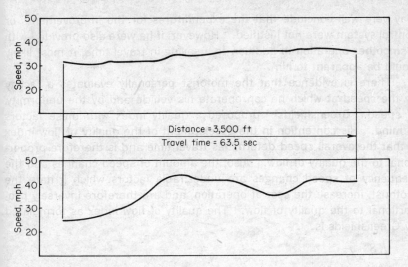

Fig. 14.2   Comparison of speed profiles of a section of freeway having identical travel times.

sarily reflected in travel time is further exemplified by Fig. 14.2. This figure represents a comparison of a section of the Gulf Freeway, 3,500 ft in length, with identical travel times but with two distinct patterns of speed. The upper chart shows relatively smooth flow; however, the lower chart illustrates rapid decelerations, which are indicative of motorist discomfort. Violent braking operations indicate dangerous conditions on the facility and add to the annoyance of the motorist; yet these maneuvers are not necessarily reflected by travel time, as is illustrated in Fig. 14.2.

The criterion of a systems approach[6] to increase the efficiency of traffic flow on a facility (or facilities) is based primarily on maximizing the system output. Since maximizing the volume output rate is equivalent to minimizing the travel time within the system, travel time is also used as the measurement of service provided to the motorist.

With a relatively large system and high volumes a small saving in travel time per vehicle, after geometric improvements have been made or during some control procedure, may result in a substantial reduction in travel time within a system. But will the reduction in travel time be appraised equally by both the motorist and the traffic engineer? For example, a saving of 2 min per vehicle within a large closed system may reflect considerable improvement to the traffic engineer. However, this saving may not be noticed by the average motorist because this may represent only a small percentage of his overall travel time. Therefore, unless a noticeable reduction of his travel time is provided, the motorist may very well conclude that the expenditures for the improvements or control system were not justified.   However, if he were also provided with a smoother operation in addition to a saving in travel time, it most likely would be apparent to him.

There is evidence that the motorist personally evaluates a facility by the speed at which he can operate his vehicle and by the uniformity of speed.   Greenshields[7,8] proposed a quality index with these factors in mind.   His contention in the development of the quality of flow index is that the overall speed determines travel time and is therefore proportional to the quality of flow.   Also, the amount of speed changes and the frequency of speed changes are undesirable factors which irritate the motorist, increase the cost of operation, and are therefore inversely proportional to the quality of flow.   The quality of flow index as formulated by Greenshields is

$$Q = \frac{KS}{\Delta s} \sqrt{f}$$

where S = average speed, mph
Δs = absolute sums of speed changes per mile
f = number of speed changes per mile
K = constant of 1,000

Platt[9] observed that traffic delays and traffic control devices are more annoying to the driver than slow-moving traffic because they cause the motorist to stop.   Also it was established that driver satisfaction and driver effort do not vary linearly with speed but vary as complex functions. For these reasons two additional terms were added by Platt to Greenshields' quality of flow index.   Some of the factors measured in the additional terms were speed change rate, steering reversal rate, accelerator reversal rate, and brake application rate.

The Highway Capacity Committee of the Highway Research Board has proposed six levels of service as a basis of design..   The basic speed-volume curve is divided arbitrarily into six levels of service as related to volumes and freedom to maneuver.

The aforementioned indices developed by Greenshields and Platt seem to possess excellent attributes for evaluating the overall efficiency of a long stretch of highway.   However, it has been established that the varying nature of the geometrics is accompanied by a varied degree of optimum speed, volume, and density for short successive sections of freeway.   If the parameter measuring the quality of flow is to be related to these known quantitative parameters, it must reflect the qualitative efficiency of small roadway sections.   The parameter must not only reflect the engineer's evaluation of level of service but, more important, that measure of service which is considered by the individual motorist.

14.2 THE PARAMETER

Because of recent research results, a traffic parameter referred to as *acceleration noise* has received attention as a possible measurement of traffic flow quality for two basic reasons.   First, it is dependent upon the three basic elements of the traffic stream, namely, (1) the driver, (2) the road, and (3) the traffic condition.   Second, it is, in effect, a measurement of the smoothness of flow in a traffic stream.   A presentation of the history, definition, and manifestations of this parameter is made in the following paragraphs.

On a given highway with traffic volume so light that it does not restrict his maneuverability, a motorist normally attempts to drive at a uniform and comfortable speed.   Unconsciously, however, he accelerates and decelerates occasionally and deviates from a uniform speed throughout his journey.   When traffic volumes have increased to a level which restricts his desirable speed, the motorist is perhaps forced to change lanes and increase speed to overcome slower-moving vehicles, resulting in higher and more frequent deviations from a uniform speed.   The accelerations during his trip can be considered random components of time, and the acceleration distributions essentially follow a normal distribution.[10]   The smoothness of his journey can be described by the amount that the individual random accelerations disperse about the mean acceleration.   This deviation is measured by determining the standard deviation (also referred to as the root-mean-square deviation) of the accelerations.

The acceleration noise (standard deviation of the accelerations) can be considered as the disturbance of the vehicle's speed from a uniform speed or can be identified as a measurement of the smoothness of traffic flow.   The term *noise* is used to indicate that disturbance of the flow, comparable to the coined phrase *video noise*, which is used to describe the fluttering of the video signal on a television set.

The accelerations of a vehicle can either be measured directly by an accelerometer or approximated from a speed-time graph of the vehicle's trip.   This latter method is fully explained later in the chapter.

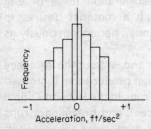

Frequency

−1   0   +1
Acceleration, ft/sec^2

Fig. 14.3 Distribution of accelerations with minor deviations.

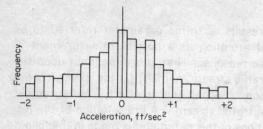

Fig. 14.4 Distribution of accelerations with larger deviations.

The distribution of the accelerations of a vehicle which has been driven at almost a uniform speed throughout its trip is similar to that shown in Fig. 14.3. Higher deviations from a uniform speed might result in an acceleration distribution similar to that presented in Fig. 14.4. Acceleration noise varies with the amount and frequency of acceleration and deceleration. The more violent and more frequent these maneuvers are, the higher is the noise. Since violent deceleration seriously affects the acceleration noise, this parameter also reflects potentially dangerous traffic conditions.

One of the most critical factors which affects the characteristics of traffic flow is the driver. However, because of the many external and internal elements which affect his decisions,[14] evaluation of his behavior in a traffic stream presents a most complex problem.

Acceleration noise seems to be a useful parameter in helping to evaluate the behavior of various drivers in a traffic stream in terms of danger potential. Since this parameter is affected by acceleration and deceleration, a reckless driver who attempts to drive faster than the stream of traffic will accelerate and decelerate violently, and perhaps frequently, and will experience a much larger acceleration noise than the motorist that is content (or patient) with present flow conditions.

It might be conjectured that if a motorist were able to operate a vehicle on a perfect roadbed without the influence of traffic, the acceleration noise would be zero. However tests[11] conducted using four drivers attempting to maintain constant speeds between 20 and 60 mph on the General Motors proving ground resulted in acceleration noise of 0.32 ft/sec^2. It is quite obvious then that although a motorist desires to maintain a constant speed with ideal conditions, he unconsciously is unsuccessful.

Other experiments[11] using the same drivers and without the influence of traffic but on roads with varied geometric design produced an increased acceleration noise. The noise determined from runs in the Holland Tunnel of The Port of New York Authority was 0.73 ft/sec^2, and preliminary runs on poorly surfaced winding country roads were even twice this amount.

The increase of noise in the tunnel was ascribed to the narrow lanes, artificial lighting, and confined conditions. Investigations on winding country roads by Jones and Potts[13] showed values between 0.79 and 1.41 ft/sec². The acceleration noise decreased with improved road design features in the latter studies.

The acceleration noise which is present on a road in the absence of traffic is called the *natural noise* of the driver on the road.[11] Based on the above findings it might be rationalized that the acceleration noise experienced on a road in the absence of traffic is some factor times the natural noise which would be experienced on a 'perfect roadbed. The factor can be ascribed basically to geometrics of the facility.

Comparison of acceleration noise on similar roadways might establish whether the intrinsic design features of the facilities have equal effects on the flow of traffic. In other words, a comparison of acceleration noise could perhaps reflect the effects to the smoothness of flow due to the difference in geometric design of similarly classed roads. One might expect a higher level of service from the facility with the lowest acceleration noise.

Of equal significance is the attribute which this parameter seems to possess in comparing the smoothness of flow before and after changes in the geometric configuration of a facility. It has been established that each roadway exhibits natural noise in an amount greater than that of a near perfect roadbed in relation to its intrinsic design features. Any improvement in design could perhaps be evaluated qualitatively by determining the amount that the acceleration noise approaches the noise of the ideal roadway. In other words, an attempt might be made to minimize acceleration noise in design procedures if this parameter shows signs of merit.

The third factor which influences the amount of acceleration noise is congestion. Results of preliminary studies[13] on suburban highways showed that the noise increased when congestion increased because of higher volumes and influence of parked cars. The amount of acceleration noise in excess of the natural noise of the facility was basically due to the existing traffic condition.

Reiterating, although stopped time or delay is a popular parameter used in the evaluation of congestion, it is not necessarily a satisfactory means of evaluating congestion. Results of studies[13] on a suburban road passing through a shopping center, which incidentally experienced a high accident rate (80 accidents per million vehicle miles), showed a stopped time of only 7 sec in an overall travel time of 128 sec. But the acceleration noise during the peak period was 1.43 compared to 0.77 ft/sec² during the off-peak period. The latter comparison no doubt gives a better indication of the degree of congestion.

14.3 MATHEMATICS OF ACCELERATION NOISE

The following is a development of an equation used by Jones and Potts[13] for approximating acceleration noise. If $v(t)$ and $a(t)$ are the speed and acceleration of a car at time t, then the average acceleration of the car for a trip of time T is

$$a_{ave} = \frac{1}{T} \int_0^T a(t_i)\, dt = \frac{1}{T} [v(T) - v(0)]$$

The standard deviation of a set of n numbers $x_1, x_2, \ldots, x_n$ is denoted by s and is defined by

$$s = \left[\frac{1}{n} \sum_{i=1}^n (x_i - \bar{x})^2 \right]^{\frac{1}{2}}$$

where $\bar{x}$ is the mean of the x's.

From Fig. 14.5 it is seen that if $a(t_i)$ denotes the acceleration of a vehicle at time t_i, the square of the difference between any acceleration and the average acceleration is denoted by

$$[a(t_i) - a_{ave}]^2$$

The summation of these differences over a time T becomes

$$\int_0^T [a(t_i) - a_{ave}]^2\, dt$$

The number of t's in the sample is equal to $\sum_{i=0}^{T} t_i = T$. Therefore the acceleration noise σ can be written

$$\sigma = \left\{ \frac{1}{T} \int_0^T [a(t_i) - a_{ave}]^2\, dt \right\}^{\frac{1}{2}} \tag{14.3.1}$$

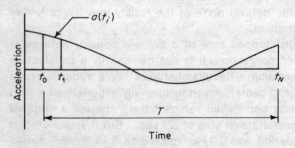

Fig. 14.5 Acceleration vs. time.

and

$$\sigma^2 = \frac{1}{T} \int_0^T [a(t_i) - a_{ave}]^2 \, dt \tag{14.3.2}$$

Expanding (14.3.2),

$$\sigma^2 = \frac{1}{T} \int_0^T [a(t_i)]^2 \, dt - \frac{1}{T} \int_0^T 2a(t_i)a_{ave} \, dt + \frac{1}{T} \int_0^T (a_{ave})^2 \, dt$$

$$\sigma^2 = \frac{1}{T} \int_0^T [a(t_i)]^2 \, dt - 2a_{ave} \frac{1}{T} \int_0^T a(t_i) \, dt - (a_{ave})^2$$

$$\sigma^2 = \frac{1}{T} \int_0^T [a(t_i)]^2 \, dt - 2(a_{ave})^2 + (a_{ave})^2$$

$$\sigma^2 = \frac{1}{T} \int_0^T [a(t_i)]^2 \, dt - (a_{ave})^2 \tag{14.3.3}$$

The value of σ^2 can be estimated by approximating Eq. (14.3.3) with

$$\sigma^2 \approx \frac{1}{T} \sum_{i=0}^{T} \left(\frac{\Delta v}{\Delta t}\right)^2 \Delta t - \left(\frac{v_T - v_0}{T}\right)^2$$

where v_0 and v_T are the initial and final velocities, respectively. If Δv is taken as a constant (say 2 mph) throughout the measurement, then

$$\sigma^2 = \frac{(\Delta v)^2}{T} \sum_{i=0}^{T} \frac{n^2}{\Delta t_i} - \left(\frac{v_T - v_0}{T}\right)^2 \tag{14.3.4}$$

where n is the number of speed changes of 2 mph in time Δt.

If the final velocity is the same as the initial velocity, the average acceleration is zero and the second term of Eq. (14.3.4) drops out. When the acceleration noise is measured on a long stretch of highway where T is very large, the second term is very small and can be ignored.[13]   Therefore for relatively long trips

$$\sigma^2 = \frac{(\Delta v)^2}{T} \sum_{i=0}^{T} \frac{n^2}{\Delta t_i}$$

If Δt is measured for each successive speed change of 2 mph, that is, $n = 1$, the approximation equation of acceleration noise as developed by Jones and Potts[13] becomes

$$\sigma = \left[\frac{(\Delta v)^2}{T} \sum_{i=0}^{T} \frac{1}{\Delta t_i}\right]^{\frac{1}{2}} \tag{14.3.5}$$

To calculate σ in units of feet per second per second (ft/sec^2), Eq. (14.3.5) becomes

$$\sigma = \left[\frac{(1.465)^2(\Delta v)^2}{T} \sum_{i=0}^{T} \frac{1}{\Delta t_i} \right]^{1/2}$$ (14.3.6)

where v is in units of miles per hour and T and t are in units of seconds.

The deviations of the accelerations when a vehicle is stopped in traffic is zero. Since this type of situation would reduce the value of σ even though it adds to the annoyance and frustration of the motorist, σ is measured only while the vehicle is in motion, and therefore T is taken as the running time of the vehicle.

If a continuous analog record is made of a vehicle's speed vs. time while traveling on a highway, the acceleration noise can be determined using Eq. (14.3.6). For every change in speed (Δv) of say 2 mph, Δt is measured from the chart, and the value $1/\Delta t$ can be accumulated up to time T.

14.4 MOMENTUM–KINETIC ENERGY

In fluid mechanics, fluids are commonly divided into liquids and gases. The chief differences between liquids and gases are that liquids are practically incompressible, whereas gases are compressible and must be so treated. A single stream of traffic offers a striking analog to the flow of a compressible gas in a constant-area duct. Both consist of discrete particles: individual molecules in the case of a fluid and individual vehicles in the case of the traffic stream.

Lighthill and Whitham[14] as well as Richards[15] applied fluid dynamic principles to various highway occurrences and concluded that discontinuities in traffic flow are propagated in a manner similar to shock waves in the theory on compressible fluids. Greenberg[16] carried the fluid dynamic analogy still further, developing functional relations for the basic interactions between vehicles. Herman and Potts[17] emphasized the macroscopic nature of the traffic quantities, flow, and concentration by referring to the somewhat analogous properties of a gas, such as pressure, volume, and temperature. Pressure can be computed from the average number of molecular collisions on the containing wall over a long interval of time. Since the relation between the pressure, volume, and temperature is called the *gas equation of state*, the relation between the traffic flow and traffic concentrations has come to be known as the *traffic equation of state*.

Because of the widespread interest in recent years in high-speed gas flow, the dimensionless parameter called the *Mach number* has become very significant in fluid dynamics. The Mach number is defined as the ratio of the actual fluid velocity to the acoustic velocity, or velocity of sound propagation, under the conditions of the flow where the velocity in question is measured. Following the logic of compressible flows wherein local sonic speed represents the condition of maximum flow per unit area,[18] the critical velocity for traffic flow corresponds to the optimum velocity u_m or the velocity of traffic at capacity q_m.

The analogy between fluid dynamics and traffic dynamics is seemingly endless. For example, just as the gas equation of state can be derived from the microscopic law of molecular interaction for two molecules, it has been shown that the traffic equation of state can be derived from the microscopic car-following law governing the motion of two cars.[19]

The speed of waves carrying continuous changes of flow through the stream of vehicles is given by the derivative q' of the q-k equation defined in (12.5.25). Discontinuities in traffic flow are propagated in a manner similar to shock waves in the theory of compressible fluids. The speed of a shock wave U is given by the slope of the chord joining the two points on the flow-concentration curve which represent the conditions ahead of and behind the shock waves. Application of the mean-value theorem suggests that the speed of the shock wave is approximately the mean of the speeds of the waves running into it from either side.

$$U = \frac{1}{2}\,(q'_1 + q'_2) \tag{14.4.1}$$

The very strong analogy between traffic flow and fluid flow suggests that the conditions of continuity of momentum and energy should be fulfilled at the surface of a traffic shock wave, just as the equations of dynamic compatibility must be fulfilled in fluid dynamics. Multiplying (12.5.6) by u and (12.5.11) by k, then adding the two equations gives

$$\frac{\partial (ku)}{\partial t} = \frac{-\partial [ku^2 + k^{n+2}c^2/(n+2)]}{\partial x} \tag{14.4.2}$$

Equation (14.4.2) is the law of conservation of momentum in the differential form as applied to traffic flow. Comparing Eqs. (12.5.6) and (14.4.2) with the classical forms in hydrodynamics, we can add to the analogy between the fluid and traffic quantities. This correspondence, which has been discussed in the Introduction and in this section, is summarized in Table 14.1.

It is apparent that the equations of continuity, motion, and momentum are identical in the two systems when the exponent of proportionality n is set equal to -1. Because, in classical systems, the conservation-of-momentum equation serves to establish the form for momentum, the

Table 14.1   Correspondence between physical systems

Characteristic	Hydrodynamic system	Traffic system
Continuum	One-dimensional compressible fluid	Single-lane traffic stream
Discrete unit	Molecule	Vehicle
Variables	Mass density ρ	Concentration k
	Velocity v	Speed u
Equations:		
Continuity	$\dfrac{\partial \rho}{\partial t} + \dfrac{\partial (\rho v)}{\partial x} = 0$	$\dfrac{\partial k}{\partial t} + \dfrac{\partial (ku)}{\partial x} = 0$
Motion	$\dfrac{\partial v}{dt} + \dfrac{c^2}{\rho}\dfrac{\partial \rho}{\partial x} = 0$	$\dfrac{du}{dt} + c^2 k^n \dfrac{\partial k}{\partial x} = 0$
Momentum	$\dfrac{\partial (\rho v)}{\partial t} + \dfrac{\partial (\rho v^2 + \rho c^2)}{\partial x} = 0$	$\dfrac{\partial (ku)}{\partial t} + \dfrac{\partial [ku^2 + k^{n+2}c^2/(n+2)]}{\partial x} = 0$
State	$P = c\rho T$	$q = ku$
Parameters	Critical velocity v_c	Critical speed u_m
	. . . .	Critical concentration k_m
	Critical flow Q_c	Capacity flow q_m
	Momentum ρv	Flow ku
	Kinetic energy $\rho v^2/2$	Kinetic energy $\alpha k u^2$
	. . . .	Optimum speed u_m'
	. . . .	Optimum concentration k_m'
	. . . .	Optimum flow q_m'
	Internal energy ϵ	Internal energy σ
	Friction factor f	Natural noise σ_n

quantity ku in Eq. (14.4.2) is defined here as the *momentum* of the traffic stream.   If in fact traffic momentum is equal to ku, it is apparent that it is also equivalent to traffic flow and that the flow-oriented parameters (u_m, k_m, and q_m) discussed in the previous section are based on optimizing this momentum.

From basic fluid mechanics, it is well known that the kinetic energy of an element of fluid is $\frac{1}{2}\rho v^2$.   However, in dealing with flow situations in open channels, the true kinetic energy of a stream is not $\frac{1}{2}\rho v^2$, since velocities vary in different parts of the stream and the contribution of velocity to kinetic energy is a nonlinear function.   For these reasons, it is necessary to use a kinetic energy correction factor α, and the kinetic energy of the traffic stream will be defined as $\alpha k u^2$, where α is a dimensionless constant.   Squaring (12.5.25) and then dividing by k yields

$$E = \alpha k u_f^2 \left[1 - 2\left(\frac{k}{k_j}\right)^{(n+1)/2} + \left(\frac{k}{k_j}\right)^{n+1} \right] \qquad n > -1 \qquad (14.4.3)$$

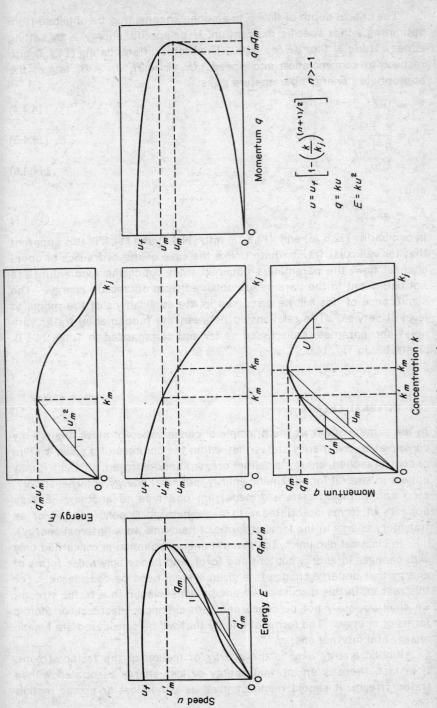

$$u = u_f \left[1 - \left(\frac{k}{k_j} \right)^{(n+1)/2} \right] \qquad n > -1$$

$$q = ku$$

$$E = ku^2$$

Fig. 14.6 Relationship between fundamental traffic variables and parameters.

The critical depth of flow y in an open channel may be obtained from optimizing either specific momentum M or specific energy E by setting either $dM/dy = 0$ or $dE/dy = 0$. Therefore, differentiating (14.4.3) with respect to concentration and speed, $dE/dk = dE/du = 0$, to get the appropriate "energy" parameters gives:

$$k'_m = (n + 2)^{-2/(n+1)} k_j \qquad n > -1 \tag{14.4.4}$$

$$u'_m = \frac{n + 1}{n + 2} u_f \qquad n > -1 \tag{14.4.5}$$

$$q'_m = k'_m u'_m \tag{14.4.6}$$

and

$$E'_m = \alpha k'_m u'^2_m \tag{14.4.7}$$

In comparing (12.5.26) and (12.5.27) with (14.4.4) and (14.4.5), it is apparent that for vehicular traffic, quite unlike the case of the hydraulics of open-channel flow, the parameters obtained from optimizing momentum are not equivalent to the parameters obtained from optimizing energy. The significance of this will be discussed in the applications of the model to level of service. The relationship between the fundamental traffic variables and parameters discussed so far and summarized in Table 14.1 is illustrated in Fig. 14.6.

14.5   INTERNAL ENERGY

In the same manner as the principle of conservation of mass, the energy conservation law merely states that within the confines of a given system or control section, energy is neither created nor destroyed, although it may appear in several forms (kinetic energy, potential energy, internal energy, etc.) and it may be transformed from one type to another. Energy appears in forms associated with a given mass (kinetic energy) or as transitory energy in the familiar forms of heat and work (internal energy).

In classical dynamics, the general energy equation is concerned only with changes in energy content and form, and hence only those forms of energy that undergo change in a given system need be considered. For this reason, in this discussion of energy conservation in a traffic stream, we shall obviously not be concerned with chemical, electrical, or atomic forms of energy. The forms of energy that will be considered are kinetic energy and internal energy.

Kinetic energy $\alpha k u^2$ is the energy of motion of the traffic stream. If in fact, there is an internal energy or lost energy associated with a traffic stream, it should manifest itself as either lost or erratic motion because of adverse geometrics and traffic interaction. In Sec. 14.1 it was seen that the measure of the jerkiness of the driving in this stream is given by acceleration noise. The units of both kinetic energy and acceleration noise are those of acceleration, which adds credibility to the hypothesis that acceleration noise represents internal energy.

The conservation of energy for the traffic stream over a section of road x is simply a case, then, of the total energy T being equal to the kinetic energy E plus the internal energy I of the traffic stream. With this notation, an energy balance, or accounting, becomes

$$T = E + I \tag{14.5.1}$$

$$T = \alpha k u^2 + \sigma \tag{14.5.2}$$

where σ represents the acceleration noise for a vehicle on a section of highway of length x.

It is common to say that there is a loss of energy due to the effects of friction in a system, whether it is a classical system or traffic system. The general energy equation indicates that there can be no loss in energy from the system if the principle of conservation of energy is to hold true. Reexamination of the energy equation will yield a clue to the true meaning of friction. Energy is not really lost; it is simply converted from one of the mechanical forms (kinetic energy) to internal energy, a thermal form of energy. One recalls from the second law of thermodynamics that the mechanical forms of energy, such as kinetic energy, are more valuable than an equivalent amount of thermal energy or internal energy. This is certainly true in the case of traffic flow. Thus one can say that the forces of friction (adverse geometrics and traffic interaction) tend to convert the desirable forms of energy (traffic motion) into the less valuable forms (traffic interaction).

Two boundary conditions become important in evaluating the parameters, T and α, in (14.5.2). As the internal energy approaches zero, kinetic energy approaches the total energy of the stream and

$$T = \alpha k'_m u'^2_m = \frac{4}{27} \alpha k_j u_f^2 \qquad n = 1 \tag{14.5.3}$$

The converse leads to an expression for total energy in terms of acceleration noise:

$$T = \sigma_{\max} \tag{14.5.4}$$

where $\sigma_{\max}$ is the maximum acceleration noise. Equating (14.5.3) and (14.5.4), we can express α as

$$\alpha = (k_j x)^{-1} \tag{14.5.5}$$

where

$$x^{-1} = \frac{27}{4} \frac{\sigma_{\max}}{u_f^2} \tag{14.5.6}$$

The units of x are feet. Intuitively this would be the length of section x over which acceleration noise should be averaged. Test runs yielding such representative values as $\sigma_{\max} \approx 2 \text{ ft/sec}^2$ $u_f \approx 60$ mph for $x = 500$ ft tended to substantiate this interpretation. The significance of α would be the reciprocal of the maximum number of vehicles possible on the section of highway x. Again, this seems logical; looking at (14.5.1) and (14.5.2), it is apparent that the internal energy of the stream is being estimated by a single average vehicle, whereas the kinetic energy is being estimated by all traffic. The parameter α serves to adjust E and I so that their sum is equal to the total energy T which is a constant, in keeping with the concept that energy cannot be created or destroyed.

Acceleration noise is the standard deviation of changes in speed. The form of $\sigma = f(u)$ may be deduced from (14.5.2) and (14.5.4)

$$\sigma = \sigma_{\max} - \alpha k u^2 \tag{14.5.7}$$

From (12.5.25) it is apparent that

$$k = k_j \left(1 - \frac{u}{u_f}\right)^{2/(n+1)} \tag{14.5.8}$$

Substituting (14.5.8) in (14.5.7) and differentiating (14.5.7) with respect to u gives the vehicular speed u'_m at which the acceleration noise or lost energy is a minimum. It is

$$u'_m = \frac{n+1}{n+2} u_f \qquad n > -1 \tag{14.5.9}$$

This is identical to the speed given in (14.4.5) that maximizes the motion or kinetic energy of the traffic stream. For example, letting $n = 1$ in (14.5.8) and substituting in (14.5.7) gives

$$\sigma = \sigma_{\max} - \alpha k_j u^2 + \alpha k_j \frac{u^3}{u_f} \tag{14.5.10}$$

for which the optimum speed would be from $d\sigma/du = 0$

$$u'_m = \tfrac{2}{3} u_f$$

Finally, $\sigma_{\min}$ would be obtained by using (14.5.9) and (14.5.7)

$$\sigma_{\min} = \sigma_{\max} - \alpha k'_m u'^2_m \tag{14.5.11}$$

We know that the total measured acceleration noise σ_T is composed of the noise due to the interaction of traffic σ_t and the natural noise σ_N

due to the geometrics of the facility. The natural noise σ_N could be obtained either by measurements made by a test vehicle under conditions when there was no traffic interaction (say at 3 A.M.) or by fitting a regression line to acceleration noise vs. speed data to obtain σ_{min}. The relations between the parameters derived here are summarized in Fig. 14.7.

The significance of the internal-energy–acceleration-noise model developed is that it provides estimates of all the parameters summarized in Table 14.1 and Fig. 14.7 based on data collected with a single test vehicle equipped with a recording speedometer. The study procedure, data reduction, and broad applications of the model are discussed in the following sections.

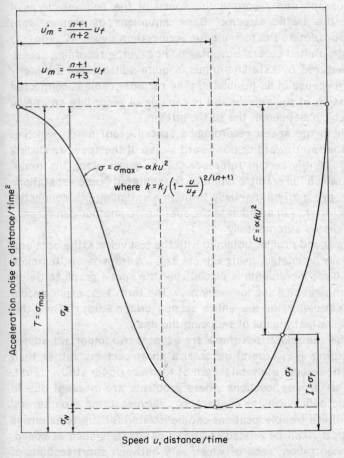

$$u'_m = \frac{n+1}{n+2} u_f$$

$$u_m = \frac{n+1}{n+3} u_f$$

$$\sigma = \sigma_{max} - \alpha k u^2$$

$$\text{where } k = k_j \left(1 - \frac{u}{u_f}\right)^{2/(n+1)}$$

$$E = \alpha k u^2$$

$T = \sigma_{max}$

σ_M

σ_N

σ_f

$I = \sigma_T$

Acceleration noise σ, distance/time^2

Speed u, distance/time

Fig. 14.7   Relations between acceleration noise parameters for internal-energy model.

14.6 STUDY METHODS

The use of the floating-car method has long been established as providing accurate determination of the average speed of traffic on a roadway. This method may also be applied in measuring acceleration noise, with the assumption made that the accelerations of a floating car also represent a good average of the accelerations of the traffic stream.

In studies of acceleration noise, two different methods of recording the necessary data to calculate the parameter have been employed. In one study[11] an accelerometer mounted in the test vehicle was photographed and σ was determined by analysis of an acceleration-time curve. Other researchers,[13] realizing that this method was too time consuming, utilized a tachograph and recorded the speed of the test vehicle as it progressed in the traffic stream. Basic equations of motion were employed to approximate the formula for acceleration noise.

The tachograph had several advantages; however, a recording speedometer manufactured by Esterline-Angus is quite suitable for the study on the freeway because of its flexibility. The recorder cable is connected to the transmission of the vehicle and is capable of recording an analog graph of the vehicle's speed in the traffic stream.

In addition to the speed recording, a special event mark, which is indicated by either an upward or downward sweep of the speed-recording pen, is used to identify certain reference-station points along the route. These stations, which for example might be the center of grade separations on the freeway, are marked manually through a switching box connected to the speed recorder. In addition to the speed pen, another pen records 100-ft distance marks automatically.

From these speed profiles obtained with the test vehicle, the acceleration noise is easily estimated, using Eq. (14.3.6). A transparent template graduated in 1.0-sec increments is placed over the speed graph to determine Δt to the nearest 0.5 sec for every Δv. The term $1/\Delta t$ may be accumulated on a calculator for the entire section under study. Thus this method provides a fast means of reducing the data.

It should be remarked that there are at least two important advantages of measuring σ in several successive short sections rather than measuring the noise over the total length of roadway under study. First, one can easily determine locations where problems are inherent due to either geometric deficiencies or congestion. By measuring σ on several short sections, these trouble locations can be isolated for future extensive study. Second, if σ can be related to known parameters such as speed, volume, or concentration, each of which vary between short sections of freeway, it is also important to measure acceleration noise in short sections.

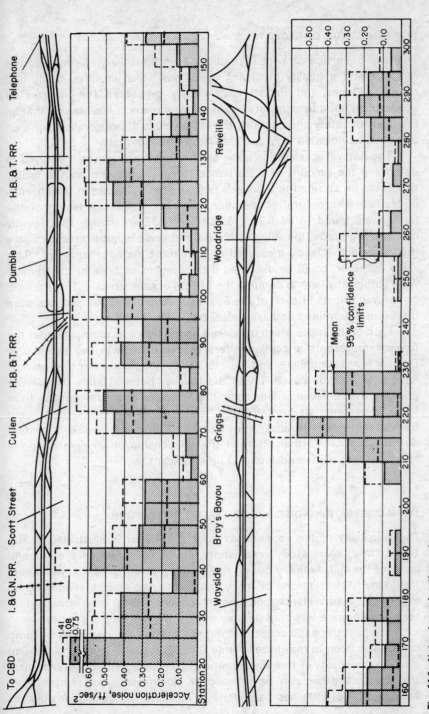

Fig. 14.8 Natural acceleration noise, σ_n.

14.7 EVALUATION OF GEOMETRICS

The level-of-service concept is predicated on the theory of providing the motorist with a quality of driving conditions which includes safety and comfort. Jones and Potts[13] have clearly presented the increase of acceleration noise because of winding country roads. It is obvious that on the basis of safety and comfort, the large values of acceleration noise on these roads indicated very poor service to the motorists.

The natural acceleration noise on the Gulf Freeway was measured to determine the amount of disturbance to a vehicle's trip on the facility that could be ascribed to a driver's natural speed changes in the absence of traffic. The natural acceleration noise for the entire length of the facility was found to be 0.38 ± 0.02 ft/sec^2. Even though the freeway has a rolling grade, the value of the natural noise is not drastically above the value (0.32 ft/sec^2) measured on an almost perfect roadbed.[11] This value, however, could be deceiving if the tangent sections of the freeway were sufficiently long to dampen the effects of geometrics. The intrinsic, or natural, effects in each of the 500-ft continuous freeway sections were therefore determined and are presented in Fig. 14.8.

Each of the roads which is marked on the figure as crossing under the freeway represents a location where grades exist, with two exceptions. Dumble is not continuous and therefore does not pass under the freeway. Also, Bray's Bayou passes under the freeway but well below the existing grade. The remaining roads constitute locations of grades ranging from 1.0 to 5.0 percent. It is noted that σ_N is approximately zero on the tangent freeway sections. However, the pattern of increased noise is evident on the grades. Further study of the relationship of acceleration noise to grades was made and found to be significant.

14.8 ACCELERATION NOISE DUE TO TRAFFIC INTERACTION

The sensitivity of the acceleration noise parameter is illustrated in Fig. 14.9. The "before study" represents normal operation on the inbound lanes of the Gulf Freeway during the morning peak. The "on-ramp control study" refers to one of many ramp-metering studies conducted on the facility's surveillance and control project. It is very obvious that acceleration noise was considerably less during the control studies.

In the past the basic concept of improving freeway operations by ramp control has been to maximize volume throughout, with little consideration given to the quality of freeway operation. It is an accepted fact that most freeway bottlenecks can be reduced or eliminated by on-ramp control techniques, but little has been done to measure the qualitative changes

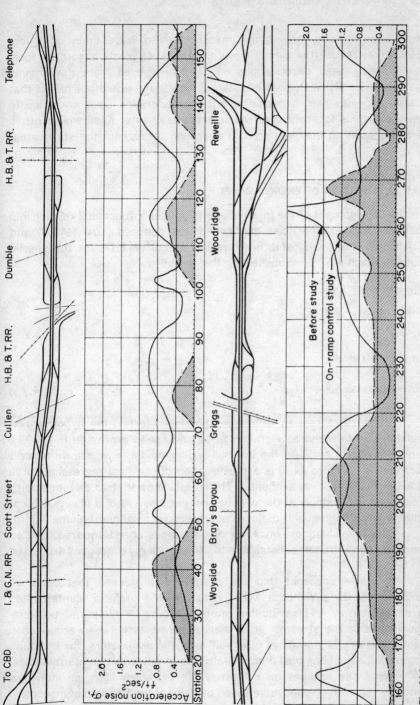

Fig. 14.9 Acceleration noise profiles, 7:30 A.M.

due to the controls, other than a measurement of travel time. It is very possible to reduce the overall travel time and yet create hazardous locations where rapid decelerations occur, far removed from the original bottlenecks. As was previously stated, travel time would not reflect that condition. However, a measurement of acceleration noise could locate any hazardous locations and would indicate whether additional controls are necessary to maintain smooth operation throughout the study area.

14.9 VERIFICATION OF ENERGY MODEL

In the development of the internal-energy model, a functional relationship between acceleration noise and speed was deduced [Eqs. (14.5.7) and (14.5.8)]. This relationship is expressed in Eq. (14.5.10) for the special case in which $n = 1$. Substituting the identities

$$\sigma_{\max} = \sigma_M + \sigma_N$$

and

$$\sigma_t = \sigma - \sigma_N$$

in (14.5.10) gives

$$\sigma_t = \sigma_M - \alpha k_j u^2 + \alpha k_j \frac{u^3}{u_f} \tag{14.9.1}$$

Standard regression techniques employed to test the hypothesized relationship between acceleration noise and speed expressed in (14.9.1), and thereby to evaluate the internal-energy model, have shown a good correlation. Figure 14.10 is a presentation of the final regression curve for 1 of the 56 freeway sections. The results show that acceleration noise increases very rapidly at the onset of congestion. Also, it is very evident that the maximum quality of flow is not realized at maximum volume output.

Similar evaluations, involving density times speed squared, plotted against speed strongly substantiate the kinetic energy concept as expressed in Sec. 14.4.[1]

It seems apparent that the maximum satisfaction that could be experienced by a motorist would result from a uniform uninterrupted freeway journey. This condition can be achieved by minimizing acceleration noise in the stream. If a relationship exists between acceleration noise and any of the above quantitative traffic parameters, the optimum operating conditions based on smoothness of flow can be determined.

When the volumes on a freeway are extremely light, an individual motorist can select a desirable speed because of the freedom of movement which he experiences. The acceleration noise of his trip is due primarily

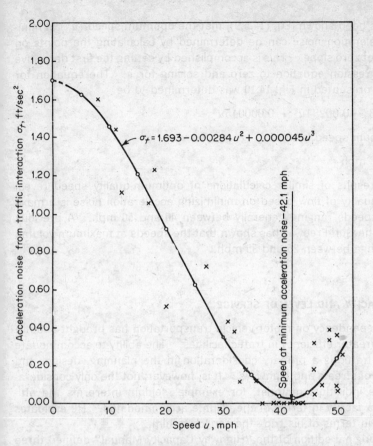

Fig. 14.10 Relationship between speed and acceleration noise due to traffic interaction, σ_t.

to his driving abilities and the geometrics of the freeway. As volumes increase, a level is reached at which an individual motorist finds it difficult to adjust laterally. Thus his speed is influenced by the vehicle in front of him, which in turn is influenced by the vehicle in front of it, and so on downstream to the lead car in the platoon. The flow at this point approaches a uniform speed and uniform headways. Consequently, the operation is smooth and the acceleration noise is at a minimum (the value of which would be dictated by the freeway geometrics). A further increase in demand is accompanied by an increase in internal friction, which promotes some instability in the stream and increases acceleration noise. It is during this unstable flow condition that driver discomfort is beginning to be realized. Additional demand increases result in inevitable freeway breakdown and a rapid decrease in driver convenience and comfort.

It has been shown [Eq. (14.5.9)] that the optimum speed (u'_m) at minimum acceleration noise can be determined by calculating the points on the curve of zero slope. This is accomplished by setting the first derivative of the regression equation to zero and solving for u. The equation for the curve presented in Fig. 14.10 was determined to be

$$\sigma_t = 1.693 - 0.00284u^2 + 0.000045u^3$$

The optimum speed is thus

$$u'_m = 42.1 \text{ mph}$$

The results of similar calculations of optimum-quality speed reveal that the quality of flow based on minimizing acceleration noise is a maximum at speeds ranging generally between 40 and 50 mph. A previous study on the Gulf Freeway has shown that the speeds at maximum volume output range between 25 and 35 mph.

14.10 CAPACITY AND LEVEL OF SERVICE

Greater dependency on motor vehicle transportation has brought about a need for greater efficiency in traffic facilities. The ability to accommodate vehicular traffic is a primary consideration in the planning, design, and operation of streets and highways. It is, however, not the only consideration. The individual motorist, for example, seldom interprets the efficiency of a facility in terms of the volume accommodated. He evaluates efficiency in terms of his trip—the service to him.

The original edition of the "Highway Capacity Manual" defined three levels of roadway capacity—*basic capacity, possible capacity,* and *practical capacity.* It was considered of prime importance that traffic volumes be accurately related to local operating conditions so that particular agencies could decide on the practical capacities for facilities within their jurisdiction. The manual recognized that practical capacity would depend on the basis of a subjective evaluation of the quality of service provided.

The present Capacity Committee of the Highway Research Board has elected, in the new edition, to define a single parameter—possible capacity—for each facility. Possible capacity is simply the maximum number of vehicles that can be handled by a particular roadway component under prevailing conditions. The practical-capacity concept has been replaced by several specific *service volumes* which are related to a group of desirable operating conditions referred to as level of service.

Ideally, all the pertinent factors—speed, travel time, traffic interruptions, freedom to maneuver, safety, comfort, convenience, and economy—should be incorporated in a level-of-service evaluation. The committee has, however, selected speed and the service-volume-to-capacity v/c ratio as the factors to be used in identifying level of service because "there are insufficient data to determine either the values or relative weight of the other factors listed."

Six levels of service, designated A through F, from best to worst, are recommended for application in describing the conditions existing under the various speed and volume conditions that may occur on any facility.   Level of service A describes a condition of free flow, level of service E describes an unstable condition at or near capacity, and level of service F gives a condition of forced flow.   Levels of service B, C, and D describe the zone of the stable flow, with the upper limit set by the zone of free flow and the lower limit defined by level of service F.   Although definitive values are assigned to these zone limits for each type of highway in the new manual, no explanation is given of how these values were obtained.   This is in no way intended as a criticism, since it is recognized that the function of any manual is essentially that of a handbook and therefore should not include a methodical discussion of the facts and principles involved and conclusions reached for every value between its covers.

Much of the traffic engineer's dilemma can be attributed to his inability to relate capacity and level of service.   If these volumes cannot be related quantitatively for an existing facility, there can be little hope for the designer to relate them for a facility that is still on the drafting board!   The acceleration-noise–momentum-energy model discussed in this chapter provides a simple, rational means for measuring freeway capacity and level of service.

If Eqs. (12.5.25) and (14.4.3) are expressed in terms of speed only, and then normalized, they become (for the special case of $n = 1$)

$$\frac{q}{q_m} = 4\left[\frac{u}{u_f} - \left(\frac{u}{u_f}\right)^2\right] \tag{14.10.1}$$

and

$$\frac{E}{T} = \frac{27}{4}\left[\left(\frac{u}{u_f}\right)^2 - \left(\frac{u}{u_f}\right)^3\right] \tag{14.10.2}$$

The curves of (14.10.1) and (14.10.2) are plotted in Fig. 14.11.   The right side of the graph is the well-known volume-speed relationship, normalized so that the abscissa is the ratio of flow to capacity (v/c) and the ordinate, the ratio of speed to free speed.   Substituting (14.5.1) in (14.10.2) gives

$$\frac{I}{T} = 1 - \frac{27}{4}\left[\left(\frac{u}{u_f}\right)^2 - \left(\frac{u}{u_f}\right)^3\right] \tag{14.10.3}$$

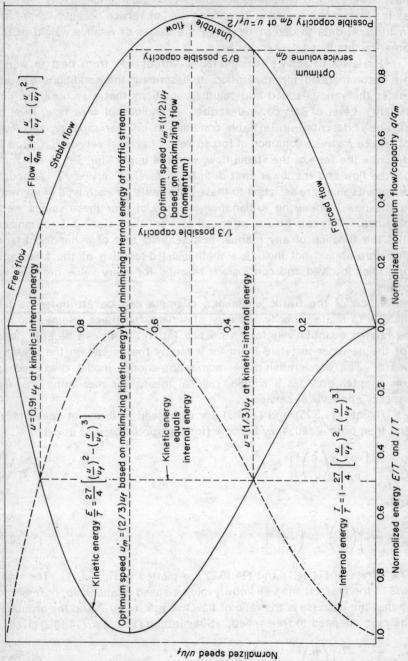

Fig. 14.11 Quantitative approach to level of service using the total energy-momentum analogy.

which is also plotted in Fig. 14.11. Equating (14.10.2) and (14.10.3) gives the two speed parameters for which the kinetic energy equals the internal energy of the traffic stream. These values are $u = \frac{1}{3}u_f$ and $u = 0.91u_f$. Proceeding to the right side of the graph, it is seen that the corresponding values for flow are $q = \frac{8}{9}q_m$ and $q = 0.328q_m$ rounded off to $q = 0.35q_m$. These two points plus the point defined as (u_m', q_m') serve to establish the four level-of-service zones defined by the 1965 "Highway Capacity Manual"—free, stable, unstable, and forced flow. In Table 14.2 the foregoing fundamental level-of-service criteria are summarized.

Table 14.2 Levels of service as established by energy-momentum concept

Level-of-service zone		Zone limits* Upper	Zone limits* Lower	Description
Free flow:	A	u_f	$0.91u_f$, $0.35q_m$	Speeds controlled by driver desires and physical roadway conditions. This is the type of service expected in rural locations
	B	$0.91u_f$, $0.35q_m$	$0.83u_f$, $0.55q_m$	Speed primarily a function of traffic density
Stable flow:	C	$0.83u_f$, $0.55q_m$	$0.75u_f$, $0.75q_m$	The conditions in this zone are acceptable for freeways in suburban locations
	D	$0.75u_f$, $0.75q_m$	u_m', q_m'	The conditions in this zone are acceptable for urban design practice. The lower limit u_m', q_m' represents the critical level of service
Unstable flow:	E_1	u_m', q_m'	u_m, q_m	A small increase in demand (flow) is accompanied by a large decrease in speed leading to high densities and internal friction
	E_2	u_m, q_m	$0.33u_f$, q_m'	This type of high-density operation cannot persist and leads inevitably to congestion
Forced flow:	F	$0.33u_f$, q_m'	0	Flows are below capacity and storage areas consisting of queues of vehicles form. Normal operation is not achieved until the storage queue is dissipated

* See Fig. 14.11.

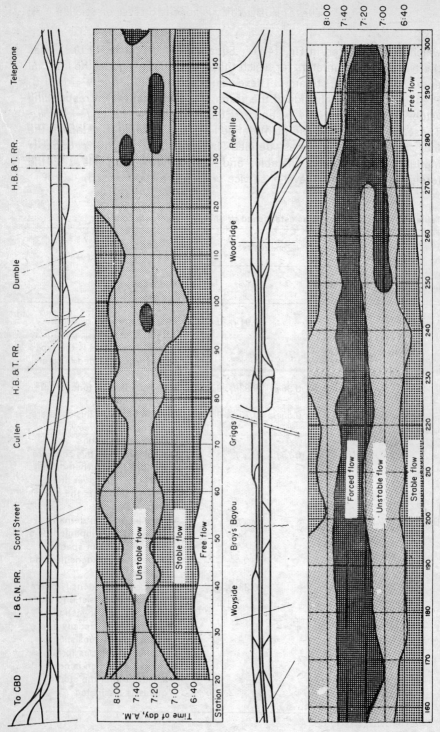

Fig. 14.12 Level-of-service contours, Tuesday, inbound.

The significance of Fig. 14.11 is that it provides a rational basis for defining level of service and relating it to the other traffic variables —speed, flow, and density.   The relationships between level of service and traffic volume (flow) are analogous to the relationships in classical hydrodynamics between energy and momentum.   Efficiency in a classical system is measured by the ratio of useful energy to total energy of E/T.   Optimum operation occurs when lost energy I is at a minimum.   In a traffic system, this concept of efficiency is manifest by maximizing the kinetic energy of the stream as a whole and minimizing the acceleration noise of the individual vehicles (internal energy).

The objective of freeways and other expressways is to provide high levels of service for high volumes of traffic.   The traffic conditions existing at maximum E/T and minimum I/T, therefore, might logically be termed *critical level of service*.   Referring to the right side of Fig. 14.11, it is seen that a small increase in demand above the volume existing at this critical level of service tends to greatly increase the density of the traffic stream, accompanied inevitably by a sharp decrease in operating speed.   That the traffic conditions k'_m and u'_m at the critical level of service are superior to those at possible capacity can be shown explicitly by dividing Eqs. (14.4.4) and (14.4.5) by (12.5.26) and (12.5.27) for $n = 1$.

$$k'_m = \tfrac{2}{3}k_m \tag{14.10.4}$$
$$u'_m = \tfrac{4}{3}u_m \tag{14.10.5}$$

The density at the critical level of service is only two-thirds the density at possible capacity, and the operating speed is one-third higher.   Of course this is accomplished at a sacrifice in the traffic volume accommodated since

$$q'_m = \tfrac{8}{9}q_m \tag{14.10.6}$$

Figure 14.12 illustrates the operation of the three inbound lanes of some 6 miles of the Gulf Freeway during the morning peak 2 hr, as obtained using the moving-vehicle study procedure described in this chapter. The figure represents a complete documentation of level of service in both time and space.   This simple procedure affords a rational means for describing the level of operation on the facility —free flow, stable flow, unstable flow, and forced flow —as established by the energy-momentum concepts illustrated in Fig. 14.11.

14.11   RELATIONSHIP OF ENERGY AND FUEL CONSUMPTION

Vehicular operating cost is an important level-of-service factor.   If the energy-momentum model for describing level of service is a valid one, there should be some relationship between fuel consumption and *energy*, since the kinetic energy of the traffic stream (volume times speed) and the internal energy of the traffic stream (acceleration noise of individual vehicles) are both assumed to describe level of service.

Capelle[20] hypothesized that the internal energy or acceleration noise measured over a segment of roadway is equal to the total fuel consumed FC minus the minimum fuel consumption FC_{min}. Thus, substituting fuel consumption for acceleration noise in the relationship in Fig. 14.7 leads to

$$FC = FC_{max} - \frac{27}{4}\left(\frac{u}{u_f}\right)^2 \sigma_{max}\left(1 - \frac{u}{u_f}\right) \tag{14.11.1}$$

This expression for fuel consumption as a function of speed appears to be realistic. For example, it is an accepted fact that the operational costs of a vehicle driven at high speeds under free-flow conditions are considerably more than those experienced when driving at a reasonable speed under free-flow conditions. It is also realistic to expect operational costs to increase at low constrained speeds when a motorist is subjected to stop-and-go conditions.

A regression analysis[20] made on data obtained during test runs between Austin Blvd. and First Avenue on the Eisenhower Expressway in Chicago using a special fuel meter showed highly significant F ratios and t values, which led to the conclusion that the data were adequately explained by the theoretical model.[20] It is seen that the energy-momentum model yields a quantitative description of yet another level-of-service factor. Capelle's work on the Eisenhower Expressway in Chicago is also significant in that it serves to validate the findings of Drew and Dudek on the Gulf Freeway in Houston, regarding the entire energy concept of traffic flow.

14.12 CHALLENGES

Based on limited data on an arterial street, Rowan[21] concludes that "the energy model which was developed to describe uninterrupted freeway traffic operation does not apply precisely to the interrupted flow conditions on major streets." However, he concedes that the model fits sufficiently to be of value in relating certain variables in the urban traffic problem.

Since in applying the model only the linear relationship between speed and density $(n = 1)$ has been tried, there is a need to investigate the nonlinear relationships $(n = -1$ and $n = 0)$ to see if a better description can be obtained. Moreover, in testing the model on city streets, perhaps more positive study techniques should be employed in order to obtain a complete range of data (for both very low and very high speeds). It is possible that use of the "maximum car" method and the "minimum car" method rather than the "average car" method employed by Rowan might provide these data, greatly improving the fit to the energy model.

Since traffic accidents are often attributed to inefficient operations because of erratic flow, an investigation should be made to determine if there is a relationship between the energy variables and accident rates.

The use of traffic control procedures to minimize freeway congestion requires continuous measurements of freeway traffic characteristics. With such a program, the moving-vehicle technique for measuring acceleration noise is obviously incompatible. An automatic sensing system for measuring acceleration noise (or an equivalent energy parameter) of the traffic stream is needed.

PROBLEMS

14.1. Given the following data from a continuous record of a vehicle's movement through a freeway section, assume the speed change is linear between the given data points. Calculate the acceleration noise for the vehicle using $\Delta v = 5$ mph.

Speed, mph	Time, min	Speed, mph	Time, min
30	0	35	16
35	1	40	18
35	3	40	19
30	4	45	20
30	5	45	22
20	6	40	24
20	8	40	25
25	9	45	26
25	10	45	28
30	12	40	29
30	13	40	32
35	14		

14.2. Determine the Greenshields quality of flow index for the vehicle in Prob. 14.1.

14.3. a. Plot a graph of normalized curves for $n = -\frac{1}{2}$, $n = 0$, $n = +\frac{1}{2}$ similar to Fig. 14.11.

b. Prepare a table similar to Table 14.2 for each of the graphs plotted in (a).

14.4. If you conducted an acceleration-noise run on a signalized arterial street at 1:00 A.M. in a city of 75,000 persons, would you expect a high or low value of acceleration noise and why?

14.5. Given the following data taken on a freeway:

Speed, mph	Density, vph
56.6	29.2
45.1	45.2
22.9	90.0
38.8	58.1
42.8	48.2
41.8	47.6
45.0	52.3
45.7	45.0
48.1	45.9

Fit the traffic flow model $u = u_f(1 - k/k_j)$ to the data to determine the parameters necessary to plot the fuel consumption vs. speed curve for the operation represented by the data set. The maximum rate of fuel consumption measured during the operation period on a test vehicle was $\frac{1}{5}$ gal per mile and the minimum was $\frac{1}{20}$ gal per mile.

14.6. The engineer's measure of efficiency of a road is often taken to be volume; and the individual motorist considers such factors as speed, number of delays, and freedom to maneuver. Discuss briefly how the energy-momentum concept allows consideration of the different views of efficiency.

REFERENCES

1. Drew, D. R., and C. J. Keese: Freeway Level of Service as Influenced by Volume and Capacity Characteristics, *Highway Res. Board Rec.* 99, 1965.
2. Dudek, C. L.: "A Study of Acceleration Noise as a Measurement of the Quality of Freeway Operation," thesis, Texas A&M University, College Station, 1965.
3. Drew, D. R., C. L. Dudek, and C. J. Keese: Freeway Level of Service as Described by an Energy-acceleration Noise Model, *Highway Res. Board Record* 162, 1966.
4. Drew, D. R.: The Energy-Momentum Concept of Traffic Flow, *Traffic Eng.*, vol. 36, no. 10, June, 1966.
5. Hall, Edward M., and Stephen George, Jr.: Travel Time: An Effective Measure of Congestion and Level of Service, *Highway Res. Board Proc.*, vol. 38, pp. 511–529, 1959.
6. Wattleworth, Joseph A.: Peak-period Control of a Freeway System: Some Theoretical Considerations, *Congress Expressway Surveillance Project Rept.* 9, August, 1963.
7. Greenshields, Bruce D.: Quality of Traffic Transmission, *Highway Res. Board Proc.*, vol. 34, pp. 508–522, 1955.
8. Greenshields, Bruce D.: Quality of Traffic Flow, "Quality and Theory of Traffic Flow," symposium, Bureau of Highway Traffic, Yale University, New Haven, Conn., 1961, pp. 3–40.
9. Platt, Fletcher N.: A Proposed Index for the Level of Traffic Service, *Traffic Eng.*, vol. 34, no. 2, pp. 21–26, November, 1963.
10. Montroll, E. W., and R. B. Potts: "Car Following and Acceleration Noise: An Introduction to Traffic Flow Theory," pp. 39–40, Highway Research Board, Washington, D.C., 1964.

11. Herman, Robert, Elliot W. Montroll, Renfrey B. Potts, and Richard W. Rothery: Traffic Dynamics: Analysis of Stability in Car Following, *Operations Res.*, vol. 7, pp. 86–106, 1959.

12. Montroll, E. W.: Acceleration Noise and Clustering Tendency of Vehicular Traffic, "Theory of Traffic Flow," symposium, General Motors Research Laboratories, 1959, pp. 147–157.

13. Jones, Trevor R., and Renfrey B. Potts: The Measurement of Acceleration Noise: A Traffic Parameter, *Operations Res.*, vol. 10, November–December, pp. 745–763, 1962.

14. Lighthill, M. J., and G. B. Whitham: On Kinematic Waves: A Theory of Traffic on Long, Crowded Roads, *Proc. Roy. Soc. (London)*, vol. 229, 1955.

15. Richards, P. I.: Shock Waves on the Highway, *Operations Res.*, vol. 7, 1959.

16. Greenberg, H.: An Analysis of Traffic Flow, *Operations Res.*, vol. 7, 1959.

17. Herman, R., and R. B. Potts: Single-lane Traffic Theory and Experiment, in "Theory of Traffic Flow," American Elsevier Publishing Company of New York, 1961.

18. Morrison, R. B.: The Traffic Flow Analogy to Compressible Fluid Flow, *Advanced Res. Eng., Bull.* M 4056-1, Ann Arbor, Mich.

19. Gazis, D. C., R. Herman, and R. B. Potts: Car-following Theory of Steady State Traffic Flow, *Operations Res.*, vol. 7, pp. 499–505, 1959.

20. Capelle, D. G.: "An Investigation of Acceleration Noise as a Measure of Freeway Level of Service," doctoral dissertation, Texas A&M University, College Station, 1966.

21. Rowan, N. J.: "An Investigation of Acceleration Noise as a Measure of the Quality of Traffic Service on Major Streets," doctoral dissertation, Texas A&M University, College Station 1967.

CHAPTER FIFTEEN
CREATIVE DESIGN

Slowly but surely humanity achieves what its wise men have dreamed.

Anatole France

15.1 INTRODUCTION

Creative system design lies in the satisfaction of mankind's needs. The creative design approach is to establish a clear, broad framework, which must be a flexible one in order to absorb future changes. This broad framework will aim at total design and will be concerned with the structural, geometric, and functional aspects of the facility in question.

The question in utilizing creative design to satisfy mankind's needs is: What are these needs? There are, for example, a number of competing interests in the freeway traffic problem which confuse the choice of viewpoint. The city dweller who works downtown and must be there at 8 A.M. 5 days a week has one viewpoint, the person who purely by chance passes through the town on a through route has another, the suburbanite who lives just outside the city limits has another, and the commuter who rides the train to the city every day has still another. On top of this are the retail merchants, who want their customers to have easy access to their stores; the pedestrians, including schoolchildren, who must sometimes walk out of their way to cross a freeway; the police, who must have quick access to the facility; the resident owners whose sleep is interrupted by the rumbling of heavy trucks; the utility companies who must lay lines beneath the facilities; and the transportation companies that compete with automobiles for customers and, as often is the case, for space.

Add to these the divergent philosophies of the architect who frowns on "unimaginative freeway aesthetics"; the city planner with his "Chinese wall" complex; the rapid transit evangelists, many of whom were dragged screaming into the twentieth century, with their plea to return to the "good old days" of the streetcar; and the highway engineer with his balanced mass-diagram obsession.

The proper viewpoint is some weighted average which gives the greatest good to the greatest number, but as some authors[1] have suggested it is not always clear just how to take this average.   For example, there is growing sentiment that the trouble with freeways lies in the fact that there are too many interchanges, causing weaving and increased traffic interaction.   True, if there were not as many interchanges, freeway operation would be improved.   However, eliminating interchanges means that a freeway driver must either go several blocks past his street and then double back or exit long before he would prefer.   On the other hand, a driver desiring to use a freeway with fewer interchanges has a much longer trip on the surface streets before he can enter the freeway.   In both cases, this adds up to increased time and distance in traffic, making the elimination of interchanges doubtful as a creative solution.

15.2 FREEWAY DESIGN

Highway design is an engineering function—not a handbook problem. The engineer is faced with the problem of predicting traffic demands in future years and providing facilities that will accommodate that traffic under a selected set of operating conditions, or levels of service.   Too often highway design has been accomplished by adopting a set of handbook standards, which has resulted in the construction of many seriously inadequate facilities.

A freeway is *not* built specifically for some date 20 years in the future. It must go to work the first day and serve efficiently all through its expected life.   And, if history is indicative, many will be serving for quite a number of years beyond some arbitrary design year.   Just as the day has gone when a freeway can be designed within the confines of two parallel right-of-way lines, so has the day gone when only the 20-year "complete system" can be considered when designing a particular facility.   Traffic projections and designs must be made on partial or incomplete systems if desirable service is to be obtained in the years before the whole system is completed. With the modern tools available, the designer should have at his disposal an accurate estimate of traffic demand for each stage of completion of the planned system.

In addition to the demand variable, something about the expected capacity of the freeway system must be known. The largest number of vehicles that can pass a given point in one lane of a multilane highway, under ideal conditions, is between 1,900 and 2,200 vph. This represents an average maximum volume per lane sustained during the period of 1 hr. Studies have found higher volumes for specific lanes or for short time periods. Where at least two lanes are provided for movement in one direction, disregarding distribution by lanes, the capacity of a freeway under ideal conditions is considered to be 2,000 vph per lane. Only in situations where the peak-period demand extends very nearly for the entire hour will this capacity value be realized. Where conditions are less than ideal because of reduced widths and sight distance, grades, commercial vehicles, etc., the capacity will be somewhat lower. Moskowitz and Newman[2] suggest some correction value to be used when these conditions are anticipated.

Finally we come to measures of effectiveness. "The problem is too many traffic jams" or "Too many accidents" or "It takes too long to drive to work," and so forth. The two most important characteristics of a measure of effectiveness are that (1) it be quantitative—capable of being expressed unequivocally in terms of a number and (2) it be efficient in a statistical sense—having a comparatively small variance and hence obtainable with reasonable accuracy. There are other desirable characteristics of the measure of effectiveness. It should be simple, where this is compatible with completeness, and should have physical meaning; there is then less chance of error through its use. Furthermore, if it has physical meaning ideal performance can easily be found and compared with the actual physical performance. It is useful to know whether the theoretical limits have been reached or whether there is still considerable room for improvement.

Thus, in the freeway design problem, the product of speed and volume, or kinetic energy as defined in previous chapters, is simple, has real physical meaning, and can be compared with ideal performance. This measure of effectiveness is simply referred to as *level of service*.

When the traffic volume equals the capacity of a freeway, operating conditions are poor. Speeds are low, with frequent stops and high delay. In order for the highway to provide an acceptable level of service to the road user, it is necessary that the service volume be lower than the capacity of the roadway.

The level-of-service approach establishing levels of operation from free flow to capacity which has been established in the 1966 "Highway Capacity Manual" is designed to allow the engineers and administrators to provide the highest level of service economically feasible. The momentum-energy analogy is an effort to explain the relationship between capacity and level of service rationally and quantitatively. It must be recognized that highway traffic represents a stochastic phenomenon. Therefore, any highway facility must be designed with the realization that from time to time demand will exceed capacity.

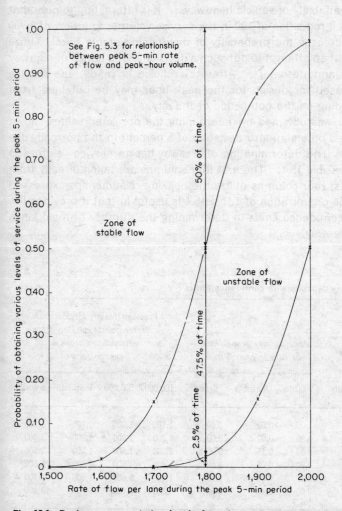

Fig. 15.1 Design curves relating level of service to flows during the peak 5 min.

Since congestion may last much longer than that interval in which demand exceeds capacity, it is important that precautions be taken to prevent this. Based on a stipulated rate of flow for a 5-min period, a designer can ensure that congestion will not occur to whatever degree of confidence desired. This can be illustrated by the design curves relating level of service to a 5-min rate of flow (Fig. 15.1). Thus, with a possible capacity of 2,000 vph per lane, a service volume of 1,800 has a 50 percent probability of guaranteeing *stable flow* during the peak 5-min period. On the other hand, there is a 50 percent chance of *unstable flow* occurring and a 2.5 percent chance of *force flow* which can lead to congestion, because of the statistical variability of vehicle headways. It is interesting to note that a relatively small reduction of 100 in the service volume, to a flow of 1,700 vph, greatly increases the probability of maintaining stable flow. These curves represent an attempt to put such a decision in the hands of highway administrators and designers. After this choice is made, the service volume to be used for design for the peak hour may be obtained from Fig. 5.3, depending on the population of the city.

Figure 15.1 was obtained by determining the predicted values shown in Fig. 5.3, based on a standard deviation of 5 percent in the normally distributed error. The determination of freeway-design service volumes is summarized in Table 15.1. The first four columns are taken directly from Fig. 15.1; the last four columns utilize the peaking relationships expressed in Fig. 5.3. The organization of Table 15.1 is useful in that it provides the designer with confidence limits in determining the number of main lanes needed on a freeway.

Table 15.1   Freeway capacity with confidence limits

Peak 5-min flow, vph	Approximate probabilities of flow in peak 5 min			Freeway-design service volume (total hourly vol./lane) for metropolitan areas with populations of			
	Stable	Unstable	Forced	100,000	500,000	1 million	5 million
1,500	1.00	0.00	0.00	1,100	1,200	1,300	1,300
1,600	0.98	0.02	0.00	1,200	1,300	1,300	1,400
1,700	0.85	0.15	0.00	1,300	1,400	1,400	1,500
1,800	0.50	0.48	0.02	1,400	1,500	1,500	1,600
1,900	0.15	0.69	0.16	1,500	1,600	1,600	1,700
2,000	0.03	0.47	0.50	1,500	1,600	1,700	1,800

15.3 THE FREEWAY SYSTEM

The performance of a system depends on all its components and cannot be determined by the analysis of any individual component. The freeway system can be broken down into at least six components: the express lanes, the entrance ramps, the exit ramps, the frontage roads, the cross streets, and the interchanges. Analysis of each will reveal important characteristics that must be considered in producing a satisfactory design. Classical systems are described in terms of a set of variables, commonly called parameters, by means of which thè system performance is described. In a freeway system the number of lanes of traffic, the number of on-ramps, the number of off-ramps, and the number of lanes per ramp are design parameters.

After deciding upon the value of a particular parameter, it is important to know its effect on the operation of that component and what effect it has on the overall system design. Thus, it might seem that an entrance ramp component should have two lanes. However, it is well known that a two-lane entrance ramp is not compatible with the freeway express lanes unless an express lane is added to make the number of lanes downstream of the ramp one more than the number upstream.

15.4 INTERCHANGE REQUIREMENTS

The full control of access features of freeways automatically requires grade separations for intersecting streets and interchanges to provide for turning movements between the intersecting facilities. Interchanges may be classified according to the number of legs, as direct or indirect, and as signalized or unsignalized. The three principal types are the cloverleaf, directional, and diamond. Fully directional and cloverleaf interchanges are unsignalized types, and therefore their capacities are essentially determined by their ramps. Diamond interchanges are always signalized in urban areas, and their capacities are therefore dependent upon the individual intersections or the coordinated system of intersections.

Although Eqs. (7.11.2) through (7.11.7) were developed for a signalized intersection, the theory can be extended to cover the capacity of signalized interchanges. The high-type intersection can be considered as an at-grade facility in which the conflicting movements are separated both by the signal phases, or *time*, and the distance between the component intersections of the interchange, or *space*. Figures 15.2 to 15.4 illustrate possible phasing arrangements for variations of the diamond interchange. The conventional diamond interchange consists of four one-way diag-

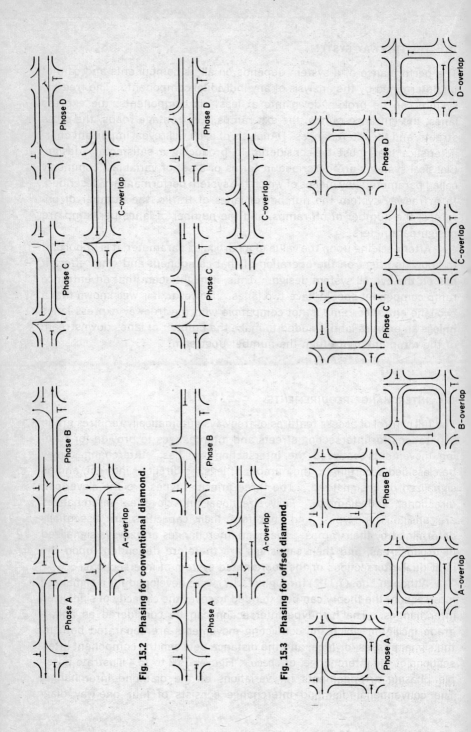

Fig. 15.2 Phasing for conventional diamond.

Fig. 15.3 Phasing for offset diamond.

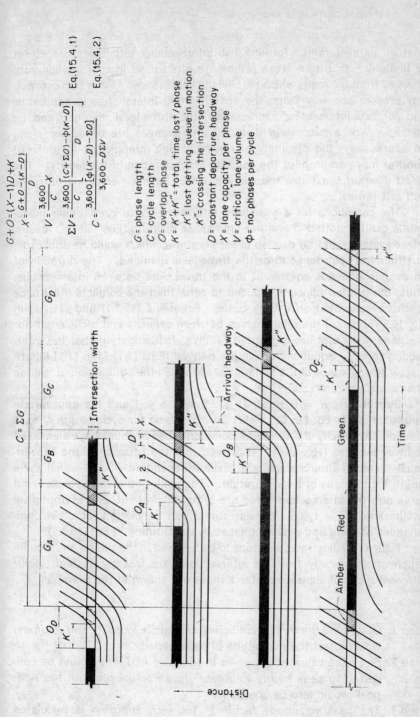

Fig. 15.5 Time-space relationship for multiphase signal systems.

$$G + O = (X - 1)D + K$$

$$X = \frac{G + O - (K - D)}{D}$$

$$V = \frac{3,600}{C} X$$

$$\Sigma V = \frac{3,600}{C} \left[\frac{(C + \Sigma O) - \phi (K - D)}{D} \right] \qquad \text{Eq.(15.4.1)}$$

$$C = \frac{3,600 \left[\phi (K - D) - \Sigma O \right]}{3,600 - D \Sigma V} \qquad \text{Eq.(15.4.2)}$$

G = phase length

C = cycle length

O = overlap phase

$K = K' + K''$ = total time lost /phase

K' = lost getting queue in motion

K'' = crossing the intersection

D = constant departure headway

X = lane capacity per phase

V = critical lane volume

ϕ = no. phases per cycle

onal or parallel ramps forming two intersections with the cross street or highway. In urban areas ramps are often used in coordination with one-way frontage roads which parallel the expressway. Where an expressway crosses one-way pairs, the split-diamond interchange is applicable; and for two intersecting expressways, the three-level diamond can be utilized. Where successive diamond interchanges are called for along an expressway, the proximity of ramps of one interchange to those of another may compromise the capacity and operation of the facility. The *offset diamond* eliminates the weaving effect between adjoining entrance and exit ramps.

The conditions for a signalized interchange with overlapping movements can be plotted in another time-space diagram (Fig. 15.5).[3] A four-phase system with four overlap phases is shown which would be applicable to either the split diamond or the three-level diamond. The duration of an overlap phase is equivalent to the travel time between intersections. Thus, the overlap phase is reduced to zero; then the conflicts must once again be separated on a time basis. Equations (15.4.1) and (15.4.2) in Fig. 15.5 represent the relationships between capacity and cycle length for the general case of any signalized facility. It is apparent that the intersection-capacity equations previously derived [Eqs. (7.11.2) to (7.11.6)] are merely special cases of the general case in which the summation of overlap phases equals zero ($\Sigma 0 = 0.0$).

Substituting $K = 6.0$ sec, $D = 2.0$ sec, $\phi = 4$ and the appropriate values for $\Sigma 0$ in Eq. (15.4.2) yields an expression for cycle length C as a function of capacity ΣV for some common types of four-phase signalized highway facilities (Fig. 15.6). It is seen that the capacity of the conventional diamond interchange is actually independent of the signal cycle length for overlaps of 8-sec duration. The split- and three-level-diamond curve and the intersection curve are symmetrial about, and asymptotical to, the line $\Sigma V = 1,800$. Similar curves can be plotted for other combinations of cycle and overlap phases by substituting in Eq. (15.4.1).

Figure 15.7 illustrates six steps to be followed in the design and signalization of any facility from an intersection to a freeway system. Four different types of signalized interchanges are shown in the example.

Step 1. Determine the three conditions (population, location, and volumes) which affect the magnitude of peak period.

Step 2. Since the volumes are given in terms of ADT, they must be converted to peak hourly volumes. In an actual problem the A.M. peak would also be checked.

Step 3. The peak-magnitude factor Y' for each approach is calculated using the regression equation.

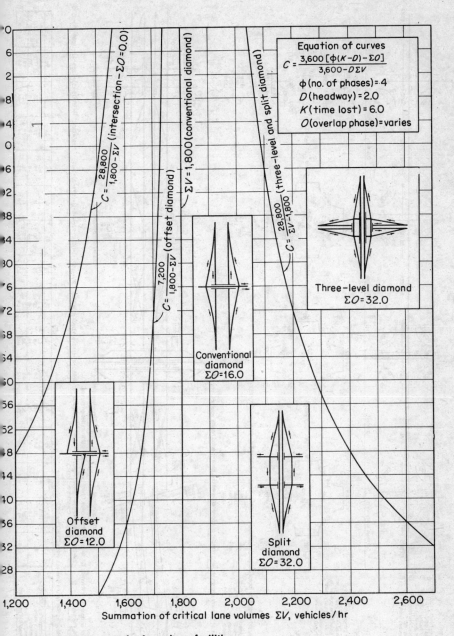

Fig. 15.6 Design curves for four-phase facilities.

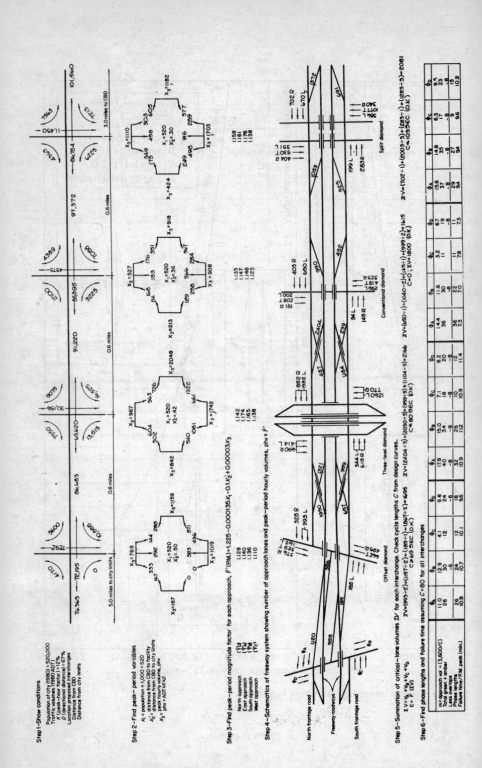

Step 1—Show conditions
Step 2—Find peak—period variables
Step 3—Find peak—period magnitude factor for each approach, P' (PM)=1.225−0.0001135X_1−0.1$X_2'^2$+0.000003X_3
Step 4—Schematics of freeway system showing number of approach lanes and peak—period hourly volumes, phv × P'
Step 5—Summation of critical—lane volumes $ΣV$ for each interchange. Check cycle lengths C from design curves.
Step 6—Find phase lengths and failure time assuming C=80 for all interchanges

Step 4. The peak-magnitude factors Y' are applied to the peak hourly approach volumes to arrive at an hourly rate of flow equivalent to the arrivals during the peak period. Where several facilities are involved, as for a freeway system, it is convenient to place the volumes and elements of the system in schematic form, as shown.

Step 5. A capacity check is made to see if the planned facilities will handle the traffic with reasonable cycle lengths. Should a facility be apparently underdesigned, additional approach lanes may be added or a higher-type facility substituted in its place.

Step 6. The average arrivals per cycle m are calculated from the critical lane volumes, and these values are used to enter the graph of Poisson curves (Fig. 7.3). The phase lengths G are tabulated for various probabilities of failure P. Any combination of $G_A + G_B + G_C + G_D$ that equals the assumed cycle length C is acceptable.

The choice of interchange at a given location will be influenced by the traffic demand at that location and by the capacity of the facility. Capacity is increased progressively from a simple intersection to the conventional diamond to the three-level diamond (Fig. 15.6). The offset diamond is predicated on an operational consideration—namely, to reduce weaving between successive ramps.

The probability of demand exceeding capacity during a cycle on an approach represents a logical design criterion. The versatility of the procedure is emphasized in step 6 (Fig. 15.7) in the many phasing combinations available. The proximity of another intersection or a ramp might dictate favoring one phase at the expense of the others. Therefore, it would be possible to prevent excessive queues leading to interference on adjacent facilities and perhaps progressive failures. It is at this point in the procedure that the engineer's judgment must be utilized.

15.5 MAIN-LANE REQUIREMENTS

After the initial determination of the number of freeway lanes using Fig. 15.1 and Table 15.1, the operating conditions at critical locations of the freeway must be investigated for the effect on capacity and level of service.[4] Unless some designated level of service is met at every point on the freeway, bottlenecks will occur and traffic operation will break down. Critical locations on a freeway are manifest by either sudden increases in traffic demand, the creation of intervehicular conflicts within the traffic stream, or a combination of both.

The traffic demand on a freeway can change only at entrance or exit ramps. Two of the most critical points on a freeway will be upstream from an exit ramp and downstream from an entrance ramp, where traffic demand will necessarily be at a maximum.   Operating conditions at exit ramps are generally similar to the operating conditions at an upgrade but can be much more severe where there is a backup from the exit ramp onto the main roadway proper.   Many exit-ramp problems could be avoided by providing for speed reduction on the ramp rather than on the shoulder lane of the freeway.   Even where long parallel deceleration lanes are provided, they are not used because of the unnatural maneuver involved.   Unfortunately, the close spacing of interchanges and use of frontage roads favor the use of short slip-type ramps.   Where a high-exit-volume slip ramp is used, definite consideration should be given to placing yield signs on the frontage roads, thus preventing backup from the exit ramp onto the freeway.

Entrance ramps may create two potential conflicts with the maintenance of the adopted level of service of a roadway section.. First, the additional ramp traffic may cause operational changes in the outside lane at the merge.   This condition, of course, will be aggravated by any adverse geometrics, such as high angle of entry, steep grades, and poor distance.   Second, the additional ramp volume may change the operating conditions across the entire roadway downstream from the on-ramp.   This is particularly true where there is a downstream bottleneck.

There are three basic procedures employed in determining the capacity of entrance ramps.   One method is based on preventing the total freeway volume upstream from the ramp plus the entrance ramp volume from exceeding the capacity of a downstream bottleneck.   A second method takes into consideration the distribution of freeway volumes per lane and then limits the ramp volume to the merging capacity (assumed here to be equal to the service volume selected in Table 15.1) less the upstream volume in the outside lane.   The third method states that the ramp capacity is limited by the number of gaps in the shoulder lane which are greater than the critical gap for acceptance.[4]   It is believed that the second method is the most practical in designing a new facility.   The first method is predicated on knowing the capacity of bottlenecks —something that is not known in the case of a new facility.   The advantage of the third approach is that it recognizes that ramp capacity and operation must be affected by the geometrics of the ramps.

The last critical location to be considered is the weaving section. Weaving sections often simplify the layout of interchanges and result in right-of-way and construction economy.   The capacity of a weaving section is dependent upon its length, number of lanes, running speed, and relative volumes of individual movements.   When large-volume weaving movements occur during peak hours, approaching the possible capacity of the section, probable results are traffic friction, reduced speeds of operation,

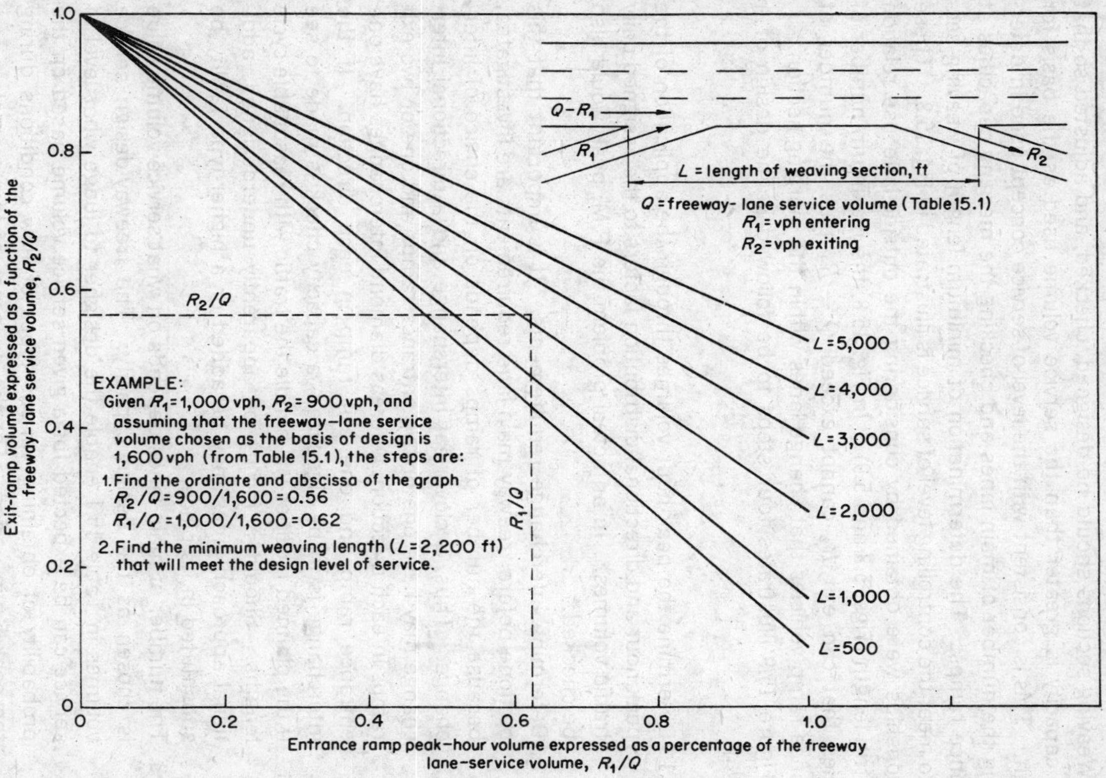
1.0
0.8
0.6
0.4
0.2
0
0.2
0.4
0.6
0.8
1.0
Exit-ramp volume expressed as a function of the freeway-lane service volume, R_2/Q
Entrance ramp peak-hour volume expressed as a percentage of the freeway lane-service volume, R_1/Q
$Q-R_1$
R_1
R_2
L = length of weaving section, ft
Q = freeway-lane service volume (Table 15.1)
R_1 = vph entering
R_2 = vph exiting
R_2/Q
R_1/Q
L=5,000
L=4,000
L=3,000
L=2,000
L=1,000
L=500
EXAMPLE:
Given R_1=1,000 vph, R_2= 900 vph, and assuming that the freeway-lane service volume chosen as the basis of design is 1,600 vph (from Table 15.1), the steps are:
1. Find the ordinate and abscissa of the graph
R_2/Q = 900/1,600 = 0.56
R_1/Q = 1,000/1,600 = 0.62
2. Find the minimum weaving length (L=2,200 ft) that will meet the design level of service.

Fig. 15.8 Determination of minimum length of weaving section to meet the design level of service.

and a lower level of service. This can sometimes be avoided by the use of additional structures to separate ramps, reversing the order of ramps to place the critical weaving volumes on frontage roads, and the use of collector-distributor roads in conjunction with cloverleaf interchanges.

Weaving sections should be designed, checked, and adjusted so that their capacity is greater than the service volume used as the basis for design. This is consistent with the level-of-service concept used in determining the number of main lanes and checking the merging capacities at entrance ramps. The determination of minimum length of weaving section to meet the controlling level of service is illustrated in Fig. 15.8. These relationships were obtained by considering the outside-lane use relation with trip length (Figs. 5.8 and 5.9). In Fig. 15.8, the maximum number of vehicles that can exit R_2 cannot exceed $Q - R_1$ plus the number of entrance ramp vehicles that change lanes within the merging section.

Figure 15.9 illustrates four steps to be followed in the design of a freeway.

Step 1. Determine the peak-hour volumes through the application of the peak hour and directional-distribution factors to the assigned daily traffic volumes. In an actual problem the P.M. peak would also be checked.

Step 2. Determine interchange requirements. It is important that this be done before freeway main-lane requirements are investigated, because the number of ramps depends on the choice of interchange. Thus, a cloverleaf interchange and a directional interchange may have one or two entrance ramps and one or two exit ramps in each direction, whereas diamond interchanges have one entrance ramp and one exit ramp in each direction. If the interchange is to be signalized, a capacity check is made to see if the planned facilities will handle the traffic with reasonable cycle lengths. Should a facility be apparently underdesigned, additional approach lanes may be added or a higher-type facility be substituted in its place.

Step 3. The number of main lanes depends on what service-volume value is chosen as the design capacity. The freeway-design service volumes in Table 15.1 enable the designer to judge what level of service can be expected for a given service volume based on the probability of obtaining various types of flow conditions during the peak 5 min period. For the purposes of this example a service volume of 1,700 vph is chosen. The operating conditions at critical locations must be checked to ensure that the designated level of service is met at every point on the freeway. The critical sections considered in this paper are merging and weaving sec-

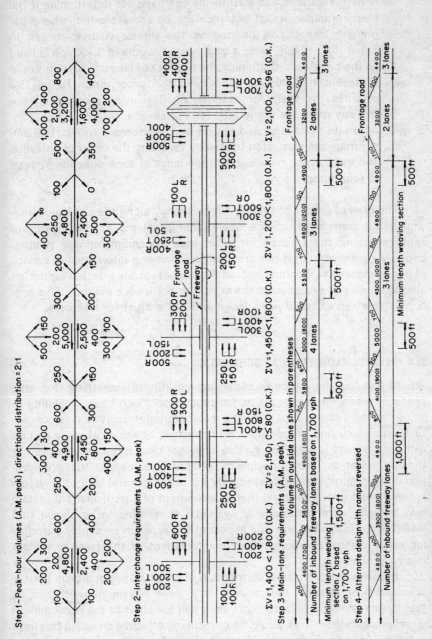

Fig. 15.9 Freeway design procedure.

tions. Figures 5.4 to 5.7 provide the basis for determining if the merging capacities at entrance ramps are exceeded, where the merging capacity is defined as the service volume chosen in Table 15.1. Thus, since a total hourly volume of 1,700 vph is used as the basis for determining the number of lanes, 1,700 vph would represent the merging capacity in this procedure. Figure 15.8 provides the basis for determining if weaving sections on the freeway meet the designated level of service.

Step 4. Alternate designs should always be considered. In Fig. 15.9 one alternative is illustrated by merely reversing the order of entrance and exit ramps, resulting in three lanes in each direction instead of four lanes.

The level of service should be "in harmony" along the stretch of freeway being considered. Since operational problems at one point are reflected along the freeway for a distance depending on the volume-capacity relationship, it is not practical to consider a lower level of service at one or more critical points; instead the level of service selected for design should be met or exceeded at the critical or bottleneck points. This concept is referred to as *balanced design* and it is a must for freeways.

15.6 REVERSE–FLOW FREEWAY SCHEMATICS

Where one has the choice of supplying too little of something (too little space between interchanges) or too much (too much space resulting in lor er surface-street trips), using a two-parameter analysis it is conceptually simple to find an optimum. In practice, however, what we end up with in this case is what the system engineer calls *suboptimization* rather than optimization. This principle of suboptimization states that optimization of each subsystem independently will not in general lead to a system optimum and, more strongly, that improvement of a particular subsystem may actually worsen the overall system. It remains to be shown that interchange spacing is only one aspect of the problem by simply introducing other system components or manifestations of the freeway system design problem.

Because the directional distribution of traffic during the peak hour is greatly out of balance, the capacity of a given pavement width can be greatly increased by devoting more than half the lanes to the predominant direction of traffic. This would be particularly effective on radial freeways in large cities carrying heavy volumes toward the CBD in the morning and outward in the evening.

This type of operation has been used rather effectively to provide traffic relief on existing major arterials and bridges. For example, a six-lane facility may be operated with four lanes in one direction and two lanes in the other to fit an unbalanced traffic flow. Since during the off peak the facility is operated with three lanes in each direction, traffic control is accomplished by (1) signing and (2) cones between, and traffic signals over, the convertible lanes.

As pointed out in the AASHO Redbook,[5] this principle, though theoretically applicable to expressways, is complicated by the median. Under some arrangements traffic on the left of the median is isolated from all exits for some distance, and the anticipated operation would not be compatible with freeway, or even expressway-type, operation.

The heart of the matter lies in the complexity of a freeway system. The freeway designer must avoid the dangers of suboptimization —in this case dealing with the problem of interchange spacing and reversible express lanes separately. What follows in the form of a typical freeway design illustration is an attempt to give the reader a feeling for the point of view which distinguishes creative freeway system design from classical geometric design.

The DHV traffic volumes and turning movements from Fig. 15.10 show that seven locations have turning movements which warrant interchanges. The problem which emerges in deciding on an interchange type is providing for the efficient movement of high ramp volumes onto the freeway. Such conventional devices as the use of two-lane entrance ramps or two separate entrance ramps per interchange to accommodate high turning movements do not really solve the problem since all vehicles are still forced to merge onto a single outside freeway lane. Rather than solutions, more often than not these tactics are no more than exercises in "pencil whipping."

Of equal concern is the high number of freeway lanes needed to handle the assigned volumes. Efficiency of operation is at its peak with three freeway lanes in one direction and drops off rapidly as four lanes of traffic, or five, in each direction are contemplated. Yet, such are the requirements in the hypothetical case illustrated in Fig. 15.10 if a conventional freeway is to be used.

As indicated previously, freeways with a separate third roadway for reverse-flow operation are feasible where peak-hour traffic volume requires an eight-lane or wider facility and traffic by directions is substantially unbalanced. Although AASHO[5] limits their discussion of the applicability of this concept to depressed freeways, there seems to be no reason why it could not be applied to elevated facilities. Reverse-flow sections of 3-2-3 lanes and 3-3-3 lanes would be theoretically equivalent in capacity during peak hours to ten- and twelve-lane facilities, respectively.

AASHO[5] further limits the reverse-flow freeway concept to situations where the traffic using the reverse-flow roadway is destined for a distant point without intermediate ingress or egress. This seems to be an unnecessary, even an unrealistic, restriction, as shall be shown.

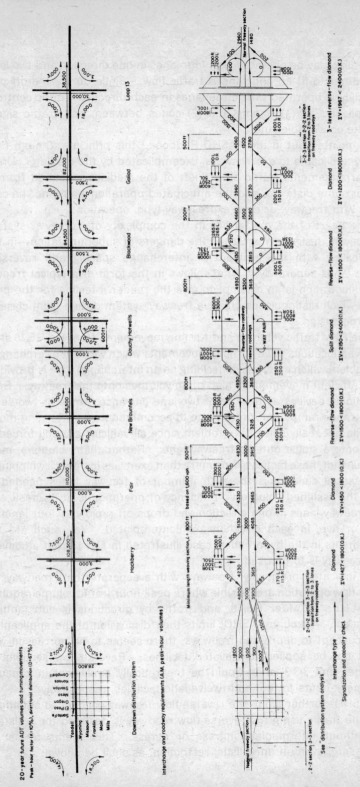

Fig. 15.10 Reverse-flow freeway schematic.

The schematic layout in Fig. 15.10 includes the relative locations and types of interchanges. The choice of interchange in each case is based on capacity limitations and operational considerations. Thus, for the three low-demand interchanges, a diamond interchange with a critical lane capacity of 1,800 vph (see Fig. 15.6) could be used, since the summations of critical lane volumes are 1,427 and 1,450 vph for the two interchanges closest to the CBD and 1,200 vph for the next to the last interchange. The 600-ft spacing between the two cross streets in the center of the schematic suggests that they be utilized as one-way pairs and a split diamond be constructed for their turning movements. The summation of critical lane volumes for this interchange, 2,150 vph, is less than the 2,400 vph capacity of the split diamond (see Fig. 15.6).

For the two interchanges on either side of the split diamond, the turning movements show that a conventional diamond suffices at both locations. The decision to use a reverse-flow diamond in each case is based on the high entrance ramp volumes of 1,400 vph and 1,500 vph, respectively. The utilization of two ramps on the center reverse-flow dual roadway spreads each of these volumes more dramatically for the last interchange, a three-level reverse-flow diamond, in which the total volume desiring to enter the freeway equals 3,000 vph. Using a conventional freeway section this would be impossible; however, it can be accomplished if the ramp movements are interchanged with the center reverse-flow dual roadway as well as the outside freeway roadways.

It is seen in Fig. 15.10 that utilization of the reverse-flow dual roadway also solves the problem of reducing the number of freeway lanes. Traditionally, access to the center roadway is provided by crossover lanes connecting with the outside roadways. However, provision of entrance and exit ramps directly to the center roadway from the intersecting cross street seems desirable from the point of view of reducing ramp volumes by increasing the number of ramps—accomplished without increasing the weaving volumes on the outside freeway lanes. Distances between weaving exit and entrance ramps shown on Fig. 15.10 are minimums obtained by entering Fig. 15.8 with the appropriate ramp volumes (based on a design service volume of 1,600 vph).

Essentially the reverse-flow diamond interchanges are a combination of two interchanges—a partial cloverleaf on the reverse-flow center roadway and a diamond on the outside freeway roadways. It should be emphasized that because the loops of the cloverleaf operate off the center dual roadway in the center of the right-of-way instead of off the freeway lanes, only slightly more right-of-way is required than would be required for a conventional diamond interchange. In fact, the freeway ramps form the diamond part of the interchange, much as they would in the conventional design.

Uniformity of pattern and its effect on operation of interchanges should always be considered. A dissimilar arrangement of exits and entrances between successive interchanges causes confusion, decelerations, shock waves, and forced weaving. It is seen in Fig. 15.10 that the driver on the outside roadways is faced with the familiar exit-entrance pattern of the conventional diamond, whereas on the center reverse-flow dual roadway the cloverleaf continuity is preserved throughout.

15.7 DOWNTOWN DISTRIBUTION SYSTEM

The daily pattern of arrivals and dispersals in the central business district is one of the most involved problems confronting designers today. All the transportation systems—the subway, the surface transit system, the pedestrian, and of course the automobile—must enter the central business district at the same time.

Probably the most critical point in the urban freeway system is where the facility crosses the central business district. The problem is one of designing the freeway to discharge and collect traffic in a relatively short distance. Leisch[6] has documented many diagrammatic schemes using lateral and parallel distributors, emphasizing that some form of distribution system is essential in conjunction with urban freeway development in order to serve downtown areas properly.

It must be realized that a new freeway skirting the downtown area may completely alter the existing circulation pattern and severely tax the capacities of the downtown streets. In order to prevent this, the entire problem must be attacked on a total system basis rather than by a piecemeal approach. Thus, the downtown distribution system should include the freeway, the interchanging ramps, the entire network of feeder streets in the downtown area affected by the freeway, and the downtown signalization system. Although these components may be analyzed separately, initially, the usual tests of system compatibility must be satisfied.

Figure 15.11 illustrates an analysis of the downtown distribution system for the hypothetical conditions established in Fig. 15.10. The existing street system consisted of two-way operation except for the one-way pair consisting of Kansas and Campbell Streets with Kansas northbound and Campbell southbound. The shaded area defines the proposed freeway system. Numbers on the streets are peak hourly volumes, based on the future design year.

Taking the components of the distribution system one by one, we consider first the freeway itself. An important characteristic of freeways is that they are big and have the same power as any big topographical

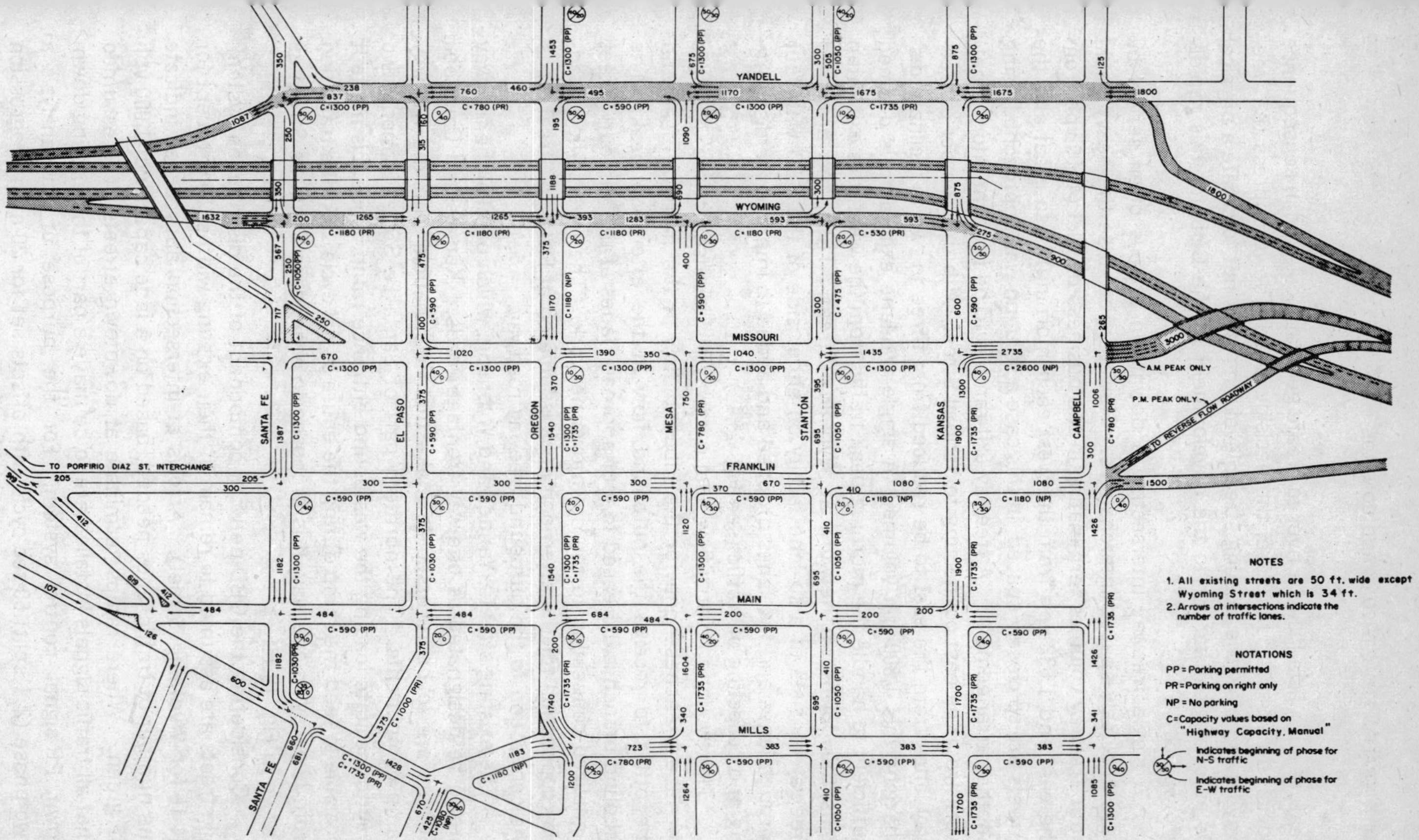

Fig. 15.11 Distribution system analysis.

feature such as a hill or a river to create geographical, and in consequence social, divisions. Therefore, the decision was made to depress the freeway lanes below the existing street level, and thus not to create a physical or psychological barrier to the development of the CBD for this rapidly growing city.

A single freeway intersecting a downtown area is often served by several closely spaced interchanges. However, because the heavy peakhour freeway volumes terminating in the business district (4,800 vph from the east and 1,632 vph from the west each morning) is so great and the streets are so closely spaced, in this case a solution was approached using another arrangement. A freeway exit ramp is to be brought into the network from the east connecting to Yandell Street, with the number of freeway lanes from the east to be reduced from three to two to facilitate discharging this 1,800-vph volume. A single two-lane exit ramp is deemed sufficient to handle the morning peak traffic from the west; this exit ramp joins the downtown network at Wyoming Street. Thus, the proposed freeway consists of a six-lane facility on either side of the CBD with four through lanes and two lanes dropped and added to improve the discharging and collecting of traffic (see Fig. 15.11).

The two-lane reverse-flow dual roadway divided itself at the CBD into two-lane branches—one for the inbound flow in the morning peak and the other to receive the outbound flow in the afternoon peak. The inbound branch is increased to four approach lanes at the Missouri Street signalized connection to the downtown network. The reverse-flow roadway connection for the afternoon peak is initiated at Franklin Street, as is the connection to the outbound freeway roadway.

The existing streets, Yandell and Wyoming, which parallel the freeway are to be designated as one-way frontage roads. Kansas and Campbell Streets are to be reversed to make them compatible with the freeway ramp components. The northern part of Santa Fe Street is made two way to facilitate loading the westbound entrance ramp, since this made it possible to feed the ramp during the entire signal cycle. At the transition from two way to one way, it is proposed to channelize Santa Fe Street as shown in Fig. 15.11.

Considering the CBD-network component of the distribution system, all streets are assumed to remain at their existing widths (50 ft except for 34 ft for Wyoming Street). Arrows at intersection approaches indicate the number of traffic lanes needed, based on a lane capacity of 1,000 vph of green. Where ramps terminate at downtown streets, it is assumed that all traffic signals installed are to be made a part of the existing downtown PR-signal control system. For the purpose of this analysis a two-phase (2:1 split) 60-sec cycle with offsets set for 20-mph progression is used. A peak-hour factor of 10 percent and a 67:33 directional distribution are assumed. Based on these assumptions, downtown peak-hour parking would be virtually unrestricted.

In addition, values recommended by the "Highway Capacity Manual" are shown for each downtown block, as an alternate form of capacity analysis. Based on these values, some parking restrictions would be necessary during the peak hour as indicated by the designations PR for "parking on right only" and NP for "no parking."

In conclusion, a complete analysis of the downtown distribution system, including the freeway components, feeder streets, downtown network, and traffic signal components, seems to suggest that the proposed geometrics fulfill two important objectives—that is, providing a freeway route close enough to the CBD to maintain its integrity, yet far enough away to reduce congestion.

15.8 GEOMETRICS OF THE REVERSE–FLOW INTERCHANGE

Having established the types of interchanges and number of lanes, it remains to dimension the roadways. A plan profile for a reverse-flow interchange is illustrated in Fig. 15.12. The typical section for this type of system design is shown in Fig. 15.13.

The choice of 12-ft lanes on the express lanes and reverse-flow lanes needs no explanation. Since the maximum number of lanes per roadway is two, a constant cross slope of $\frac{1}{8}$ in./ft is sufficient. Provision for future widening was made on the inside of the express lanes; no provision was made on the reversible throughout the limits of interchanges. However, the 100 ft center to center between the expressways and the reverse flow allow plenty of latitude. It should be added that 200 ft between opposing roadways should eliminate glare from high beams at night, since the operating hours for the reversible would normally be during the daylight hours for most seasons of the year.

Standard shoulder widths of 10 ft on the right and 6 ft on the left would be employed. Shoulders adjacent to speed-change lanes should be carried 10 ft wide, measured from the edge of the through lane, until a minimum width of 6 ft from the edge of surfacing for speed-change lanes is attained. The shoulders would then be continued for a uniform 6-ft width on the ramp. All shoulders would receive surface treatments. The slope across the shoulders would normally be $\frac{1}{4}$ in./ft on the inside and $\frac{1}{2}$ in./ft on the outside giving a difference in slope with the pavement of 3 percent in both cases.

Back slopes for ditches in cut sections would be 4:1. Embankment slopes would vary with the heights or fill, from 8:1 for fills 5 ft or less to

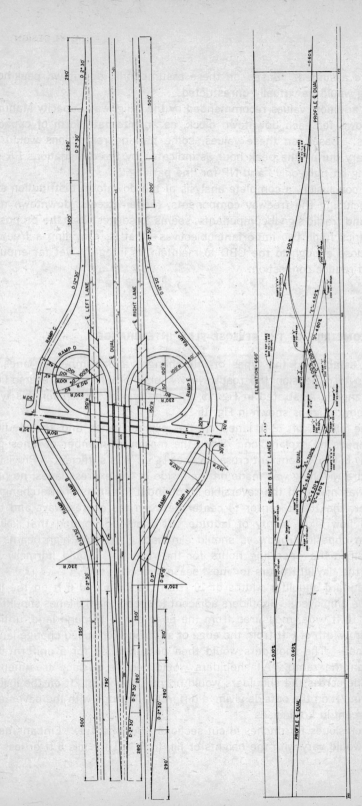

Fig. 15.12 Geometrics of reverse-flow diamond interchange.

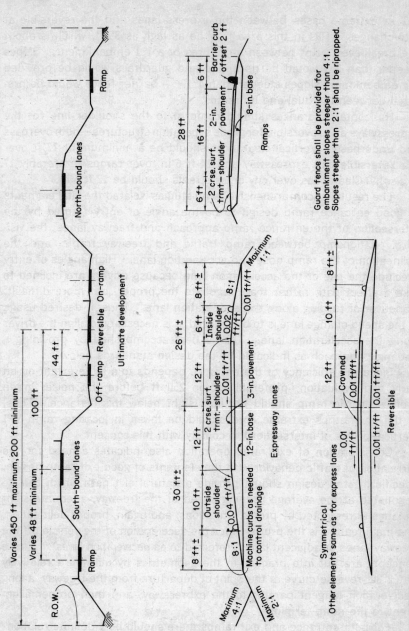

Fig. 15.13 Typical sections.

Guard fence shall be provided for
embankment slopes steeper than 4:1.
Slopes steeper than 2:1 shall be riprapped.

Ramps

Barrier curb
offset 2 ft

28 ft

6 ft   16 ft   6 ft

2 crse. surf.   2-in.
trmt–shoulder   pavement

8-in. base

Ultimate development

Off-ramp   Reversible   On-ramp

44 ft

North-bound lanes

Ramp

R.O.W.

Varies 450 ft maximum, 200 ft minimum

100 ft

Varies 48 ft minimum

South-bound lanes

Ramp

Expressway lanes

30 ft±

10 ft   12 ft   12 ft

Outside
shoulder

2 crse. surf.
trmt–shoulder

Machine curbs as needed
to control drainage

0.04 ft/ft   –0.01 ft/ft

12-in. base   3-in. pavement

8:1

Maximum
4:1

Maximum
2:1

26 ft±

6 ft   8 ft±

Inside
shoulder

0.02
ft/ft

8:1

Maximum
1:1

0.01 ft/ft

Reversible

12 ft   10 ft   8 ft±

Crowned

0.01   0.01
ft/ft   ft/ft

0.01 ft/ft   0.01 ft/ft

Symmetrical
Other elements same as for express lanes

1:1 in extreme cases between the express lanes and the reversible at interchanges.   Fills in this area may be as high as 35 ft, which is about the break-even point between structures or a 1:1 embankment.   Slopes steeper than 2:1 would be riprapped and guardrails would be provided for embankment slopes steeper than 4:1.   A barrier fence would be provided between the dual and its ramps.

Horizontal clearances should conform to the shoulder line for the expressways and reversible for grade separation structures—both overpass and underpass.   Vertical clearances should be a minimum of 17 ft over the reversible and expressway and 14 ft 6 in. over ramps (preferably 17 ft also).   Clearances over city cross streets should be 12 ft 6 in.

The results of comprehensive ramp studies isolated the vital elements of good entrance ramp design to be the angle of entry formed by the intersection of the entrance ramp approach and freeway lanes, the visibility relationship between ramp traffic and freeway traffic, and the delineation of the ramp nose and acceleration lane.   High angles of entry preclude the use of the acceleration lane because drivers are inclined to take a direct path rather than negotiate the proper, but more difficult, maneuver of turning along the acceleration lane.   Thus, if desired usage of the speed-change lane is to be realized, it is necessary to align the driver along the acceleration lane.   This is best accomplished by providing a flat-angle approach as indicated by the design standards below.

Since the efficiency of the entrance depends to a large extent on an early gap evaluation, preferably some 250 ft before the noise, grade profiles on the ramp should not limit sight below this distance.   Thus, in the geometrics extreme care should be taken in locating ramp VPI (vertical points of intersection) to comply with this concept.

Consideration of exit ramp operation also indicates a need for the correlation of traffic behavior with the elements of good exit ramp design. Specifically, the design should provide a natural exit path which can be negotiated at the average running speed of the freeway, adequate sight distance unrestricted by profile limitations, and again, proper delineation. The main feature is the provision for the deceleration of the vehicle off the freeway lanes or adjacent lanes (referred to as *deceleration lanes*).   These concepts are put into practice in the geometrics by utilizing a relatively flat 2°30′ reverse curve at the point of departure from the freeway, a long deceleration tangent parallel to the expressway, and then proper alignment to the ramp terminal.

For both entrance and exit ramps there should be a definite relationship between the design speed and the running speeds at each end of the ramp.   Guide values for ramp design speeds in relation to the highway design speed take on less significance because of the geometrics proposed in the previous discussion; namely, the concept of speed adjust-

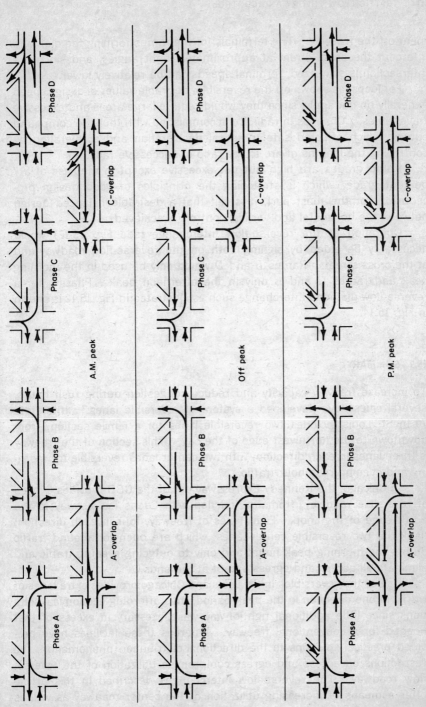

Fig. 15.14 Phasing for reverse-flow diamond interchange.

ment off the freeway. The terminals for the ramps forming the diamond intercept the cross street at approximately right angles, and since the intersection is signalized, terminal speeds can be relatively lower.

For loops, as shown off the reversible, desirable values of design speed generally do not apply, since they would require large areas and excessive travel times. Changes in radii are accomplished with multiple compound curves, which effected safe, practical deceleration and acceleration by controlling the lengths of arc to the ratio of successive radii.

Ramp grades are high but not excessive except in the case of a 7 percent grade, which is stretching the principles of good design procedure. For the most part, it is felt that a desirable balanced design between the horizontal and vertical controls is achieved.

Operation of the reverse-flow interchange (see Fig. 15.12) would necessarily be aided by signing both on the reverse-flow roadway and at the cross street. Ramps B and D would only be used in the morning peak and ramps G and E only in the afternoon peak. Phasing for a reverse-flow diamond interchange such as illustrated in Fig. 15.12 is shown in Fig. 15.14.

15.9 SUMMARY

To increase freeway capacity and reduce congestion during rush hours, several cities have developed a system of reversible lanes.[7] Interstate 70 in St. Louis includes two reversible lanes for a 6-mile section, from downtown to the northwest edge of the city. This section of the freeway is three lanes in each direction, with two center lanes reversible for morning and evening rush-hour traffic.

In Chicago, the Kennedy Expressway out of the CBD provides 6 miles of reversible roadway, from the junction with Edens Expressway to the north edge of the Loop. Eight lanes of freeway, four in each direction, flank the two reversible center lanes, which are open to inbound traffic during the morning peak hours and only to outgoing evening traffic and eliminate all access and egress except at the ends.

Use of the reversibles in St. Louis and Chicago provides extra lanes of traffic in one direction in the peak period, thus affording economical solutions since little additional right-of-way was .iecessary in excess of that needed for a conventional freeway. Whereas these facilities have provided practical solutions to the directional-distribution phenomenon, the restrictions on access and egress compromise utilization of the reverse-flow roadway. The reverse-flow interchanges described in this article offer a means of increasing utilization of the center roadway as well as improving operation of the outside freeway roadways by reducing weaving volumes.

The reverse-flow freeway offers a versatile planning tool.   A freeway is not built specifically for some date 20 years in the future; it must go to work the first day and serve efficiently all through its expected life.   Traffic projections and designs must be made on partial or incomplete systems if desirable service is to be obtained in the years before the whole system is completed.

Looking ahead to 1975, Seattle found that it could best satisfy its anticipated traffic volume requirements with a four-lane reversible roadway, integrated with the 16.5-mile eight-lane Seattle Freeway.   Downtown, the freeway expands from eight lanes to twelve, the center four reversible for morning and evening peak-hour demands.   The outer freeway lanes will be utilized while the reversible roadway is under construction.

Two general techniques of stage construction—the completion of a highway section to something less than the ultimate planned improvement but to a stage where it can be used for traffic operation—can be utilized in a reversible freeway system.   In one method, the route can be progressively constructed by sections; first from the CBD to some intermediate point, etc.   In the second method, a first-stage improvement can be made consisting of the construction of the outside roadways—the construction of the reverse-flow center roadway to bring it up to the ultimate design being made later.

PROBLEMS

15.1. A freeway can be considered as a system with components, six of which might be:
  a. Express (through) lanes
  b. Entrance ramps
  c. Exit ramps
  d. Frontage roads
  e. Cross streets
  f. Interchanges
A system exists only when a functional interdependence exists among the components of the system.   Discuss the identification of a freeway as a system with respect to the design and operation of the above listed components.

15.2. A diamond interchange is to be constructed 5 miles from the central business district of a city of 2 million persons.   During the A.M. peak hour the freeway outbound is expected to be 2,000 vph and the freeway inbound, 5,200 vph.   Parallel one-way frontage roads will be used, and the arterial street will be bridged over the freeway.   At the intersections of the frontage roads with the arterial the expected peak-hour flows are as tabulated on the following page.

Movement	Flow, vph
Arterial through each intersection with outbound frontage road	1,000
Arterial right turn onto outbound frontage road	300
Outbound frontage road right onto arterial	100
Outbound frontage road through each intersection with arterial	200
Outbound frontage road left onto arterial	250
Arterial through each intersection with inbound frontage road	900
Arterial right turn onto inbound frontage road	350
Inbound frontage road right onto arterial	300
Inbound frontage road through each intersection with arterial	450
Inbound frontage road left onto arterial	200

Prepare a written report to include sketches and calculations indicating lane configuration and traffic control of the interchange. For any signalization determine phasing and cycle lengths.

15.3. A proposed elevated freeway will pass through the edge of the central business district of a city of 1 million persons for a distance of 2 miles. As much as possible the freeway will be aligned between parallel one-way streets spaced on a 300-ft grid. All streets in the central business district are one way (direction of flow alternates at consecutive streets) on a 300-ft grid spacing. The freeway will have five lanes each way. The land use adjacent to the proposed route consists of hotels, high-rise apartments, multistory office buildings, and general commercial development. Employ your creativeness to recommend how the following problems should be approached:

a. How is drainage of surface water from rainstorms to be removed?

b. What utilization is to be made of the land under the elevated freeway?

c. How would you connect the surface streets of the central business district to the elevated freeway?

15.4. If you want the probability of stable flow during a 5-min period to be equal to or greater than 0.7, estimate the upper limit on the design service volume (per lane) in an urban area with a population of 5 million.

15.5. A depressed freeway intersects parallel major arterials $\frac{1}{4}$ mile apart. During the morning rush, as one moves toward the center of the city, at the first arterial a heavy on-ramp volume is followed by a high-volume exit at the second arterial. Indicate possible solutions to reduce the weaving interference on the freeway in the form of brief line diagrams with accompanying descriptions.

15.6. A freeway lane service volume of 1,700 vph is selected. An on-ramp with a flow of 800 vph is followed by an off-ramp with a flow of 300 vph. How long should the weaving section be?

15.7. You are a member of the traffic and transportation engineering staff of a major metropolitan area of 3 million persons. Projected urban area growth, potential development of large-scale high-rise apartment dwellings, concentrated employment in the central business district, and large outlying low-density housing areas created the need for expanded freeway and public transit. The freeways will either be depressed or elevated, and the transit will either be subway or elevated rail. List the considerations you would investigate to present an evaluation of alternatives from a design and operations viewpoint.

REFERENCES

1. Goode, H. H., and R. E. Machol: "System Engineering," McGraw-Hill Book Company, New York, 1957.
2. Moskowitz, Karl, and Leonard Newman: Notes on Freeway Capacity, *Highway Res. Board Rec.* 27, 1963.
3. Drew, D. R.: Design and Signalization of High-type Facilities, *Traffic Eng.,* vol. 33, no. 10, July, 1963.
4. Drew, D. R., and C. J. Keese: Freeway Level of Service as Influenced by Volume and Capacity Characteristics, *Highway Res. Board Rec.* 99, December, 1965.
5. "A Policy on Arterial Highways in Urban Areas," American Association of State Highway Officials, Washington, D.C., 1957.
6. Leisch, J. E.: "Interchange Design," DeLeuw, Cather and Company, Chicago, 1962.
7. "What Freeways Mean to Your City," Automotive Safety Foundation, 200 Ring Building, Washington, D.C., 1964.

CHAPTER SIXTEEN
SURVEILLANCE AND CONTROL

Control is achieved most efficiently, of course, not by the inspection operation itself, but by getting at causes.

Dodge and Romig

16.1 INTRODUCTION

Fundamentally, the capacity of any system can be increased by either physical expansion or by better control of the existing plant. Historically, both methods have been applied to the surface-street traffic problem, and the first method has been applied to freeway facilities. Whereas early freeways were four-lane facilities, many new freeways are 10 lanes wide. It only remains for the second method —the control of existing freeways —to be explored.

16.2 FREEWAY OPERATIONS

The motorist on a freeway expects to have his needs anticipated and fulfilled to a much higher degree than on conventional roads. This expectation can sometimes be fulfilled by the application of capacity considerations to rational geometric design. More often than not, however, actual traffic and travel patterns differ from the projected values, mal ng constant freeway operational attention after construction a must.

Congestion occurs on a freeway section when the demand exceeds the capacity of that section for some period of time. Because congestion can lower the rate of flow, a short period during which the capacity is lower than the capacity of adjacent sections is called a *bottleneck.* Bottlenecks can be caused by changes in the freeway alignment (horizontal or vertical), reduction in lane widths, the presence of an entrance ramp, etc. Accidents, disabled vehicles, and maintenance or law enforcement operations can also cause temporary bottlenecks by reducing the effective capacity or level of service provided.

Freeway design does not always eliminate the need for sound traffic regulation. A reasonably homogeneous traffic stream, particularly with respect to speed, is essential for efficient freeway operations. Pedestrians, bicycles, animals, and animal-drawn vehicles are excluded from freeways. Motor scooters, non-highway (farm and construction) vehicles, and processions such as funerals are also generally prohibited from the freeway. Towed vehicles, wide loads, or other vehicle combinations such as trailers drawn by passenger vehicles which impede the normal movement of traffic may be barred during the peak-traffic hours or during inclement weather.

Minimum speed limits are being increasingly used and have been found of great benefit, particularly on high-volume sections. The effect of this type of control is to reduce the number of major accident-potential lane-change maneuvers. The effects of slow-moving vehicles on both capacity and accident experience are so pronounced that a greater use of minimum limits appears probable. There is a need to eliminate all vehicles incapable of compatible freeway operation.

Increasing attention has been given to the possibility of, and need for, using variable speed control on urban freeway sections as a means of easing the accordion effects in a traffic stream as congestion develops. Drastic speed variations might be dampened by automatically adjusted speed message signs in advance of bottlenecks.

A properly designed entrance ramp with provision for adequate acceleration should allow the entering driver adequate distance to select a gap and enter the outside lane of the freeway at the speed of traffic in the lane. These merging areas operate best when there is a mutual adjustment between vehicles from both approaches. Yield signs impose drastic speed restrictions under the laws of a number of states, thus causing operational problems, and are no longer mandatory on the Interstate System. It is generally felt that any speed restriction or arbitrary assignment of right-of-way should be avoided unless inadequacies in the design make it imperative.

Almost any engineering problem including freeway operation may be described as a systematic attempt to resolve a capacity-demand relationship at an acceptable level of service. We try to build enough strength into the materials of a layered pavement system, for example, to withstand shear stresses due to anticipated loads. However, the mere fact that the strength (capacity) exceeds the load stresses (demand) does not guarantee an acceptable level of service. The deflection, smoothness, texture, and color contrast also affect the driver's ride and, as such, are level-of-service factors that must be considered.

Traffic engineering is the science of measuring traffic characteristics and applying this information to the design and operation of traffic systems. The traffic engineer's basic problem of resolving a capacity-demand relationship is similar to that of any other engineer: He must be able to either measure the parameters defining capacity and demand very accurately, or he must be able to control them. Returning to the pavement design analogy, we see that although the strength of the materials in a pavement can probably not be estimated as accurately as the capacity of a freeway lane, the pavement designer knows that the loads (demand) on the facility are controlled, and in most states limited, by law. If urban freeways are to operate at the levels of service for which they were designed, the demand on these facilities must also be regulated. Such regulation is the subject of the next section.

16.3   FREEWAY SURVEILLANCE AND CONTROL

It is generally agreed that one key to significant progress in operation of urban freeways lies in improved surveillance techniques. The term *surveillance* has developed in the highway terminology primarily in the last decade and denotes the observation of conditions in time and space. Initially, urban freeway surveillance was limited to moving police patrols. Recently, helicopters have been used for freeway surveillance in many metropolitan areas. Efficient operation of high-density freeways is, however, more than knowing the location of stranded vehicles or the qualitative description of the degree of congestion by high-flying disk jockeys.

Television surveillance became an operational reality in the late 1950s both in the United States and Europe. The Port of New York Authority utilized closed-circuit television for monitoring traffic in the Hudson River tunnels. In Germany, a well-publicized TV system was developed to monitor traffic at a major, complex urban intersection.

Experimentation with closed-circuit television as a freeway surveillance tool was initiated on the John C. Lodge Freeway in Detroit. This offered the opportunity of seeing a long area of highway in a short, almost instantaneous period of time, made possible by spacing cameras along the entire section of roadway.   The system was put into use in the summer of 1961.[1]

During the period in which the television system was being made operational, the staff of the Detroit Project considered the requirements and specifications of an automatic detection system.   To improve surveillance, classification and speed sensors with the necessary relay racks, analog computers, and display panels were added.   In the summer of 1962, the John C. Lodge Freeway became a control project as well as a surveillance project when lane and off-ramp signals and variable speed signs were initiated.   Six months later, on-ramp closure signals became operational.   Thus, in four short years an entirely new concept in freeway traffic control was developed, based on a closed-circuit television surveillance system and a traffic detection and measuring system, with capabilities of transmitting control messages to matrix variable speed signs, lane controls, and ramp controls.

In the case of television surveillance, alone, evaluation of freeway operation depends mostly on the visual interpretation of the observers. Many traffic people believe that this is not enough.   The Chicago Surveillance Project maintains that even trained observers offer no uniform objectivity.   In other words, the quality of operating conditions must be detected automatically by monitoring operating characteristics.   When the characteristics are in certain predetermined levels of operation, certain previously designated courses of action may be taken.

The Chicago Area Expressway Surveillance Project was established in 1961 as a research program of the Illinois Division of Highways, financed by Federal, state, county, and city funds.   Its objective was to develop, operate, and evaluate a pilot network and control system to reduce travel time and increase traffic flow over the Congress (now the Eisenhower) Expressway.   Thus, unlike many projects which end their responsibility for traffic at the edge of the road, this project was equally concerned with the arterial subsystems that feed the expressway or receive vehicles from it.

As indicated before, some of the most significant advances in freeway surveillance and control have been obtained on research facilities other than freeways.   Most notable is the work of The Port of New York Authority in the Holland Tunnel.   In November, 1963, a new tunnel traffic control system was authorized for the south tube of the Lincoln Tunnel.   Anticipated benefits were more effective use of tunnel capacity and more efficient patrolling with fewer police officers.   Features include a traffic surveillance system of vehicle detectors with traffic-condition display and control devices, a special computer which receives traffic information and activates controls to prevent congestion from occurring inside the tunnel, closed-circuit television, two-way radio communication between control centers and police in the tunnel, and a monorail for speedily moving tunnel police to any point in the tunnel.[2]

16.4   THE GULF FREEWAY PROJECT

In most cases, the primary use of television has been for operational, rather than research, purposes.   The National Proving Ground for Freeway Surveillance, Control, and Electronic Traffic Aids on the John C. Lodge Freeway served as the first major installation which was planned for, and now continues, a program of research per se.   It has provided technical knowledge and experience to several new television surveillance projects at the Baytown Tunnel in Houston, the Gulf Freeway in Houston, and the Interstate Freeway in Seattle.   The Baytown Tunnel and Seattle Projects are primarily operational; the Gulf Freeway Surveillance and Control Project is a combined research and operational project.

In September, 1963, the Texas Transportation Institute was authorized to initiate studies in the area of freeway surveillance and control in order to increase the efficiency of existing urban freeways and to determine how to improve the level of service of future facilities.   A research project sponsored by the Texas Highway Department and the U.S. Bureau of Public Roads was formulated with the basic objective of developing criteria for the design and operation of automatic surveillance and control systems which would permit the attainment of acceptable levels of service on heavily traveled urban freeways during peak periods of demand.

This research project was conceived as centering on a study of the Gulf Freeway in Houston.   It was reasoned that the development of a system of automatic surveillance and control for this facility would furnish an excellent pilot study from which technology could be developed that would be applicable to similar systems throughout the state and nation. The research work was interpreted as consisting of three steps:

1. To develop basic research on the characteristics of operation on both the freeway and connecting at-grade arterial system in order to determine the requirements of the proposed surveillance and control system
2. To experiment with detection and sensing equipment and develop prototypes of eventual automatic controllers that could anticipate the buildup of congestion and react to prevent it
3. To install and evaluate the final automatic surveillance and control system

16.5   SURVEILLANCE PROJECTS AS RESEARCH FACILITIES

Inherent in the research process are theory formulation, experimentation, and evaluation.   Theory formulation includes establishing the criteria for system optimization and then formulating a mathematical model.   Experi-

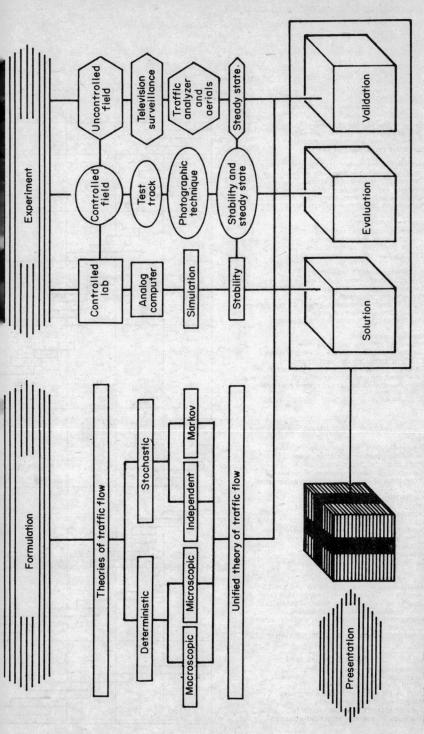

Fig. 16.1 Traffic research procedure.

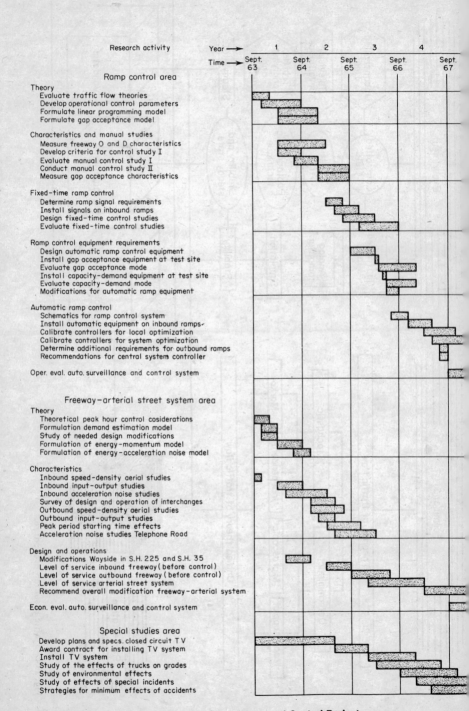

Fig. 16.2 Work plan for Gulf Freeway Surveillance and Control Project.

ment can be of a controlled or an uncontrolled nature. Control experiments, in traffic research, may be either in the laboratory or in the field. Examples of the former are simulation, both analog and digital; examples of the latter are test tracks such as the one operated by General Motors and, to some extent, the tunnel and freeway surveillance and control projects identified in the previous section.

The steps in the traffic research process are diagramed in Fig. 16.1. The implication is that controlled laboratory experiments are an attempt at solution where other modes fail. The controlled field experiments of the test track are useful in evaluating theories, but actual validation must be based on real-world traffic. The last step and probably the most important is presentation.

The procedure described in Fig. 16.1 served as a guide in the research conducted on the Gulf Freeway Surveillance and Control Project. In order to reach the three broad objectives enumerated in the previous section, it has been necessary to proceed through numerous studies involving different aspects of the total research problem. This is illustrated in Fig. 16.2, in which the three principal areas of concern listed are ramp control, the freeway–arterial street system, and special studies. The various activities listed in the figure indicate broad research steps that have been and will be taken in these three areas.

The organization of the activities of Fig. 16.2 under theory, characteristics, etc., clearly indicates deference to the first two axioms, formulation and experiment, of the research philosophy depicted in Fig. 16.1. The adherence to the third axiom, presentation, is evidenced by 14 formal project research reports[3 to 16] in the first three years duration of the project. In this way, the research effort and results were to be fed back to the planners and designers of the next generation of freeway facilities.

16.6 THE EVOLUTION OF RAMP CONTROL CRITERIA

When demand exceeds or sometimes even approaches the capacity of a system, there is a self-aggravating deterioration of operation and buildup of congestion. Classical control systems are employed either to make the facility flexible enough to accommodate fluctuations in demand or to reduce the magnitude of the demand fluctuations. Freeway surveillance and control projects are necessarily limited to the latter. One approach, pioneered by the Detroit Project, is to inform the motorist of traffic conditions by utilizing lane controls and variable speed messages. A second, and more positive, approach is exercised at the point or points of ingress (the entrance in the case of tunnel control, on-ramps in the case of freeway control).

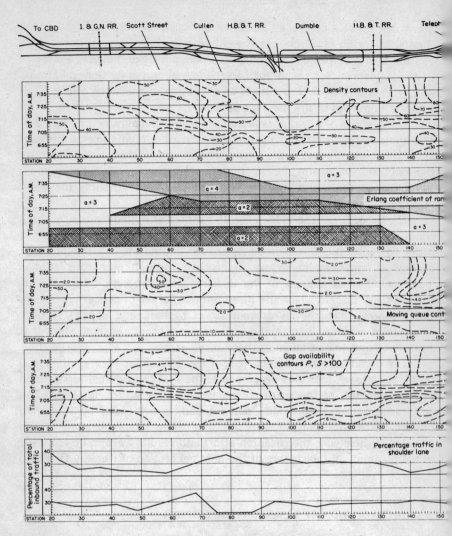

Fig. 16.3 Inbound distributional characteristics on shoulder lane.

Metering, the process of controlling the amount of entering traffic to prevent congestion, was developed by The Port of New York Authority. The first step was the identification of the bottleneck at the foot of the tunnel upgrade. Second, a mathematical model[17] was formulated to describe the behavior of vehicular traffic in the tunnel. The significant feature of the model was its prediction of shock waves upstream from the bottleneck. The remedy consisted of metering traffic at the entrance of the tunnel to prevent the development of instability by (1) keeping traffic

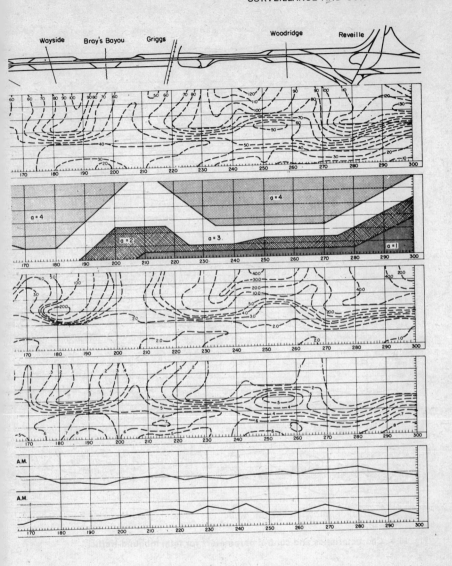

density below some critical value and (2) keeping the traffic demand below the bottleneck capacity.

Based on the success of metering in the tunnel, a similar control plan was formulated for the Eisenhower Expressway by the staff of the Chicago Project. Two bottlenecks on the outbound facility were identified within the study area.[18] The one farthest upstream was caused by a reduction in the number of lanes from four to three without a corresponding reduction in traffic demand. The second bottleneck, further downstream

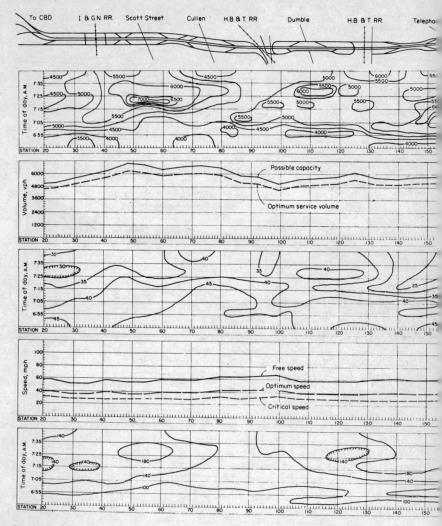

Fig. 16.4 Characteristics and control parameters for total inbound traffic.

and the last bottleneck on the outbound expressway, was caused by fairly heavy on-ramp traffic and was located at the top of an approximately 1,000-ft, 3 percent upgrade.

Two metering techniques were developed. One technique utilized a point density, or occupancy, measurement on the freeway just upstream of the entrance ramp to be metered; the other utilized a volume measurement on the freeway about ½ mile in advance of the entrance ramp and an exit ramp volume between the freeway volume measurement and the entrance ramp. After further study, the technique based on occupancy of the

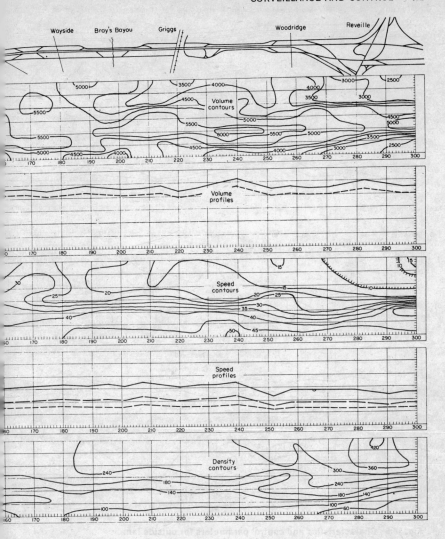

center lane was used as a control parameter for initiating metering.   From a relation established between the center-lane occupancy and the maximum safe ramp volume, a metering rate was established for various levels of occupancy.

Some researchers who followed the Chicago experiments were more impressed by the use of a freeway capacity-demand relationship as a control parameter for ramp metering.   Wattleworth[10] has championed "capacity-demand" criteria in which an individual ramp would be metered according to the difference between the upstream freeway demand and

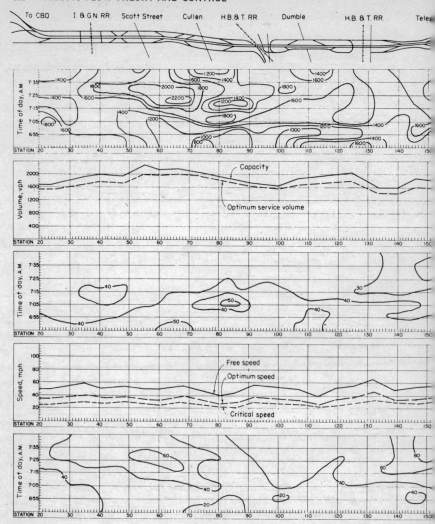

Fig. 16.5 Characteristics and control parameters for outside lane.

downstream freeway bottleneck capacity.   He has also developed a linear programming model in which several entrance ramps in a freeway system would be metered to maximize the output of the system, subject to constraints assuring that the demand will not exceed the total directional capacity at each freeway bottleneck.[19]

In a paper presented in 1963, Drew[20] describes a *moving queues* model based on coordinating ramp metering with the detection of *acceptable gaps* in the outside freeway lane.   An acceptable gap is defined as one equal to or larger than the critical gap (that gap for which an equal percentage of

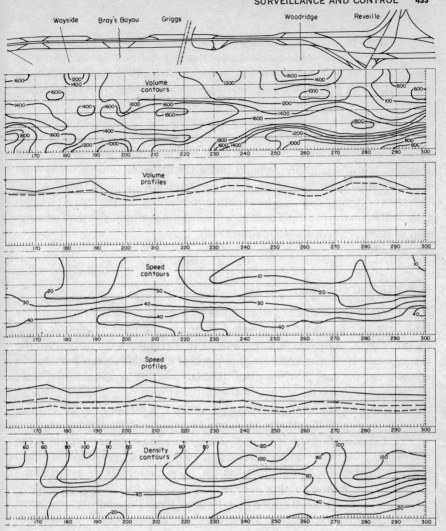

ramp traffic will accept a smaller gap as will reject a larger one) for a merging ramp vehicle. Moving queues or platoons occur when the time headway, or gap, between successive vehicles is less than an *arbitrary queueing headway.* Since the arbitrary queueing headway is taken as the critical gap, the number of ramp vehicles to be metered in some time constant equals the number of moving queues detected. The average number of vehicles per moving queue, as the reciprocal of the probability of a gap larger than the critical gap, provides a rational index of freeway operation.

An interesting aspect of the model is the flexibility of metering a single ramp vehicle per available acceptable gap on the freeway (hereafter referred to as the *gap acceptance mode*) or metering ramp vehicles in bunches. The latter—the *bulk-service* technique—is described in a project report.[15]

Figures 16.3 to 16.5 illustrate summaries of the freeway characteristics utilized in three promising ramp metering philosophies. The figures apply to the gap acceptance mode, the capacity-demand mode for the total freeway, and the capacity-demand mode for the outside lane. The data were obtained from time-lapse aerial photos of the inbound Gulf Freeway during the morning peak. The various *critical* and *optimum* control parameters plotted on the profiles are applications of the *energy momentum* description of level of service described in Chap. 14.

16.7 MERGING CONTROL

In order to make the most effective use of freeway entrance ramps under moderate to high traffic demands, a maximum number of vehicles must merge safely, without causing significant disruptions to traffic on the through lanes. Indications are that a major limitation in ramp operation stems from the fact that the driver is unable either to estimate the sizes of the gaps in the outside lane of the traffic stream or to predict when an acceptable gap will arrive at the point where the merge is to take place. Indeed, it is doubtful that one driver in one hundred knows what an acceptable gap for merging is, as evidenced by the large number of ramp accidents. Certainly, an acceptable gap defies description in terms of any simply stated rule of thumb, for example, "one car length for each 10 mph of speed in following another vehicle."

When the volume of traffic on a freeway begins to approach capacity, the merging driver is placed in an extremely difficult position. The number of acceptable gaps in the freeway stream decreases sharply as the freeway volume increases. At these higher volumes, the merging driver cannot always defer his decision to merge until he is on the acceleration lane. Rather he must detect the location of gaps in the oncoming stream before he reaches the acceleration lane. Operating this way, he must then project the location of a gap onto the acceleration lane in order to decide whether it will be available to him. This in turn requires that he estimate his own speed and acceleration as well as the speed of the gap in order to decide whether there will be sufficient space for the merging maneuver to be completed successfully within the limit of the acceleration lane.

It is apparent that under these circumstances, the merging driver's task becomes a formidable one. Mathematically, however, the problem is quite simple, requiring only a knowledge of the location of the freeway gaps and their speed, the accelerating capability of the ramp vehicle, the length of the ramp, and the length of the acceleration lane.

In order to markedly improve merging performance and overall freeway conditions, it is necessary to supply the information that the driver needs to coordinate in time and space his entrance onto the acceleration lane, his speed change, and his point of merge. Since a perfectly determinate solution is possible, and instrumentation to carry out these operations is well within the grasp of existing electronic technology, some sort of merging control system seems feasible.

In present freeway operation, further complications are caused by simultaneous arrival in the merging area of a platoon of vehicles from the ramp and a platoon of vehicles in the outside lane of the freeway. It follows, therefore, that the ramp vehicle arrivals to the merging zone should be controlled, volumetrically and timewise, depending on the size of gap available and the capacity of the merging zone and downstream freeway.

16.8 MERGING CONTROL SYSTEM

To effect a synchronized merge, one must be able to (1) detect acceptable gaps on the outside freeway lane, (2) predict when these acceptable gaps

Fig. 16.6 Ramp control signal.

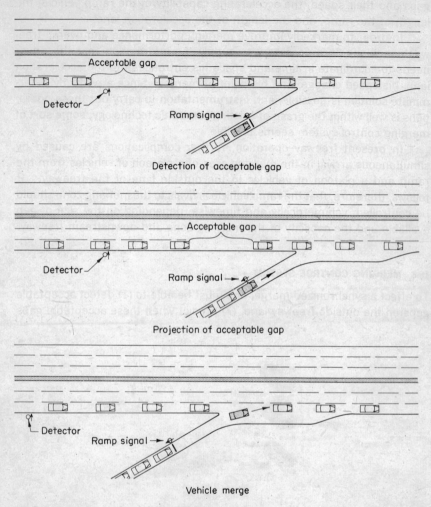

Fig. 16.7 Illustration of automatic ramp controller operation.

will reach the merging point, and (3) arrange for the speed adjustment of the merging ramp vehicle so it hits the gap at the merge point.   A sensor to measure the time interval between successive vehicles (gap detection) and vehicular speed (gap projection) presents no problem.   However, as evidenced in Chap. 9, there is an interaction between the speed of the ramp vehicle and the acceptable gap which must be resolved.

The easiest way to standardize the required speed adjustment on the part of ramp drivers is to stop all ramp vehicles at some point on the ramp far enough from the merge point to allow these vehicles to accelerate to

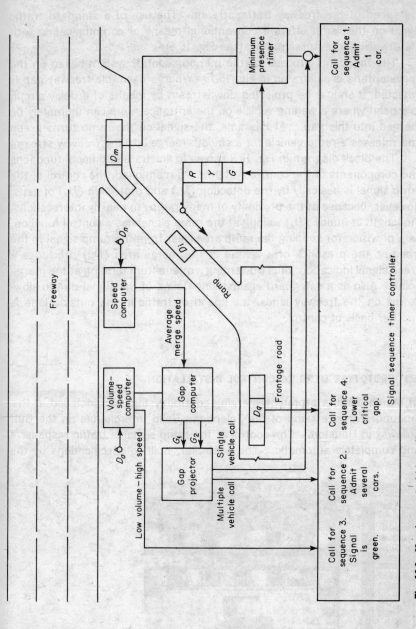

Fig. 16.8 Merging control block diagram.

the speed of the freeway traffic stream. The use of a standard traffic signal on the ramp offers a conventional means of communicating with the driver in accomplishing this (see Fig. 16.6).

The functional process for the merging control system based on the above criteria is illustrated in Fig. 16.7. When a desirable freeway gap is detected, it should be projected downstream by means of a delay circuit to a point where a waiting vehicle on the entrance ramp can ultimately be merged into this gap. At this time, the signal on the ramp turns green and releases a ramp vehicle for a smooth merge into the freeway stream.

The block diagram in Fig. 16.8 serves to illustrate additional functions and components of the control system. As mentioned, the control of the ramp signal is basically by the detection (D_a) and projection (D_n) of gaps. However, because of the proximity of many ramps to nearby intersections, the length of queue (D_q) waiting at the ramp signal has a control function. As a provision for keeping the ramp area clear from the ramp signal to the freeway, the presence of a vehicle in the merge area (D_m) precludes a green signal indication, thus preventing a queue from forming at the merge point. Also as a safeguard against long delays of the signal due to slow-downs on the freeway lanes, the speed of traffic in the outside lane is another basis of control.

16.9 PROTOTYPE MERGING CONTROL INSTALLATION

In March, 1966, a prototype merging control system was installed at the inbound entrance ramp of the Telephone Road Interchange on the Gulf Freeway in Houston. The control of the ramp signal is traffic responsive and completely automatic. The controller, built to specifications by the

Fig. 16.9 Merging controller in freeway surveillance center.

Automatic Signal Division, Laboratory for Electronics, Inc., is housed in the Surveillance Center of the Gulf Freeway Surveillance and Control Project in Houston (see Fig. 16.9). Control is designed for either single-vehicle or multivehicle entry (designated G_1 or G_2). A description of the function of the prototype merging control system is explained in the following paragraphs.[21] The time-space diagram in Fig. 16.10 has been prepared to complement this description.

Gap projection A sonic detector is mounted in a side-fire position on a luminaire standard about 950 ft upstream of the ramp nose (see note 1 in Fig. 16.10). The detector measures all gaps in the outside lane and calculates the speed of traffic flow (see note 2 in Fig. 16.10). When a gap is detected that is equal to or greater than the designated acceptable gap size, it is projected in the signal controller at a rate defined by the vehicle speed in the outside lane. If a ramp vehicle is waiting at the ramp signal, a call for the green signal is made when the projected gap reaches the position in time, designated the *decision point,* at which the travel time of the gap to the merge area is the same as the travel time of the ramp vehicle from the signal to the merge area (see note 3 in Fig. 16.10). However, the green signal will not be called if there is a ramp vehicle over the merge detector (see note 6 in Fig. 16.10).

If the gap is equal to or greater than the designated acceptable gap size for more than one vehicle, the controller holds the green signal until the gap passes the decision point (see note 7 in Fig. 16.10).

Speed of outside-lane traffic A sonic detector is mounted in a side-fire position on a luminaire standard at the nose of the entrance ramp. The detector measures the speed of traffic flow, which is used to select the size of the acceptable gap. When speeds in this area drop below a preset speed, a background cycle rate, set on the fixed rate control, is put in effect; the signal continues to release vehicles when acceptable gaps are available but also releases vehicles after a specified waiting time. The fixed rate setting is usually in the range of 150 to 200 vph.

Length of queue A loop detector is placed in the pavement of the left lane of the inbound frontage road near the Telephone Road intersection. If the queue at the ramp signal is greater than 14 or 15 vehicles, the detector is actuated and a lower critical-gap setting is put in effect. This new gap stays in effect only as long as the queueing detector is timed out.

Occupance of the merge area A loop detector is placed in the pavement of the ramp just upstream of the merge area. All vehicles entering the freeway from Telephone ramp will actuate the detector. If a vehicle stops on the ramp in this area, blocking the entrance to the freeway, the detector will time out and the signal controller will hold on red until the detector is cleared (see note 6 in Fig. 16.10).

The distinguishing feature of the merging control system discussed in this paper is that it is the only form of ramp control that aids the ramp driver in the merging maneuver.   The gap-oriented system installed at the Telephone Road interchange locates freeway gaps and their speeds; it compares these gaps to a critical gap, which is the size gap required by the ramp drivers for merging; it takes into account the accelerating capability

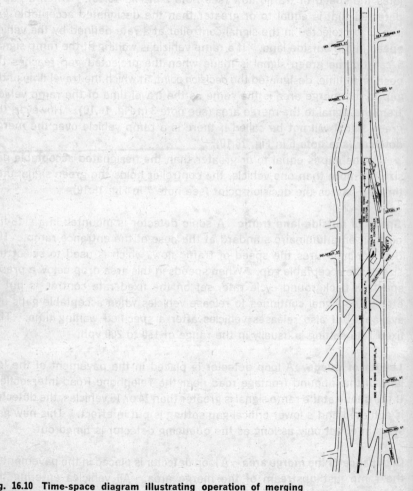

Fig. 16.10 Time-space diagram illustrating operation of merging control system (right and far right).

Plan view

of the ramp vehicle and the length of the ramp and acceleration lane; and it solves the equations of motion automatically before the ramp signal is actuated to allow a vehicle, or vehicles, to make a smooth merge such as the one pictured in the sequence in Fig. 16.11.

In addition to increased efficiency, other benefits are greater safety and higher ramp capacity for a comparatively low cost of installation. Considering safety, any speed differential at a point in the traffic stream in either a longitudinal or transverse direction is dangerous. A vehicle that stops in a travel lane is in particular danger; it is a safety hazard to the

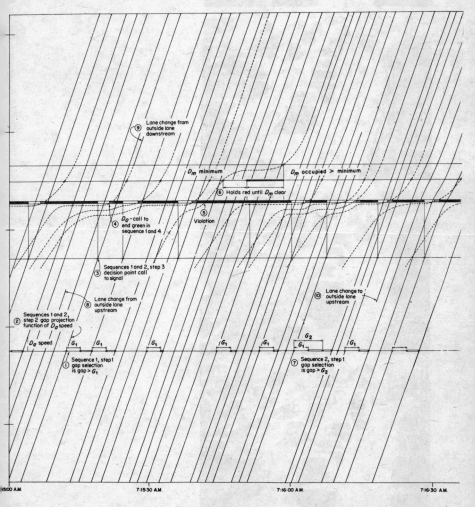

Fig. 16.11. Time sequence of a con-trolled merge.

Fig. 16.11 Time sequence of a controlled merge.

remaining traffic, to its driver, and to its occupants.   This is indicated by the high percentage of accidents that are of the rear-end type occurring at induced stop and yield locations such as in the freeway merging area. The merging control system virtually eliminates ramp vehicles stopped in the merging area, thereby contributing greatly to safe operation.

The merging control system affords more opportunity for increased ramp capacity than other metering models.   In systems which meter ramp vehicles one at a time, the ramp capacity is obviously a function of the ramp cycle length.   Since it takes about 6 sec to go through the ramp signal cycle, the maximum metering rate is one vehicle every 6 sec, or 600 vph. The merging control system can meter at a faster rate because it has the flexibility to meter more than one ramp vehicle whenever large freeway gaps are detected.

The proposed gap-oriented merging control system is relatively inexpensive because only the outside lane of the freeway is sensed for characteristics rather than all the freeway lanes.   This enables an appreciable reduction in the number of detectors.   Moreover, since time is one of the simplest variables to measure, the detection of acceptable gaps can be obtained with a comparatively simple analog device.

Lastly, the system provides the merging driver with the necessary information to know that a sufficient gap is available.   Yet, because of its nature, it is automatically a metering system.   Such a dynamic merge-aiding and metering technique appears to be a very attractive and inexpensive way of maintaining high efficiency of flow on a freeway, at the same time obtaining maximum ramp capacity and merging safety.

16.10 APPLICATION TO BUS RAPID TRANSIT

With the continued growth of large urban centers, the problems of intra-urban circulation are becoming more and more serious.   Traffic congestion which was once limited to peak hours of relatively short duration is reaching more severe proportions over longer periods of time in many cities.

Proponents of rapid transit are demanding that a "moratorium" be called on freeway construction in large urban areas and that rail rapid transit facilities be substituted.   Such proposals are usually supported by studies "demonstrating the increase in movement by people by the addition of mass transportation facilities."   Thus a study by the American Transit Association maintains that the substitution of a streetcar line for one lane of autos will carry three times the number of seated passengers and four times the number if standees are counted.[22]

Fortunately, few transportation people view the streetcar or rail rapid transit as a creative solution to the urban transportation problem.   After all, the existing urban centers have developed based on auto-oriented transportation. This development has manifested itself in much lower density configurations than is generally considered necessary to sustain rail rapid transit.

Bus rapid transit would appear to be better suited than rail rapid transit as the means of mass transportation, especially in presently developed areas. This stems from the observation that the bus is a very flexible mode, as compared to rail transit, and might more adequately provide service between existing residential areas and the major traffic attractors within the urban areas. Bus service can of course subsist on considerably lower volumes of corridor movement than can rail since they share the street system with the automobile and therefore need not amortize the entire facility alone.

Freeway passenger capacity could be raised by allocating more space to buses during peak hours. Three freeway lanes will carry about 10,000 passengers per hour since the average commuter-hour load is from 1.5 to 1.9 persons per automobile. A single freeway lane reserved for buses has a possible capacity of three times that number, assuming there is sufficient demand and it is possible to designate an exclusive lane for buses. One possibility would be based on utilization of the reverse-flow freeway design suggested in Chap. 15, with the reverse-flow roadway in the center devoted to exclusive transit use.

The disadvantage inherent in these schemes is the underutilization of highway capacity since no practical bus frequency can provide enough demand to fill a freeway lane. There are two additional possibilities which could avoid these losses and improve the performance of express bus systems. One procedure would be to control the entry of all vehicles including buses and thus keep the freeway lanes operating at an optimum level of service (Chap. 14). A control system with this capability is now being tested on the Gulf Freeway in Houston, although no attempt has been made to implement bus rapid transit. It has been designated mode 2 to contrast it with the merging control system described in Secs. 16.7 to 16.9.

This second mode of automatic ramp metering which is being evaluated at the Telephone Road entrance ramp is the capacity-demand mode. Many facets of its operations are similar to those of mode 1. The same basic signal operation is used. Two loop detectors on the ramp indicate (1) the presence of a vehicle waiting at the signal and (2) the passage of a vehicle past the signal. The controller preempts also operate in the same manner in mode 2 as in mode 1. That is, (1) when a ramp vehicle stops in the merging area, the ramp signal remains red until the vehicle merges, (2) a background cycle which defines the maximum waiting time at the signal is also used, and (3) when the queue of vehicles at the ramp signal

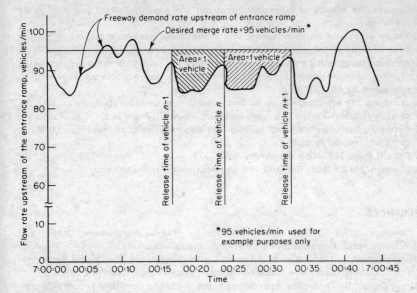

Fig. 16.12 Relationships among freeway demand rate, desired merge rate, and release times of vehicles from the ramp signal.

has a length of 14 or 15 vehicles, a preset high metering rate is used. Under mode 2 control, the metering rate on the ramp is adjusted according to the freeway demand rate to keep the total merging rate less than capacity, or service volume (see Fig. 16.12). In this way, freeway congestion can theoretically be prevented since demand is kept less than capacity.

The second alternative which might be instituted is a direct application of the merging control philosophy discussed in Secs. 16.7 to 16.9. Under this plan controlled entrance ramps would have two lanes with one for cars and trucks and the other for buses. Each lane would operate in conjunction with a merging controller based on appropriate gap settings. Wohl[23] refers to this as the *head-of-the-line* or *priority merge*. The bus ramp lane could be on the right to facilitate its use for bus stops. For existing facilities, some minor design modifications would obviously be needed. The feasibility of bus rapid transit operations on urban freeways using traffic surveillance and control is now being studied by the Texas Transportation Institute.

PROBLEMS

16.1. Discuss the effects and consequences of gap stability and lane changing on the *gap acceptance* mode of ramp metering.

16.2. Discuss the effects of lane distribution on the *capacity-demand* mode of ramp metering.

16.3. For an eight-lane divided freeway with a directional peak-hour volume of 2,000 in each of three of the lanes and 1,200 in the outside lane at a point upstream of an entrance ramp with a *critical gap* of 4.0 sec, find the peak-hour ramp capacity using (a) the capacity-demand approach and (b) the gap acceptance approach as expressed by Eq. (9.18.3).

16.4. What are the effects on capacity and operation in small errors in measuring the system variables for the two ramp-metering modes in Prob. 16.3?

16.5. Assuming the entrance ramp in Prob. 16.3 was redesigned to lower its *critical gap* to 3.0 sec, repeat Prob. 16.3.

16.6. Repeat Prob. 16.5 using the curves in Fig. 10.8.

16.7. Compare the capacity-demand and gap acceptance philosophies.

REFERENCES

1. "Development of National Proving Grounds for Freeway Surveillance, Control and Electronic Aids," Detroit Surveillance Project, unpublished.
2. Foote, Robert S.: Installation of a Tunnel Traffic Surveillance and Control System, in "Traffic Control Theory and Instrumentation," Plenum Press, New York, 1965.
3. Drew, Donald R.: Theoretical Approaches to the Study and Control of Freeway Congestion, *Texas Transportation Inst. Res. Rept.* 24-I, 1964.
4. Pinnell, Charles: Optimum Distribution of Traffic over a Capacitated Street Network, *Texas Transportation Inst. Res. Rept.* 24-2, 1964.
5. Drew, Donald R., and Charles J. Keese: Freeway Level of Service as Influenced by Volume Capacity Characteristics, *Texas Transportation Inst. Res. Rept.* 24-3, 1964.
6. Drew, Donald R.: Deterministic Aspects of Freeway Operations and Control, *Texas Transportation Inst. Res. Rept.* 24-4, 1964.
7. Drew, Donald R.: Stochastic Considerations in Freeway Operations and Control, *Texas Transportation Inst. Res. Rept.* 24-5, 1964.
8. Haynes, Dr. John J.: Some Considerations of Vehicular Density on Urban Freeways, *Texas Transportation Inst. Res. Rept.* 24-6, 1964.
9. McCasland, William R.: Traffic Characteristics of the Westbound Freeway Interchange Traffic on the Gulf Freeway, *Texas Transportation Inst. Res. Rept.* 24-7, 1964.
10. Wattleworth, Joseph A.: System Demand-Capacity Analysis on the Inbound Gulf Freeway, *Texas Transportation Inst. Res. Rept.* 24-8, 1964.
11. McCasland, William R.: Capacity-Demand Analysis of the Wayside Interchange on the Gulf Freeway, *Texas Transportation Inst. Res. Rept.* 24-9, Texas A&M University, College Station, 1965.
12. Pinnell, Charles, Donald R. Drew, William R. McCasland, and Joseph A. Wattleworth: Inbound Gulf Freeway Ramp Control Study I, *Texas Transportation Inst. Res. Rept.* 24-10, Texas A&M University, College Station, 1965.
13. Drew, Donald R., and Conrad L. Dudek: Investigation of an Internal Energy Model for Evaluating Freeway Level of Service, *Texas Transportation Inst. Res. Rept.* 24-11, Texas A&M University, College Station, 1965.
14. Drew, Donald R.: Gap Acceptance Characteristics for Ramp-Freeway Surveillance and Control, *Texas Transportation Inst. Res. Rept.* 24-12, Texas A&M University, College Station, 1965.
15. Pinnell, Charles, Donald R. Drew, William R. McCasland, and Joseph A. Wattleworth: Inbound Gulf Freeway Ramp Control Study II, *Texas Transportation Inst. Res. Rept.* 24-13, Texas A&M University, College Station, 1965.

16. Wattleworth, Joseph A.: Peak-period Analysis and Control of a Freeway System, *Texas Transportation Inst. Res. Rept.* 24-15, Texas A&M University, College Station, 1966.
17. Greenberg, H.: An Analysis of Traffic Flow, *Operations Res.*, vol. 7, 1959.
18. May, A. D., P. Athol, W. Parker, and J. B. Rudden: Development and Evaluation of Congress Street Expressway Pilot Detection System, *Highway Res. Board Rec.* 21, 1963.
19. Wattleworth, J. A.: "Peak-period Control of Freeway System: Some Theoretical Considerations," doctoral dissertation, Northwestern University, Evanston, Ill., 1963.
20. Drew, Donald R., and Charles Pinnell: Some Theoretical Considerations of Peak-hour Control for Arterial Street Systems in "Traffic Control Theory and Instrumentation," Plenum Press, New York, 1965.
21. McCasland, William R.: Operation of Gap Projection Controller, unpublished.
22. "Moving People in the Modern City," American Transit Association, New York, 1966.
23. Wohl, Martin: "Improvement of Bus Service between Washington and Portions of Northern Virginia," unpublished.

NAME INDEX

Numbers in parentheses are Reference numbers and are followed by the page(s) on which they are cited in text.

Alligaier, Earl, (11) 27
Anderson, A. A., (4) 26; (5) 26
Athol, P., (4) 303; (18) 429

Barnett, J., (18) 65
Bartlett, Neil R., (15) 28, 29
Bartz, Albert E., (15) 28, 29
Bergami, D., (1a) 258
Berry, D. S., (19) 65
Blunden, W. R., (3) 178; (6) 248
Braunstein, M.L., (49) 37
Brethorton, M. H., (5) 130
Brody, Leon, (10) 27; (33) 33
Brown, R. G., (3) 334
Buckley, D. J., (21) 216; (3) 242
Buhr, J. H., (17) 204; (18) 209, 212, 213; (19) 211, 213; (20) 213, 218; (12) 286
Bull, A. W., (17) 17

Capelle, D. G., (20) 384
Carmichael, T. J., (25) 21
Chandler, R. E., (14) 342
Chapanis, A., (30) 32
Charlesworth, G., (10) 312
Claffey, P. J., (14) 56
Clissold, C. M., (3) 178
Clough, D. J., (4) 244
Conger, John J., (3) 26; (12) 27
Conrad, L. E., (4) 11
Cook, J. S., (30) 32
Cozan, L. W., (40) 33, 34, 35

Dart, O. K., (9) 283
Dawson, R., (9) 190
Dobbins, D. A., (4) 26; (5) 26
Doebelin, E. O., (1) 331, 333
Domey, Richard C., (31) 32, 33

Drew, D. R., (38) 33; (24) 72; (6) 92; (7) 92, 96, 98; (7) 135; (7) 168; (4) 178, 184, 188, 189, 192; (17) 204; (18) 209, 211, 213; (19) 211, 213; (20) 213, 218; (1) 254; (12) 286; (9) 310; (18) 325; (10) 340; (15) 346; (1) 355, 376; (3) 355; (4) 355; (3) 396; (4) 400; (3) 427; (5) 427; (6) 427; (7) 427; (12) 427; (13) 427; (14) 427; (15) 427, 434; (20) 432
Dudek, C. L., (2) 355; (3) 355; (13) 427

Edie, L. C., (5) 247; (3) 303; (16) 322
Eisner, S., (2) 81
Erickson, E. L., (1) 177, 196

Feller, William, (6) 132; (7) 181, 182
Finney, D. J., (12) 193; (13) 193
Firey, J. C., (15) 56
Fisher, R. B., (3) 178
Foote, Robert S., (2) 423
Forbes, T. W., (17) 29; (25) 30, 31; (26) 30
Fosberry, R. A. C., (32) 32
Fox, B. H., (46) 37; (50) 37; (53) 37
Fox, M. L., (16) 17
Fox, M. W., (50) 37
Foxworth, De V. M., (19) 18
Froehlich, W. R., (2) 8
Fry, Glenn A., (19) 29
Fry, Thorton C., (3) 156

Gallion, A. B., (2) 81
Garwood, F., (8) 188
Gaskill, Herbert S., (12) 27
Gazis, D. C., (57) 40; (11) 341; (13) 341, 342; (19) 365
George, Stephen, Jr., (5) 356
Gerig, F., (18) 209, 211, 213
Gerlough, D. L., (3) 129; (8) 118; (6) 274, 281
Gilgour, T. R., (20) 30
Gillis, C. M., (10) 54
Glad, Donald D., (12) 27
Goldstein, L. G., (9) 27
Goode, H. H., (5) 264; (11) 285; (4) 335; (1) 389
Greenberg, H., (8) 306; (16) 364; (17) 428
Greenshields, B. D., (12) 103; (13) 103; (1) 177, 196; (2) 300, 310; (7) 358; (8) 358

Haile, E. R., (27) 31
Hald, A., (6) 167

Hall, A. D., (1) 112
Hall, Edward M., (5) 356
Harr, M. E., (6) 305
Hart, J. W., (2) 177, 178, 180, 196
Hart, William M., (23) 30
Hawkins, D. L., (23) 70
Haynes, J. J., (1) 299; (8) 427
Heath, Earl Davis, (13) 27
Herman, R., (57) 40; (10) 190, 192; (11) 341; (13) 341, 342; (14) 342; (11) 360, 361, 372, 374; (17) 364; (19) 365
Holstein, W. K., (1) 256
Homburger, W. S., (1) 150
Hulbert, Slade F., (44) 37; (45) 37; (47) 37
Hurd, F. W., (15) 197
Hurst, P. M., (37) 33; (39) 33; (43) 36

Johannesson, S., (11) 103
Johnson, A. E., (5) 54
Johnson, A. N., (10) 103
Jones, Trevor R., (13) 361, 362, 363, 372, 374

Katz, M. S., (26) 30
Keese, C. J., (36) 33; (12) 54; (24) 72; (7) 92, 96, 98; (1) 355, 376; (3) 355; (4) 400; (5) 427
Kell, J. H., (1) 150
Keller, Leo, (12) 27
Kendall, D. G., (9) 169; (2) 240
Kennedy, N., (1) 150
Kobayashi, M., (52) 37
Kometani, E., (9) 338

LaMotte, L. R., (17) 204
Lang, C. H., (11) 54
Lauer, A. R., (6) 27
Lee, Clyde R., (28) 31
Leisch, J. E., (6) 408
Leonards, G. E., (6) 305
Lewis, R. M., (7) 281
Lighthill, M. J., (11) 314; (14) 360, 364
Lipscomb, J. L., (8) 92, 96
Loutzenheiser, D. W., (27) 31
Lund, M. W., (30) 32

McCasland, W. R., (18) 325; (9) 427; (11) 427; (12) 427; (15) 427, 434; (21) 439
McConnell, W. A., (3) 8; (17) 62
McCormick, E. J., (1) 25; (16) 28; (56) 40

McCoy, P. T., (1) 50
McFarland, R. A., (22) 30; (31) 32, 33; (34) 33
Machol, R. E., (5) 264; (4) 335; (1) 389
Major, N. G., (21) 216; (3) 242
Malfetti, James L., (2) 25
Maradudin, A. A., (57) 40; (11) 190
Mathewson, J. H., (12) 16; (45) 37
Matson, T. M., (15) 197
Matsunaga, T., (52) 37
May, A. D., (4) 303; (18) 429
Mayer, R. A., (5) 12, 21; (6) 12; (19) 65
Mayne, A. J., (6) 180
Meese, G. E., (21) 30
Meserole, T. C., (12) 286
Michael, H. L., (7) 281
Michaels, R. M., (40) 33, 34, 35; (48) 37; (16) 202
Miller, A. J., (14) 319
Mills, B. C., (32) 32
Molina, E. C., (2) 125, 130
Montroll, E. W., (14) 342; (10) 359; (11) 360, 361, 372, 374; (12) 360
Moore, B. C., (3) 261
Morgan, C. T., (30) 32
Morgenthaler, G. W., (10) 284
Moroney, M. J., (1) 124
Morrison, R. B., (5) 303; (18) 365
Morse, P. M., (7) 230, 248
Moskowitz, Karl, (6) 54; (12) 316; (2) 390

Neal, H. E., (54) 39
Newell, G. F., (15) 319; (12) 341
Newman, Leonard, (12) 316; (2) 390
Nilsson, J. W., (3) 334
Noble, C. M., (21) 65
Normann, O. K., (8) 14

Olmstead, F. R., (55) 39
Olley, M., (15) 17
Olson, P. L., (58) 41
Olson, R. M., (23) 70
Oppenlander, J. C., (14) 192, 194, 196

Parker, W., (4) 303; (18) 429
Pawley, E. R., (4) 11
Pearson, Karl, (4) 166; (5) 166
Peckham, Robert H., (23) 30
Perchnok, (35) 33; (37) 33; (39) 33; (42) 36
Pinnell, C., (12) 54; (6) 92; (7) 135; (7) 168; (1) 254; (18) 325; (4) 427; (12) 427; (15) 427, 434; (20) 432

Pipes, L. A., (8) 338
Platt, F. N., (11) 15; (9) 358
Potts, R. B., (13) 341, 342; (10) 359; (11) 360, 361, 372, 374; (13) 361, 362, 363, 372, 374; (19) 365

Raff, M. S., (2) 177, 178, 180, 196
Rainey, Robert V., (3) 26; (12) 27
Raven, F. H., (7) 337
Reddy, M. S., (7) 305
Rekoff, M. G., (2) 119; (15) 346
Richards, Oscar W., (24) 30
Richards, P. I., (15) 364
Roos, Daniel, (25) 75
Roper, Val J., (21) 30
Rothery, R., (58) 45; (11) 341; (11) 360, 361, 372, 374
Rowan, N. J., (36) 33; (1) 50; (23) 70; (21) 384
Rudden, J. B., (4) 303; (18) 429

Saal, C. C., (13) 55
Sandefur, G. G., (8) 281
Sasaki, T., (9) 338
Sawery, William L., (12) 27
Sawhill, R. B., (15) 56
Schmidt, R. E., (7) 11
Schuhl, Andre, (2) 155
Segal, L., (14) 17
Selling, L. S., (7) 27
Sequein, E. L., (39) 33
Severy, D. M., (12) 16
Shapiro, D., (1) 177, 196
Silver, C. A., (51) 37
Simonson, Wilbur H., (7) 54
Skordahl, D. M., (4) 26; (5) 26
Smith, W. S., (15) 197
Solberg, Per, (14) 192, 194, 196
Sowkup, W. R., (1) 256
Stephens, B. W., (48) 37
Stonex, K. A., (1) 8; (13) 17; (21) 19; (29) 31; (21) 65
Stuart, A., (9) 169
Sugarman, R. C., (49) 37
Suhr, Virtus W., (6) 27

Tanner, J. C., (5) 180
Taragin, A., (41) 35; (20) 65, 67
Thompson, D., (16) 61
Tippett, L. H., (4) 263

Tocher, K. D., (2) 258, 262, 273
Truxal, J. G., (6) 336
Turrell, Eugene S., (12) 27

Uhlaner, J. E., (9) 27

Van Lennep, D. J., (8) 27
Van Steenberg, W., (9) 27

Wait, John V., (15) 28, 29
Wallace, Ralph, (14) 27
Walsmith, Charles R., (3) 26; (12) 27
Wardrop, J. G., (10) 312
Wattleworth, J. A., (17) 204; (18) 209, 211, 213;
   (19) 211, 213; (17) 324; (18) 325; (19) 326;
   (6) 357; (10) 431; (12) 427; (15) 427, 434; (16)
   427; (19) 432

Webb, George M., (6) 54
Webster, F. V., (13) 319
Weingarten, H., (16) 202
Weintraub, Sol, (4) 130
Weiss, G., (10) 190, 191; (11) 190
White, W. J., (49) 37
Whitham, G. B., (11) 314; (14) 360, 364
Whitson, R. H., (20) 213, 218
Willey, W. E., (23) 21
Williamson, Eric, (5) 130
Wilson, W. E., (5) 335, 336
Wilts, C. H., (2) 331, 334
Wohl, Martin, (23) 445
Wojcik, Charles, (44) 37; (47) 37
Wolf, Ernst, (22) 30

Young, J. C., (18) 17

Zigler, Michael, (22) 30

SUBJECT INDEX

AASHO (American Association of State Highway Officials), 50, 51, 54
AASHO Redbook, 405
Abilene, Texas, 41
Abutments, 31
Acceleration, 8
  centrifugal, 17
  centripetal, 63
  nonuniform theory, 8–10
  uniform theory, 10–12
Acceleration distance, 211, 212
Acceleration distribution, 359, 360
Acceleration lane, 174, 211–214, 219
  definition, 174
  length of, 211–214
  parallel-type, 219
  tapered, 219
  use of, 174
Acceleration noise, 267, 359–364, 369, 370, 372, 374–377, 383, 384
Acceleration-time curve, 372
Acceleration-time relationship, 199

Accelerometer, 359, 372
Acceptable gap, 432
Accepted-gap number, 213, 214
Access, 49, 81, 108
Access control, 84
Accident analyses, 14, 87
  collisions, 15, 17
  noncollisions, 20
Accident statistics, 3
Acuity, visual, 28, 40
ADT (see Average daily traffic)
Aerial photography, time lapse, 350, 351, 434
Air resistance, 12
Alignment, horizontal, 64, 65, 67–69
  circular curves, 67
  compound curves, 67
  spiral curves, 65, 67
  superelevation, 65, 67, 68
  vertical, 55, 64
    maximum grades, 55, 56
    vertical curves, 55, 56, 58–61
Amber signal, 40

American Association of State Highway Officials, development of standards, 50, 51, 54
American Transit Association, 443
Analog graph, 372
Angle, of convergence, 209, 210, 214, 218
of entry, 414
merging, 209, 210, 214, 218
of incline, 13
visual, 35
Angular velocity, 34
Angular velocity model, 202
Arbitrary queuing headway, 433
Arrivals, analysis of, 128–131
distribution of, 128–131, 135, 224, 226, 227
observed, 136
Poisson, 135, 139, 224, 226, 250
predicted, 135, 136
uniform, 135, 139
Arterials, 81, 84, 108
Assignment, traffic, 100–102
Assignment network, 102
Asynchronous counting, definition, 166
Attitude, driver, 27
Average acceleration, 362, 363
Average car study technique, 385
Average daily traffic (ADT), 86, 88, 90, 92, 135, 396
Average daily traffic volume, 90
Average delay, 319
Average driver, 40, 103
Average speed, 372
(See also Mean speed)
Average waiting time, 187

Balanced design for freeways, 404
Barrier curbs, 53
Barriers, 54, 75, 414
Bay Town Tunnell, Houston, 426
Bernoulli experiment, 237, 238
Beta distribution, 161–164, 167, 269
derivation of, 167
estimation of parameters, 163
generalization of, 162
generating function for, 162
mean of, 162
utility as traffic model, example of, 163, 164
variance of, 163
Beta function, 161, 167
incomplete, relation to cumulative binomial, 165

Binary number system, 264, 265
Binomial coefficients, 122
Binomial distribution, 121–123, 126, 128, 130, 132, 133, 149, 193, 272
derivation of, 122
estimation of parameters, 130
mean of, 123, 126
moment generating function for, 149
moments, 123, 128
probability generating function for, 126
variance of, 123, 126, 193
Binomial response, 192
Binomial series, 122
Binomial theorem, 122
Biological assays, 193
Block diagram, 278
Bottleneck control approach, 315–319
Bottlenecks, 105–107, 316, 317, 324, 325, 374, 376, 421
definition, 421
volume density relationships, 316, 317
Boundary conditions, 299, 302–304
Brake fade, definition, 14
Braking distance, 13
Bray's Bayou, 374
Bridge rails, 55
Buffon needle problem, 257, 258
Bulk service technique, 434
Bureau of Public Roads, U.S., 69, 70, 103, 111, 424
Bus rapid transit, use of freeway, surveillance and control for, 443, 445

Capacity, 51, 90, 102–104, 106–108, 114, 115, 134, 216, 217, 311, 378
basic, 378
bottleneck, 106, 107, 316
definition, 102, 103
design (see Service volume)
entrance ramp, 400
freeway, 390, 392
highway, 103
lane, 103
merging, 216, 217, 404
possible, 105–107, 378
practical, 106, 378
signalized intersection, 134, 135
Capacity analysis, 138
Capacity-demand analysis, 95
Capacity-demand relationship, 134, 137, 226, 318

Capacity-demand relationship, of freeway, 431
Capacity-design procedure, 140–144
Capacity equations, 139
Capacity restraint, 101
Capacity restraint model, 101
Car-following laws, 344
Car-following models, linear, 337–340, 346
 nonlinear, 341, 342
Car-following variables, 350–352
Car manuvering, 342–345
Central limit theorem, application, 272, 273
Centrifugal force, 17
Centripetal force, 18
Centroid, 100
Chaining array, 289
Chaining technique, 289
Channel configuration, 224
Channelization, 87
Characteristic equation, 334
Chi-square test, 125, 129, 169
 definition, 125
 interpretation of, 125
Chicago Project, 423, 429
Classification counts, 87–89
Clearance, 50, 414
 horizontal, 35, 414
 vertical, 63, 414
Closed loop system, 117, 330, 331
Cloverleaf interchange, 134, 393, 407
Collector street, 80–82, 108
Collisions, deformations developed in, 15
 head-on, 16
Color contrast, 47, 53
Commercial vehicles (see Truck)
Comparison of hypothesized and observed
  distributions (see Fitting distributions to
  observed data)
Compressible-fluid flow analogy, 117
Computer, 278
 analog, 256, 346, 347, 349, 350
 digital, 256, 262, 265, 278, 289
Conditional probability, definition, 182
Conductivity, 304
Confidence limits, 168, 301, 392
Congestion, 3, 105, 106, 303, 323, 361, 376, 392, 421
Congress Expressway, Chicago, 423
Congruent numbers, 265
Conservation, of energy, law of, 368
 of mass, principle of, 368

Consistency, 50
Constant of proportionality, 311, 341, 342, 343, 350
Continuity, equations of, 306–308, 314, 365, 366
Continuous distributions, definitions, 147–149, 166
 moment generating function of, 149
 (See also Beta distribution; Erlang distribution; Gamma distribution; Log-normal function; Negative exponential distribution; Normal distribution)
Continuous variables, 147
Contour maps, 280, 323
Contours, density, 323
 level of service, 382
 speed, 323
Control, horizontal, 52
 vertical, 52
Control counts, 86
Control devices, 69
Control stations, minor 86, 87
Control systems, feedback, 330–332
 open loop, 330
Control theory, 72
Controlled access, 84
Cordon area, 88
Cordon counts, 86, 88
Cornering force, 17
Countdown signal, 41
Counters, 88
Counting distributions, 166
 (See also Discrete distribution)
Counting program, 86, 87
Counts, 86–88
Crash test, 70
Creative design, 388
Crest vertical curve, 56, 58–61
Critical depth of flow, 368
Critical flow, 366
Critical gap, 177–179, 187–190, 218, 219, 432
 definition, 432
 determination of, 178, 179
 distribution of, 187–190
 for merging, 179
Critical lag, 177
 definition, 177
Critical lane volume, 138
Critical level of service, 383
Critical path methods, 72, 113
Critical speed, 323
Critical velocity, 365
Cross section, 50, 52–55

Cross section elements, 52
Cross slope, 53
Culs-de-sac, 84
Cumulative probability distribution, 139, 148, 267, 271
Curbs, 53, 54
Curvature, 50, 66
    flat, 67
    horizontal, 52
    maximum, 65
    radius of, 18, 66
Curve fitting, 229–302
Curves, for cumulative Erlang, 159, 160
    for cumulative frequency, 150, 151
    for cumulative Poisson, 139
    horizontal, compound, 67, 416
        operating characteristics on, 56
        spiral, 65, 67
        (See also Horizontal curves)
    for Poisson arrivals, 135, 136
    for uniform arrivals, 135, 136
    vertical, 55, 56, 58–64
        crest, 56, 58–62
        design controls for, 56
        parabolic, 56, 57, 60
        sag, 56, 61–63
        sharp, 50
        sight distance on, 56
Cycle failure, 139
Cycle length, 135
    versus capacity for signalized, 140–144, 396–399
    determination of, 138, 139

Daily time pattern, 90
Deceleration, human tolerance to, 15
    intersection, 41
Deceleration lanes, 400, 414
Decimal number system, 264, 265
Decision point, 439
Degree of saturation, 318
Degrees of freedom for t tests, 300
Delay, 114
    at merging areas, 180, 181
Delay flow relationship, 319
Delay models, 180–187
Delays, starting, 104
Delineation, 48
Demand, traffic, 50, 90, 91, 100, 114, 134, 135, 316
Demand-capacity ratio, 108

Density, 239, 240, 299, 342, 358
    defintion, 153, 321, 322
    jam, 309
    mean, 299
    optimum, 303, 309
Density function, probability, 148, 269
Density-speed relationship, exponential
    model, 301, 311
    linear model, 301, 311
    parabolic model, 301, 311
Density-speed-volume relationship, deriva-
    tion of, 299
    boundary condition approach to, 302–304
    fluid-flow analogy, 305–311
    regression, 300–302
Density-volume relationship, 316, 317
Design, 72, 92
    freeway, 113
    geometric (see Geometric design)
    highway, 104, 105, 107
    sign, 69, 70
    system, 73
Design curves diagram, 391, 397
Design designation, 53
Design hourly volume (DHV), 90, 92, 405, 406
Design level of service, 106, 107
Design speed, 48, 51, 52, 56, 63, 65, 75, 414
Design vehicles, 51
Design volume, 107
Design year, 90
Destination studies, origin and, 88, 90, 245
Detectors, 88, 89
    gap, 196, 439
    infrared, 89
    loop, 439, 444
    queue, 439
    radar, 89
    road tube, 88
    sonic, 89, 439
    speed, 439
Deterministic approach to traffic theory, 299, 304
Deterministic theory pyramid, 117
Detroit Project, 427
Deviation, standard, 121, 359, 362
DHV (design hourly volume), 90, 92, 405, 406
Diamond interchange, conventional, 393, 394
    split, 396, 407
    three-level, 394
Dilemma zone, 40–42
Diode function, generation and multipliers, 350

Directional counts, 87
Directional distribution, 86, 90
Directional interchange, 393
Discrete distribution, definition, 147
  moments of, 123, 126, 128
  (See also Binomial distribution; Geometric
    distribution; Negative binomial distri-
    bution; Poisson distribution; Uniform
    distribution)
Displacement, lateral, 34
Distance-logic flow diagram, 290
Distribution relationships, 164–166
Diversion, 101
Downtown distribution system, 408–411
Drift, 67
Driver behavior, 33
Driver comfort, 65, 377
Driver education, 25–27
Driver error, 36
Driver eye height, 31, 40
Driver licenses, 26
Driver performance, 114
Driver response, 34
Driver simulation, 36
Drivers, 48, 50
  perception-reaction time for, 37–40
  physical characteristics of, 28–33
  psychological characteristics of, 25–28
Dumble interchange, Gulf Freeway, Hous-
  ton, Texas, 374
Duration (parameter), 92
Dynamic programming, 72
Dynamics, fluid, application of, 305, 364, 365
  vehicle, 8

Efficiency, 383
Eisenhower Expressway, 384, 429
Elements of traffic flow, 5, 112
Energy-momentum concept, 115, 117, 277, 381
Energy transfer, 22
  enforcement, 25–27
Engine resistance, 12
Entrance ramp volume, 96
Entrance ramps, 97, 106, 108, 208, 240–244, 293,
  400, 405
Environment, 112
Equation, of car-following (general), 341
  of continuity, 306–308, 314, 365, 366
  of diffusion, 304, 305
  of energy, 368, 369
  of momentum, 365, 366

Equation, of motion, 306, 311, 343, 365, 366
  generalized, 308, 310
  of moving vehicle method (fundamental),
    312, 313
  of one-dimentional wave motion (char-
    acteristic), 314
  of state, 343
    fundamental, 302, 311, 343
    gas, 364, 365
    traffic, 364
  of traffic stream, 299, 342
  of wave velocity, 303
Erlang distribution, 158–161, 180, 181, 218, 238,
  239, 248, 250, 269, 270, 273
  applied to gap acceptance, 181
  queueing, 238, 239
  merging delay, 185
Esterline, Angus, 372
Estimation, parameter, maximum likelihood, 169
  method of moments, 158, 186
Euler method of analysis, 319
Event scan technique of simulation, 274
Exit ramps, 97, 106–108
Expected values, 226, 227, 229, 232, 233, 235–
  242, 250
Experimental design, 283
Exponential distribution (see Negative
  exponential distribution)
Exponential model, 301, 311
Exponential series, 124
Expressway, 80–82, 85, 405

Feedback control system, 330–332, 335
Figures of merit, 139, 276
Finite queues, 229, 230
First come, first served (see Queue
  discipline)
Fit, goodness of (see Chi-square test)
Fitting, continuous distributions to observed
  data, 157, 158, 163, 164
  discrete distributions to observed data,
  130
Fixed objects, 55
Floating-car method, 372
Flow, 391
  definition, 321, 322
  forced, 317, 379, 381, 383, 391, 392
  free, 379, 381, 383
  stable, 379, 381, 383, 391, 392

Flow, unstable, 377, 379, 381, 391, 392
Flow concentration curve, 365
Flow diagrams, 271, 272, 288, 290, 291
Flow index, quality of, 358
Flow rate (see Volume)
Fluid-flow analogy, 305–311
Fluid mechanics, 306
Force, centrifugal, 62, 64, 65
  gravitational, 62
Fortran IV, 280
Free speed, 309
Freeway, 80, 82, 84, 85, 92, 94, 96, 98, 99, 108,
  176, 389–392, 402–404, 422–424
  capacity, 390, 392
  in city street plan, 408
  control of access, 84, 108
  design, 113, 389–392, 396, 402–404
  elements, 174
  merging, 115, 176, 286, 295
  surveillance, 85, 422–424
  system, 393
Freeway total volume, 96, 98, 99
Freeway volume, 94
Frequency, 125
Frequency distribution, 121, 150
  spot speed, 150, 151
  (See also Probability distribution)
Friction, 369
  coefficients of, 13, 57
  kinetic, 13
  static, 13
  for vehicle braking, 13–15
  for vehicle turning, 18
Frictional resistance, 48
Frontage roads, 84, 85, 174
Fuel, consumption, 8, 21, 383, 384
  economy, 8, 21
Functional classification, 80
Functional design, 47
Functions, 121, 126, 166, 167
  cumulative distribution, 148
  density, 269
  generating, 127
  moment generating, 127, 148, 149
  probability, 147
  probability generating, 126
  of sheet systems, 83
  time, 331
  transfer, 331
Fundamental volume-speed-density
  surface, 300

Game theory, 72
Games of chance, 121
Gamma distribution, 155–158, 161, 167, 242,
  269
  deviation of, 167
  estimating parameters, 158
  generalization of, 156
  generating function for, 156
  mean of, 157
  special cases of, 161
  variance of, 157
Gamma function, 156, 164, 166
  incomplete, relation to cumulative
    Poisson distribution, 165
Gap, 174, 196, 197, 439
  accepted, 432
  critical (see Critical gap)
  definition, 153, 174
  detection of, 196, 439
  ideal, 198, 220, 221
  projection, 439
  rejected, 178
  space, 174
  time, 174
Gap acceptance, 173, 190
  characteristics, 209–214
  distributions, 87
  functions, 190–193
Gap acceptance mode, 434
Gap availability, 240
  General Motors, 116, 227, 360
Generalized Pearson distributions, type I,
    162–164
  (See also Beta distribution)
  type III, 156–158
  (See also Gamma distribution)
Generating function, 126, 127, 227
  definition, 127
  moment, 127, 148, 149
  probability, 126, 227, 228
Generators, 89, 92, 100
  Erlang distribution, 249
  (See also Erlang distribution)
  hyperexponential, 249
  pulse, 89
  random number, 263–266
  traffic, 100
Geometric design, 50, 51, 79, 87, 90, 92
  ADT, use of, 90
  alignment, horizontal, 64
    vertical, 55
  channelization, 87

Geometric design, classification of high-
    ways, 80, 82
  classification counts, 87
  cross-section, 52
    elements, 52
  curbs, 53
  curves, horizontal, 52
    vertical, 54, 60
  definition, 7, 47
  design designations, 82
  design speed, 51, 52
  directional distribution, 92
  divided highways, 54
  grades, 51
  highway capacity (see Capacity)
  interchanges (see Interchanges)
  intersection (see Intersection)
  land widths, 53
  medians, 54
  relation to vehicle design, 20
  right-of-way, 51
  safe speed, 51
  separation strip (see Medians)
  shoulders, 52, 53
  sight distance, 31, 50, 52
  slopes, 20, 53
  system design, 72
Geometric distribution, 131, 132, 180
  definition, 131
  generating functions for, 132, 149
  mean of, 226
  queue length probabilities, 225, 228
Geometric progression, 131
Geometric series, 131, 339
Glare, 29
Goodness-of-fit test (see Chi-square test)
Grade differential space, 54
Grade line, 55
Grade profile, 414
Grade separation, 174, 372, 393
Grades, 51, 55, 56, 82, 84, 374
  balancing of, 51
  effect on stopping distance, 13
Gradient resistance, 12
Green phase, 134, 135, 318
  apportionment of, 134
Gridiron pattern, 81, 84
Griggs Street overpass (station 220), 323
Guardrails, 17, 54, 55, 75, 414
Gulf Freeway, 179, 240, 323, 324, 356, 357, 374,
    378, 383, 384, 424, 426, 427, 436, 438, 444

Hand tally, 88
Harmonic mean, 322
Head-of-the-line merge, 445
Headlamps, design of, 30
  height of, 62
  polarized, 42
Headlight glare, 29
Headways, 104, 135, 138, 174
  average minimum, 103, 135, 138
  arbitrary queueing, 433
  constant, 135
  definition, 153, 174
  departure, 104, 135
  safe, 303, 337, 338
  space, 156, 174, 303, 342
  starting, 248, 340
  time, 157, 174, 299
Heat flow analogy, 117, 304, 305
High-type facility, 134
High-type intersection, design of, 140
  signalization of, 140
Highway capacity (see Capacity)
Highway Capacity Committee, 104, 358, 378
"Highway Capacity Manual," 92, 105, 107, 278,
    381, 391, 411
Highway classifications, 80, 82
  planning, 86
Highway design, 104, 105, 107
Highway Research Board, 104, 111, 358
Hills, 410
Histogram, 262
Holland Tunnel of The Port of New York
    Authority, 360
Hollywood Freeway, 3
Horizontal alignment, 64–69
Horizontal clearance, 35, 414
Horizontal curves, 52, 64
  circular, 65
  compound, 67, 416
  friction on, 65
  minimum ratio for, 66
  spiral, 65, 67
  transitional, 65
Horizontal sight distance, 31
Houston, Texas, 41
Human factors, 32
Human servomechanism, 334, 335
Hydrodynamic analogy (see Fluid-flow
    analogy)
Hypothesis testing, chi-square test, 125
  t test, 300

Ice problems, 66
ICES (Integrated Civil Engineering System), 75
Icy conditions, 56
Illinois Division of Highways, 423
Illumination, 29, 38, 48, 49
Impact, force at, 15
  tests, 16
Incomplete beta function, relationship to cumulative binomial distribution, 165
Incomplete gamma function, relationship to cumulative Poisson distribution, 165
Independence of events, 114, 166
Information matrix, 281
Input function, 224
Input source, 224
Institute of Traffic Engineers Technical Committee 7-G, 88
Interchanges, 393–399, 402, 405–408, 411–416
  cloverleaf, 134, 393, 407
  diamond, 393, 394, 396, 406, 407
  directional, 393
  reverse flow, 411–416
  signalized, 134, 393, 396
  types, 393
  unsignalized, 393
  weaving section, 400–402, 404
Internal energy, 368–371
Intersection, 87, 92, 103, 108, 137, 317, 318
  capacity, 103, 108, 137
  counts, 87
  signalization, 134, 393
Intersection-capacity equations, 396
Interstate Freeway, Seattle, 424
Interstate system, 2, 421
Inversion, 266–269
Islands, traffic, 56

Jam density, 309
John C. Lodge Freeway, Detroit, 422–424

Key counts, 87
Key stations, 87
Kinematic waves analogy, 117
Kinetic energy, of traffic stream, 364–370
  of vehicle, 15

Ladder effect, 50
Lag, 174, 196, 197

Lagrange's interpolation formula, 345
Lagrangian method of analysis, 319
Land use, 47, 80
Lane change, 99, 174
Lane distribution, 94, 98
Lane width, 8, 35, 53
Lanes, number of, determination of, 391, 392
Laplace equation, 259, 261
Laplace transforms, 331, 332, 338
Lateral clearance, 34, 35
Law, of conservation of energy, 368
  of conservation of momentum as applied to traffic flow, 365
  of molecular interaction, 365
  of thermodynamics, second, 369
Laws of chance, 121
Learning, driver, 25
Least squares analysis, 300, 301
Left-turn lanes, 52
Left-turning movements, 104
Level of service, 51, 53, 73, 95, 104–108, 115, 355, 356, 374, 378–383, 390, 391, 401, 402, 404
  definition, 105
Lighting, 47–50
"Lights on" study, 98
Lincoln Tunnel, 423
Linear car-following, 337–340
Linear programming, 113
Linear programming model, freeway control, 72, 324, 326, 432
Link, 100
Local street, 80–82, 87, 108
Lodge Freeway, Detroit, 422–424
Log-normal function, 195, 196
Long Island Expressway, 3
Loop detector, 439, 444
Loop street, 84
Loops, 85, 416
Lost time for starting headways, 138
Luminaire, 49, 50, 439
  definition, 49
  height of, 50

Machine counters, 88
Mach number, 364, 365
Magnitude, 86, 92
Major arterial, 80–82, 108
Maneuver, fundamental, 174
Maneuvers, basic, 7
"Manual on Uniform Traffic Control Devices," 69

Manual counters, 88
Manual traffic counts, 88
Markings, 69
Markov process, 114, 115, 244–246
Mass transportation, 443
Maximum car study technique, 385
Maximum likelihood estimators, 169
Mean, 121, 123
  arithmetic, 322
  of continuous distribution, 147
  of discrete distribution, 126
  harmonic, 322
Mean critical gap, 178
Mean delay, 184, 319
Mean queue length, multiple channel, 235
  single channel, 226. 227
Mean speed, 299
  space, 315, 322
  time, 322
Mean-value theorem, 257, 365
Mean waiting time, 187, 232
"Measuring Traffic Volumes," 86
Mechanical traffic counters, 88
Median, 2, 67, 150
Median critical gap, 178
Median critical lag, 178
Median gap, 194
Median lanes, 405
Median width, 54
Medians, 48, 54
  barrier, 54
  deterring, 54
  traversable, 54
Mercury vapor lamp, 49
Merge, ideal, 198, 199
Merging, 115, 174, 293, 294
Merging capacity, 216, 404
Merging control, 434, 435
Merging control system, 435–443
Merging maneuvers, 177
Merging parameters, 177
Merging priority, 445
Metering traffic flow, 428—434
Method of least squares, 300, 301
Minimum-variance estimators, 169
Mobility, 3, 80
Mode, 150
Model calibration, 283, 284
Models, 113–115
  acceleration-noise–momentum-energy
    model, 379

Models, analog, 114
  angular velocity, 202–204
  computer, 256
  definition, 113
  deterministic, 114, 115, 301, 311, 349, 350
  internal energy, verification of, 376–378
  internal-energy–acceleration-noise, 371, 379
    of traffic flow, comparison of, 311, 343
  mathematical, 113, 114, 333, 350
  moving queues, 432
  multiple sense, 348, 349
  reciprocal spacing, 342
  single sense, 347
  stochastic, 114, 115, 256, 299
  system, 324–326
Moment of inertia, 123
Moment-generating function, 127, 148, 149
Moments, 123, 148, 149
Momentum, 364–368
  of traffic stream, 366
Monte Carlo methods, 256–258, 260, 262, 284
Motivation, driver, 26
Motor vehicle (see Vehicles)
Mountable curbs, 53
Moving queues, 236–240, 433
Moving-vehicle study technique, 312–314
Multilane freeway counts, 88
Multiple channel queueing, 233–235
Multiple entries, 197, 198
Multiple regression analysis applied to
  intersection demand, 92

National Committee on Urban Transporta-
  tion, 86
National Safety Council, 2
Natural acceleration noise, 361, 370, 371, 373, 374
Negative binomial distribution, 127–130, 132
Negative exponential distribution, 153–155, 216, 267, 268, 273
Network theory, 72
Newton's binomial identity, 182
Night driving, 29, 30, 47
Night visibility, 48
Node, 100
Nomograph, 96, 114
Nonexponential distributions in queueing
  process, 248–251
Normal distribution, 150, 152, 247, 269, 272, 273, 359

North Central Expressway, Dallas, 98
Null hypothesis, 301

Obstructions, 31, 35
Occupancy counts, 88
Off-tracking, 17
Operational calculus, 332
  (See also Laplace transforms)
Operational level of service, 106
Optimization, 72, 298, 299
Optimum density, 303, 309, 311
Optimum speed, 304, 311, 358, 370
Optimum velocity, 309, 365
Origin, and destination studies, 88, 90, 245
  and termination distribution of weaving
    maneuvers, 245
Outside lane, 95–99
Overhang, 49
Overlap phase, 396
Overpasses, 84
Overtaking and passing (see Sight distance)
Overtaking distance (see Sight distance)

Parabolic distribution, 273
Parking, 52, 236
  curb, 236
Parking control, 103
Parking regulations, 411
Parking spaces, 8
Parkways, 85
Passenger car, 51
  design vehicle, 51
Passing sight distance (see Sight distance)
Paved shoulders, 53
Pavements, color of, 48
  markings for, 69
  slope, 53
  surface, 48
  types, 48
  width of, 8
Peak-hour factor, 92
Peak hourly volume, 86, 90–93, 95, 101, 106
Peak-magnitude factor, 139
Peak period rate of flow, 92, 93
Pedestrian counts, 88
Pedestrian islands, 54
Perception-reaction time, 37
Periodic scan, 274, 281, 287
Permanent counters, 88

Permanent treadle-type electric contact
  detector, 89
Phase, amber, 40
  length, 87, 399
  shift, 89
Planning, 86
Plastic hinge, 70
Platoons, 236, 240, 433
Pneumatic detector (see Road tube)
Point distribution method, 256, 257, 266, 269,
  270
Poisson arrivals, 135, 139, 224, 226, 250, 276
Poisson curves, 141, 143, 399
  cumulative, 139, 140
Poisson distribution, 124–126, 128–130, 132,
  133, 135, 180, 247, 272, 274, 275
  approximation to binomial, 132–134
  cumulative, 139
  definition, 124
  estimation of parameters, 130
  implies negative exponential gaps, 55
  mean, 124–126, 128
  moment generating function, 149
  probability generating function, 126
  representing incomplete gamma function,
    156
  variance, 125, 126, 128
Polarized headlamps, 42
Port of New York Authority, 422, 423, 428
Portable counters, 88
Possible capacity, 105, 106
Power residue method, 265, 266
Power residues, 266
Power systems, vehicular, 21
Practical capacity, 105
Pressure potential, 305
Primary roads, 80
Principle of conservation of mass, 306
Probability, 120, 121, 182
  conditional, 182
  definition, 120
  laws of, 121
  mathematical, 120
  statistical, 120
Probability distribution, 121, 123, 124
Probability-generating function, 126, 227, 228
Probability scale, 150, 195
Probit analysis, 193–197, 204–208
Profile, grade, 414
Profile map, 280
Pseudo-random numbers, 263–266, 268
Pulse generator, 89

Queue, definition, 224
  moving, 236, 240
  service times, 248
  waiting time in (*see* Waiting time)
Queue discipline, 224–230, 237–239, 439
Queue length, 224
  definition, 224
  expected, 226–229, 233, 235
  finite, 229, 230
  infinite, 227
Queueing configurations, 224
  multiple channels, 233, 235
    in parallel, 250
    in series, 249
  single channels, 224–227, 233
Queueing distributions, 132, 224, 226
Queueing service rate, significance of,
  246–248
Queueing system, 224
Queueing theory, 72, 113, 223, 251

Radar, 89
Radial-circumferential design, 81
Radial loop, 85
Radius of curvature, 18
Rail rapid transit, 443, 444
Raised surface, 47
Ramp angle of entry (*see* Angle, of
  convergence)
Ramp control criteria, 427–434
Ramp design, 215, 414
Ramp vehicle, 176
Ramps, 84, 94–98, 106–108, 174, 215, 230, 395,
  399, 400, 414–416
  alignment, 414
  capacity, 400
  definition, 174
  demand, 230
  entrance, 97, 106–108, 240–244, 293, 399, 400,
    414
  exit, 97, 106–108, 399, 400, 405, 414
  maximum service volume, 243, 244
  nose, 97
  sight distance, 414
  slip, 400
  trades, 416
  volume, 94
  weaving sections, 400–402
Random arrivals, 268
Random deviate, 266–268, 272–274

Random numbers, 256, 257
  generation of, 263–266
  table of, 263
Random variable, 156, 230
Random walks, 258–261
  weighted, 261
Rear-end collision, 337
Recorders, automatic, 372
Recording speedometer, 372
Rectangular distribution, 149, 150
Red phase, 317
Reflectivity, 48
Reflectorization, 69
Refuge islands, 54
Regression, 300–302
  curvillinear, 301
  exponential, 301
  linear, 301
Regression curve, 302, 376
Relative speed, 177, 179, 202, 211, 277, 291,
  292, 341, 344
Renewal theory, 190
Replication, 283, 313
Research, 50, 111, 117, 118, 144, 173, 286, 322,
  326, 359
Research pyramid, 116
Response time for braking, 39
Reveille interchange, 323
Reverse flow, freeway schematics, 404–408
Reversible lanes, 405, 416
Right-of-way, 51, 82
Right-of-way width, 55
Right turning movement at intersections,
  103, 104
Road classification, 80, 82
Road color (*see* Color contrast)
Road margin, 53
Road Research Laboratory, Traffic and
  Safety Division, 312
Road surface, 47
Road tube, 88, 89
Road user characteristics (*see* Drivers)
Roadway design, access control, 81, 84
  channelization, 87
  cross section, 52
  curbs, 53
  design vehicle types, 51
  frontage roads, 84
  highway types, 82
  horizontal alignment, 64–69
  intersections (*see* Intersections)
  land width, 53, 82

Roadway design, medians, 54
  ramps (see Ramps)
  reversible lanes, 417
  shoulders, 48, 53
  sight distance (see Sight distance)
  system classifications, 72
  vertical alignment, 55–63
Roadway lighting, 47–50
  discernment factors, 29
  glare effects, 30
  luminaire distribution patterns, 49
  luminaire positioning, 49
  reflection characteristics, 48
  system design, 72, 73
Roadway surface, 47, 65
Roadways, capacity, 103, 107
  classification, 80, 82
Roughness, 47, 48
Route-assignment (see Assignment, traffic)
Rule of thumb, 338, 434
  car-following, 338
Rural arterials, 81
Rural roads, 53, 80, 105

Safe headway, 199, 337, 338
Safety, 53, 56, 75, 80, 89, 105, 111, 117, 278, 355, 374
Sag curves, 56, 61–63
Sample size, 168, 169, 283
Sampling, 86, 121, 262, 270, 273
Scanning techniques for simulation, 274, 275
  event, 274
  periodic, 275, 287
Seat belts, design of, 15
Secondary roads, 82
Sensitivity coefficient, 341, 345
Service rate, definition, 247
Service time distributions, 224, 226, 227, 242, 248, 249, 251
Service time moments, 242
Service volume, 107, 378, 390, 392, 402, 404
Shifted negative exponential distribution, 155, 268
Shock waves, 115, 314, 315, 364, 365, 408
Shoulders, 48, 52, 53
  contrast, 48
  cross section, 53
  widths, 8, 35
Side friction, 18, 65
Side slopes, 20

Sight distance, 50, 52, 56–59, 66
  definition, 56
  design standards, 57, 58
  head light, 61
  horizontal, 31
  passing, 56–58, 61
  stopping, 56–58, 61, 66
  for underpasses, 63
  vertical, 31
Sign supports, 17, 70–72
Sign visibility, 48, 49
Signal control, 134
Signal phases, 134
  amber, 40
  countdown, 41
  green, 134, 135
Signal reflector, 89
Signalization, 134–136, 140
Signalized interchange, 134, 140, 393, 396
Signalized intersection, 40, 134, 317, 318, 393
Signals, 38, 69, 139
Significance level (see Confidence limits)
Signs, 69, 70
  functional requirements, 31, 38
  location, 35, 70
Simulated sampling, 261–263
Simulation, 113, 115, 255, 256, 261
  analog, 116
  chaining, 281, 282
  definition, 255, 256
  digital, 116
  driver, 36
  freeway merging, 286–295
  reasons for, 284–286
  steps in, 276–278
Simulation program, 278–282
Simulator, analog, 346–350
Single channel queueing, 224–227, 233
Sinusoids, damped, 334
  undamped, 334
Skid marks, use of, in speed computations, 14
Skid resistance, 47
Skidding, 14
Slide-rule, 114
Slip angle, 17
Snow plows, 89
Sonic detector, 89
Space gap, 299
Space headway, 174
Space lag, 174
Space mean speed, 322

Space time relationships, 137, 138, 198–201
Speed, 377
  critical; 323
  free, 309
  optimum, 304, 370, 378
    of optimum service volume, 323
    of outside lane traffic, 439
  space mean, 315, 322
  time mean, 322
  wave, 365
  weaving, 292
Speed-change lanes (see Acceleration lane)
Speed-change maneuver, 12
Speed-density relationship, 300–302, 306, 311
  exponential, 301, 311
  linear, 301, 311
  parabolic, 301, 311
Speed-density-volume relationships, 299
Speed distribution, 150, 151
Speed-time graph, 359
Speed-volume relationships, 108, 379, 380
Spot illumination, 49
Spot speed, 317
  distribution of, 150, 151
    cumulative, 151
Stability, 332, 333
Stability factor, 19–20
Standard deviation (error), 121, 359, 362
State of the queue, 244
State Roads Commission of Maryland, 103
Statistical analysis, 112
  steps to follow, 168
Statistical tests for randomness, 125
Statistics, 86, 120, 121, 262
Steady-state equations, multiple channel
    queueing, 234, 235
Steady-state flow, 117, 341
Steady-state probability, 225
Steady-state temperature, 260
Stimulus-response, driver, 33–35, 38
Stochastic, 114, 115, 391
Stopping distance (see Sight distance)
Stream measurements, 319–323
Street classification, 82
Street cleaners, 89
Street counts, 87
Street system, 83
Streetcar, 443
Structural design, 47, 87, 88
Suboptimization, 73, 404
Subsystems, 73

Superelevation, 52, 65–68
  attainment of, 65
  definition, 18
  maximum, 65, 66
Surface color, 47
Surface friction, 18, 65
Surveillance, freeway, 422–424
  tunnel, 423
Synchronous counting, definition, 166
System design, 72, 73
System models, 324–326
System stability, 332–334
Systems engineering, 112

Tachograph, 372
Tangents, 56, 64, 414
Terminals, ramps, 416
Test tracks, 116, 427
Test vehicle, 312, 371, 372
Tests of hypotheses, 168, 169, 300
Texas Highway Department, 70–73, 424
Texas Transportation Institute, 50, 70, 424,
    445
Texas Transportation Researcher, 72
Theory, of convolutions, 273
  of traffic flow, 306
Theory pyramid, 116, 117
Thirtieth-highest hour, 90
Time headway, 299
Time lag, 117, 118, 336, 337, 341
Time-lapse aerial photography, 350, 351
Time-lapse photographic technique, 103
Time mean speed, 322
Time ratio, 101
Time-space diagrams, 57, 152, 175, 199, 200,
    280, 294, 295, 313, 338, 340
Time-temperature curve, 336
Timing signal, 87
Tire chains, 89
Tire pressure, 14
Tire thread, 14
Tire wear, 47
Toll booth, 226, 227, 233, 247
Topography, 52, 54, 80
Total travel time, 361, 376
Track width, 19
Tractive effort, 12
Tractive resistance, 47
Traffic assignment, 100–102
Traffic capacity, 67, 90, 103
  (See also Capacity)

Traffic characteristics, 53, 277, 305, 360
Traffic composition, 50, 80, 94
Traffic control, 103
Traffic counts, 86, 87
Traffic demand, 50, 90, 91, 100, 114
Traffic density, 304–315
Traffic design, 79
Traffic diversion curve, 101
Traffic engineer, 114, 357
Traffic engineering, 103, 112
  definition, 5, 422
Traffic equation of state, 364, 365
Traffic estimates, 90
Traffic flow, elements of, 5
  theory of, 306, 307, 331
Traffic flow variables, 114, 299
Traffic forecasting, 100
Traffic generators, 100
Traffic interaction, 374–377
Traffic lane width, 51
Traffic models, 115, 311
Traffic momentum, 365, 366
Traffic movement, 108
Traffic parameters, 355
Traffic patterns, 90
Traffic research, 111–118
Traffic seperators, 53
Traffic signals, 38, 69, 139
Traffic signs (see Signs)
Traffic stream, kinetic energy of, 365
Traffic stream flow, general equation of, 299
Traffic studies, origin-destination, 88, 90
  speed, 50
  volume, 85–89
Traffic variables, 114, 152, 305, 367, 368
Traffic volume, 50, 80, 85, 86, 90, 105
Trafficability, 305
Transfer function, 331, 332, 334, 347, 348
  human, 335
Transformation, 195, 196, 301, 339
Transformed variables, 152, 153, 301
Transit, 3, 4
Transition matrix, 245, 246
Transitional probabilities, 244
Translated distributions, 155, 156, 162, 268
  (See also specific types)
Transportation lag, 336
Travel time, 277, 312, 314, 355–357, 376
Treatment combinations, 283
Tree, 100
Trends, height of eye, 31
  vehicle sizes, 8

Triangular distribution, 273
Trip characteristic, 88
Trip distribution, 100
Trip generation, 100
Trip length, 98, 99
Truck, 51, 94
  combinations, 51
  design vehicles, 51
  effects of, 35, 37
  operating characteristics on grades, 55, 56
  traffic, 94
Tunnels, 423, 428
Turning movements, 88
Turning paths, 17–18
Turning radii, 17–19
Type I distribution, Pearson, 156, 161–164
Type III distribution, Pearson, 155–158, 242
Typical section, 411, 413

Unbalanced (unstable) flow, 377, 379, 381
Underpasses, 84
  sight distance for, 63, 64
Understeer, 17
Uniform distribution, 149, 150, 266–268, 272, 273
U.S. Bureau of Public Roads, 69, 70, 103, 111, 424
University of Maryland, 103
Upgrades, 400
Urban street, characteristics and functions of, 180
Urban traffic surveys, 90, 91
User, road, 47
Utilization factor, queueing, 318

Validation, 284
Vandals, 89
Variables, continuous, 147
  discrete, 147
Variance, of beta distribution, 163
  of binomial distribution, 123, 126, 128
  of delay distribution, 184
  of discrete distribution, 126
  of distribution of service time, 251
  of Erlang distribution, 250
  of gamma distribution, 157
  of negative binomial distribution, 130
  of negative exponential distribution, 128, 155

Variance, of normal distribution, 155
  of Poisson distribution, 125, 126, 128
Vehicles, 286–295
  center of gravity, 17, 19
  characteristics of, 287, 289
  control of, 17
  deceleration of, 287, 288, 293, 340
  design of, 16, 32, 51
  dimensions of, 8
  energy absorbing qualities of, 16
  height of, 31
  operating characteristics of, 21
  stability of, 19–20
  trends in size of, 8
  turning radii of, 17–19
  weaving, 287–289, 291, 292
  weight of, 13
Vehicular density, 105
Vehicular operating costs, 8
Vehicular sensitivity, 341
Vehicular speed, 247, 345
Vehicular velocity, 153, 299, 307
Velocity, definition, 321, 322
Vertical alignment, 55, 56, 58, 60
Vertical clearance, 63, 414
Vertical curves (see Curves, vertical)
Vertical sight distance, 31
Video noise, 360
Visibility, 48–50, 68
Vision, color, 28
  eye movement in, 38
  night, 29–31
  peripheral, 28, 29
  pupillary response, 30
Visual acuity, dynamic, 28, 40

Volume, 85, 86, 88–90, 92–95, 105–107, 153,
  358
Volume characteristics, 86, 94
  composition of traffic, 94
  directional distribution, 94
  lane distribution, 94
  peak period, 90–94
Volume counting, 86–88
Volume-speed-density relationships, 300,
  322, 323
  speed-volume, 379
Volume survey, 88

Waiting time, 230–232
  density function, 230–232
Wave, shock (see Shock waves)
Wave speed, 365
Wave velocity, 314, 315
Wave velocity equation, 303, 311
Weaving, 174, 292, 293, 400, 402, 404
Weaving section, 400–402, 408
Weaving speed, 292
Wet pavements, 57
Widening, 55
Width, lane, 8, 35
Windshields, design of, 30, 33
Windshield wipers, 32, 33
Word length, digital computer, 265

Yield signs, 400

Zones, 100, 303, 379, 381